Transforming Business Management with AI, BI, and Data-Driven Decision-Making

In today's rapidly evolving business landscape, the convergence of technology and finance is inevitable. This book discusses how artificial intelligence (AI), business intelligence (BI), and data-driven decision-making are transforming traditional business management practices and leading to unparalleled innovation and efficiency.

Transforming Business Management with AI, BI, and Data-Driven Decision-Making presents readers with valuable insights into how these fields connect. By covering practical uses and case studies that illustrate real-world applications of AI and BI as decision-making tools, this book makes complex theories more accessible and understandable. Readers will also find practical tips on leveraging AI and BI to enhance operational workflows, customer communication, and competitive positioning in today's dynamic business landscape. Practical implementation steps and a framework for successfully applying AI and BI technologies are included, addressing common challenges organizations may face. Additionally, strategies for achieving data-driven decision-making at the highest level are also offered.

By bridging the gap between theoretical concepts and practical data analysis issues, this book serves as a valuable resource for professionals in the business environment, industrial and manufacturing engineering, business analytics, and related fields.

Manufacturing Design and Technology Series

Series Editor: J. Paulo Davim, University of Aveiro, Portugal

This series will publish high quality references and advanced textbooks in the broad area of manufacturing design and technology, with a special focus on sustainability in manufacturing. Books in the series should find a balance between academic research and industrial application. This series targets academics and practicing engineers working on topics in materials science, mechanical engineering, industrial engineering, systems engineering, and environmental engineering as related to manufacturing systems, as well as professions in manufacturing design.

Emotional Intelligence and Neuro-Linguistic Programming: New Insights for Managers and Engineers
Edited by Carolina Machado and J. Paulo Davim

Business Intelligence and Analytics in Small and Medium Enterprises
Edited by Pedro Novo Melo and Carolina Machado

Enabling Technologies for the Successful Deployment of Industry 4.0
Edited by Antonio Sartal, Diego Carou, and J. Paulo Davim

Industry 4.0: Challenges, Trends, and Solutions in Management and Engineering
Edited by Carolina Machado and J. Paulo Davim

Sustainable Manufacturing for Industry 4.0: An Augmented Approach
Edited by K. Jayakrishna, K.E.K. Vimal, S. Aravind Raj, Asela K. Kulatunga, M.T.H. Sultan, and J. Paulo Davim

Functional and Smart Materials
Edited by Chander Prakash, Sunpreet Singh, and J. Paulo Davim

Advanced Manufacturing and Processing Technology
Edited by Chander Prakash, Sunpreet Singh, and J. Paulo Davim

Characterization, Testing, Measurement, and Metrology
Edited by Chander Prakash, Sunpreet Singh, and J. Paulo Davim

Non-Conventional Hybrid Machining Processes: Theory and Practice
Edited by Rupinder Singh and J. Paulo Davim

Digital Transformation and Enterprise Information Systems
Edited by Adelaide Martins and Carolina Machado

Transforming Business Management with AI, BI, and Data-Driven Decision-Making
Edited by Swati Gupta, Sanjay Taneja, and J. Paulo Davim

For more information about this series, please visit: https://www.crcpress.com/Manufacturing-Design-and-Technology/book-series/CRCMANDESTEC

Transforming Business Management with AI, BI, and Data-Driven Decision-Making

Edited by
Swati Gupta, Sanjay Taneja, and J. Paulo Davim

CRC Press
Taylor & Francis Group
Boca Raton London New York

CRC Press is an imprint of the
Taylor & Francis Group, an **informa** business

First edition published 2026
by CRC Press
2385 NW Executive Center Drive, Suite 320, Boca Raton FL 33431

and by CRC Press
4 Park Square, Milton Park, Abingdon, Oxon, OX14 4RN

CRC Press is an imprint of Taylor & Francis Group, LLC

ISBN: 978-1-032-91091-8 (hbk)
ISBN: 978-1-032-93153-1 (pbk)
ISBN: 978-1-003-56462-1 (ebk)

DOI: 10.1201/9781003564621

Typeset in Times
by codeMantra

Contents

Foreword

The modern business operation transforms itself due to the emerging artificial intelligence (AI) and business intelligence (BI) technologies, as well as data-driven decision-making. Emerging technologies have evolved from pure imagination to directly revolutionize business operations and corporate strategies, enabling new possibilities for growth through innovation. Organizations that gain insights from AI and BI for decision-making enhance their market standing, execute decisions more quickly and precisely, and sustain better adaptability.

Transforming Business Management with AI, BI, and Data-Driven Decision-Making is a timely and insightful contribution to the growing discourse on the intersection of technology and business strategy. This book presents extensive research on the integration of AI and BI in banking and healthcare domains, along with examinations of crisis management, sustainable business operations, and marketing.

This book explores modern cutting-edge research addressing financial market deepfake fraud while studying the supremacy of algorithms in economic development among other important topics.

The work stands out by successfully uniting conceptual knowledge with day-to-day usage examples. This book contains scholarly works by experts from both academia and industry, offering empirical research findings, case studies, and bibliometric analyses suitable for academic scholars and industry professionals alike. The extensive research in this text will guide business leaders, policymakers, and researchers who aim to use AI, BI, and data analytics to advance their specific domains.

The strategies developed from the information presented in this book will be fundamental for companies to obtain success in data-intensive operations. I commend the editors and contributors for compiling this remarkable body of work, which is sure to inspire innovation and foster a deeper understanding of AI and BI in business management.

Dr. Md. Motahar Hossain
Head, Department of Business Administration
Khwaja Yunus Ali University
Enayetpur, Sirajganj, Bangladesh

Preface

The rapid advancement of artificial intelligence (AI) and business intelligence (BI), along with data-driven decision-making, is transforming business practices. These technologies provide predictive analytics and automation features, which simultaneously improve operational efficiency and delivering data-driven wise decisions to organizations. *Transforming Business Management with AI, BI, and Data-Driven Decision-Making* is a comprehensive exploration of how these technologies are shaping the future of various industries.

The authors in this book processed several viewpoints and academic studies contributed by business experts, academic scholars, and industry practitioners. They examine the influence of AI and BI on company operations through the lens of data analytics. This book examines diverse practical uses which span crisis responses together with financial access enhancement as well as AI applications for healthcare systems and marketing methods and bank operations. This book also examines three main aspects regarding AI-driven solutions: their economic impact, deepfake technology effects, and AI adoption ethical difficulties. This book presents vital information about business AI and BI use to create a complete understanding of sustainable organizational growth strategies.

The documentation contains published research alongside case studies and theoretical evaluations that deliver readers comprehensive views about real-world implementation of AI and BI solutions while illustrating both positive outcomes and difficulties. This book serves scholars and business leaders as well as policymakers along with students to explain AI and BI's effects on strategic business operations.

Data intelligence along with automation has become indispensable because we are entering an era driven by AI. This book exists to help business professionals leverage AI and BI technologies through decision-fostering innovation.

Our deep appreciation goes to the authors who provided their expertise, which resulted in producing this work through their dedication. Research combined with expert insights from this book will generate inspirational ideas that propel advanced AI business management developments.

Introduction

The fast-changing business environment of today utilizes artificial intelligence (AI) and business intelligence (BI) together with data-driven solutions, which transform numerous business sectors worldwide. The effective use of these technologies drives organizations to produce new levels of operational efficiency and transformational innovation and strategic development. The book *Transforming Business Management with AI, BI, and Data-Driven Decision-Making* explores deep insights into AI and BI effects on banking alongside healthcare and marketing with crisis management documentation.

The compilation of research-based findings and case examples demonstrates the significant business effects of AI and BI on strategic planning together with decision-making along with digital transformations. The research investigates how algorithmic superiority transforms economic growth and deepfake schemes in financial operations and AI healthcare applications and marketing operations alongside sustainable business strategies.

This book offers a complete viewpoint about how technology meets business intelligence with managerial innovation through its examination of multiple topics. This book includes professional insights from expert academics and business professionals who demonstrate real-world AI and BI implementations along with analyses regarding deepfake technology and AI financial inclusion and the moral dimensions of AI adoption. This book presents important findings useful for executive leaders together with research experts and government officials who seek to use data-based approaches for durable business expansion.

As AI continues to disrupt traditional business models, this book serves as a guide to understanding and implementing AI- and BI-driven solutions in various industries. This book teaches essential navigation strategies about future intelligent business management across all sectors including banking education and healthcare.

Editors

Swati Gupta is Assistant Professor at Chitkara Business School, Chitkara University, Punjab, India. She earned a doctoral degree in management and has over 9 years of professional experience across various educational institutions. She specializes in teaching subjects such as consumer behavior, market research, marketing management, organizational behavior, sales and distribution management, services marketing, and retail management. Having served as the editor for various books under renowned publishing houses as well as a chapter reviewer, she has contributed significantly to the academic field.

Sanjay Taneja is Associate Professor in the Department of Management Studies, Graphic Era (deemed to be a university), India. His areas of interest include accounting and finance (fin tech, green finance), risks, insurance management, green economics, and management of innovation in insurance. He earned a master's degree in management with a specialization in finance and marketing. He earned a PG degree in management (Gold Medalist) at Chaudhary Devi University, Sirsa, India, in 2012 and earned a PhD (sponsored by ICSSR) in 2016 at Chaudhary Devi University, Sirsa, India. He earned a post-doctoral degree at the Faculty of Social Sciences, Department of Banking and Insurance, Usak University, Turkey, in 2023. Dr. Taneja has published research papers in reputed journals and has more than 50 publications to his credit. He also has 2 ebooks, 1 textbook, 20 edited books, and 4 book chapters, all Scopus-indexed. He has attended or organized more than 50 national and international conferences and has been invited as session chair, resource person, keynote speaker, and judge at various conferences in India and internationally. He is also a reviewer for several journals and has received several academic awards and recognitions. His motto in life is to grow and help others to grow.

J. Paulo Davim is Professor at the University of Aveiro, Portugal. He is also distinguished as an honorary Professor at several universities, colleges, and institutes in China, India, and Spain. He earned a PhD in mechanical engineering, an MSc in mechanical engineering (materials and manufacturing processes), and a mechanical engineering degree (5 years) at the University of Porto (FEUP), the Aggregate title (Full Habilitation) at the University of Coimbra, and the DSc (Higher Doctorate) at the London Metropolitan University. He is a Fellow (FIET) of IET-London and Eur Ing by Engineers Europe FEANI-Brussels. He is also a Senior Chartered Engineer by the Portuguese Institution of Engineers with an MBA and specialist titles in engineering and industrial management as well as in metrology. He has more than 35 years of teaching and research experience in manufacturing, materials, mechanical, and industrial engineering, with special emphasis in machining, tribology, and surface engineering. Dr. Davim is also interested in design, management, sustainability, artificial intelligence, engineering education, and higher education for sustainability. He has guided large numbers of postdoc, PhD, and master's students as well as coordinated and participated in several financed research projects. He has received several scientific awards and honors and has worked as an evaluator of projects for the ERC-European Research Council and other international research agencies and as an examiner of PhD theses for many universities in different countries. He is the Editor-in-Chief of several international journals, a guest editor of journals, a book editor, a book series editor, and a scientific adviser for many international journals and conferences.

Contributors

Vikas Agrawal
Vivekananda Institute of Professional Studies
New Delhi, India

Ramneek Ahluwalia
University School of Business
Chandigarh University
Mohali, India

Diksha Ahuja
University School of Business
Chandigarh University
Mohali, India

Şenol Babuşcu
Department of International Finance and Banking
Başkent University
Ankara, Türkiye

Ankita Bansal
Vivekananda Institute of Professional Studies
New Delhi, India

Mukul Bhatnagar
Department of Commerce
Graphic Era Deemed to be University
Dehradun, India

Shantharuby Buvanendra
University of Colombo
Colombo, Sri Lanka

Bhupinderpal Singh Chahal
Yorkville University
Toronto, Canada

Cyril Crasto
Balaji Institute of Management and Human Resource Management
Shri Balaji University
Pune, India

Kavita Dahiya
Center of Online Learning
Manipal University
Manipal, India

Harshi Garg
IIMT University
Meerut, India

Sachin Ghai
Department of Management Studies
Graphic Era Deemed to be University
Dehradun, India

Ashulekha Gupta
Department of Management Studies
Graphic Era Deemed to be University
Dehradun, India

Swati Gupta
Chitkara Business School
Chitkara University
Rajpura, India

Adaley Hazar
Department of International Finance and Banking
Başkent University
Ankara, Türkiye

Sanjeev Jain
Chandigarh University
Mohali, India

Kashish Jindal
Department of Management Studies
Graphic Era Deemed to be University
Dehradun, India

Kavya Jindal
Department of Commerce
Graphic Era Deemed to be University
Dehradun, India

Amar Johri
College of Administrative and Financial Sciences
Saudi Electronic University
Riyadh, Saudi Arabia

Mohammad Kashif
Department of Management Studies
Graphic Era Deemed to be University
Dehradun, India

Arvinder Kaur
Department of Commerce
Chandigarh University
Mohali, India

Rupa Khanna
Department of Commerce
Graphic Era Deemed to be University
Dehradun, India

Mohit Kukreti
University of Technology and Applied Science
Muscat, Oman

Dhananjay Kulkarni
Balaji Institute of Management and Human Resource Management
Shri Balaji University
Pune, India

Pawan Kumar
Department of Management Studies
Graphic Era Deemed to be University
Dehradun, India

Sanjeet Kumar
Department of Business Administration
Chaudhary Devi Lal University
Sirsa, India

Elif Hasret Kumcu
Aksaray University
Aksaray, Turkey

Larisa Mistrean
Academy of Economic Studies of Moldova
Chişinău, Moldova

Soni Mitali
Delhi University
New Delhi, India

Timilehin Olasoji Olubiyi
West Midlands Open University
Nigeria

Ercan Ozen
Department of Banking and Finance
Usak University
Uşak, Turkey

Rajendra Rajaram
UNISA School of Business Leadership
Pretoria, South Africa

Amol Ranadive
Balaji Institute of Management and Human Resource Management
Shri Balaji University
Pune, India

Keerthi Rao
Department of Physiotherapy
Chandigarh University
Mohali, India

Reepu
University School of Business
Chandigarh University
Mohali, India

Sabina Sehajpal
University Institute of Nursing
Chandigarh University
Mohali, India

Zelhuda Shamsuddin
Universiti Sultan Zainal Abidin
Kuala Terengganu, Malaysia

Priyank Sharma
IIMT University
Meerut, India

Rasna Sharma
University School of Business
Chandigarh University
Mohali, India

Muhammad Rehan Shaukat
University of Technology and Applied Sciences
Muscat, Oman

Ashwini Nikhil Shinde
Balaji Institute of Management and Human Resource Management
Shri Balaji University
Pune, India

Muskan Singh
Department of Management Studies
Graphic Era Deemed to be University
Dehradun, India

Raj Kumar Singh
School of Management
Graphic Era Hill University
Dehradun, India

Vivek Anand Singh
Teerthanker Mahaveer University
Moradabad, India

Ashish Kumar Srivastava
Chandigarh University
Mohali, India

Sanjay Taneja
Department of Management Studies
Graphic Era Deemed to be University
Dehradun, India

Cem Sabutay Turkdonmez
Department of International Finance and Banking
Başkent University
Ankara, Türkiye

Sami Ullah
University of Central Punjab
Lahore, Pakistan

Luan Vardari
Universiteti "Ukshin Hoti" Prizren
Prizren, Kosovo

Amar Yedake
Balaji Institute of Management and Human Resource Management
Shri Balaji University
Pune, India

1 The Impact of Deepfake Technology on Crisis Management in Business through AI and BI

Rajendra Rajaram, Pawan Kumar, and Mukul Bhatnagar

1.1 INTRODUCTION

Crisis management is a critical function within organizations, requiring swift, effective responses to mitigate damage and restore normalcy in times of adversity. As organizations face increasingly complex crises, the need for innovative tools and strategies to improve decision-making and enhance organizational preparedness has never been more pronounced. Among these innovations, deepfake technology—commonly associated with virtual avatars and artificial intelligence (AI)-generated simulations—has emerged as a powerful tool for refining crisis management training and decision-making processes (Abbas et al., 2022). Deepfake-powered simulations offer organizations the ability to create highly realistic crisis scenarios, enabling executives, employees, and stakeholders to engage with realistic crises in a controlled environment (Khan et al., 2023).

This study investigates the role of deepfake technology in crisis management simulations, explicitly examining its influence on crisis communication preparedness, resource allocation, employee engagement, and long-term recovery. By leveraging virtual avatars and realistic scenario simulations, organizations can better prepare their teams for real-world challenges, improving both decision-making and overall crisis response effectiveness (Tandon et al., 2023). However, the deployment of deepfake technology in crisis management is not without its challenges, particularly regarding the ethical implications of manipulating reality for training purposes and concerns about technological reliability and stakeholder trust.

The research delves into how deepfake technology enhances crisis communication strategies by simulating complex communication channels, decision-making processes, and the real-time engagement of employees in crisis scenarios. Moreover, it explores the broader implications of this technology on organizational resilience, employee preparedness, and stakeholder satisfaction during crises (Westerlund, 2019). As organizations grapple with the evolving landscape of crisis management, the ability to integrate cutting-edge technologies such as deepfake simulations offers a promising avenue for improving the speed, clarity, and efficiency of crisis responses (Choi et al., 2022). This study aims to contribute valuable insights into the practical

DOI: 10.1201/9781003564621-1

application of AI-powered tools in crisis management, providing a nuanced understanding of their potential benefits and limitations.

1.2 LITERATURE REVIEW

The integration of technology into crisis management has been increasingly recognized as an essential tool for enhancing organizational preparedness and decision-making. In particular, the use of AI and virtual simulations, including deepfake technology, offers significant promise in improving the realism and effectiveness of crisis training. Previous studies have emphasized the importance of crisis communication in mitigating the effects of crises, with a focus on the speed, clarity, and consistency of information dissemination (Kaushik, 2022). Deepfake technology, through the use of AI-generated virtual avatars, allows for the creation of highly realistic crisis scenarios that can enhance training by mimicking real-world crises more accurately than traditional methods (Clarke et al., 2023).

The literature suggests that effective crisis management is not only reliant on rapid decision-making but also on the ability to allocate resources efficiently, engage employees, and ensure stakeholder satisfaction. As crisis scenarios become increasingly complex, organizations must adopt innovative strategies to simulate real-time decision-making processes and evaluate the efficiency of their communication systems (Liu et al., 2024). Moreover, studies on the ethical implications of AI in crisis management have raised concerns about the manipulation of reality, urging organizations to balance technological innovation with transparency and ethical responsibility. Despite these challenges, deepfake technology presents a valuable opportunity for organizations to refine their crisis management frameworks, ensuring they are equipped to respond effectively to unpredictable, high-pressure situations (Kukreti et al., 2025; Sehajpal & Kaur, 2025).

1.3 RESEARCH METHODOLOGY

This research adopts a quantitative approach, utilizing multiple regression analysis to examine the influence of deepfake technology on various aspects of crisis management, including crisis communication preparedness, leadership decision-making, resource allocation, and employee engagement. The study involves a sample of 273 respondents from organizations that have implemented deepfake-powered simulations in their crisis management training. Data was collected using structured questionnaires, with Likert scale responses ranging from 1 (very ineffective) to 5 (very effective). The analysis examines the relationships between the independent variables (e.g., use of deepfake technology, crisis communication efficiency) and the dependent variables (e.g., crisis management effectiveness, employee engagement). Diagnostic tests such as ANOVA and hypothesis testing were used to ensure the validity and reliability of the results.

1.3.1 Data Analysis

Table 1.1 outlines the research instruments, categorizing various variables into dependent, mediating, and independent types, and presenting specific question headings along with corresponding Likert scale descriptions. The table reveals how

TABLE 1.1
Research Instruments

Type of Variable	Variable Name	Question Heading	Question Statement	Likert Scale Description
Dependent	Crisis management effectiveness	Crisis management response	How effective was your organization's response during a crisis?	1: Very ineffective, 2: Ineffective, 3: Neutral, 4: Effective, 5: Very effective
Dependent	Crisis management effectiveness	Leadership decision-making	How quickly did leadership make decisions during the crisis?	1: Very slow, 2: Slow, 3: Neutral, 4: Quick, 5: Very quick
Dependent	Crisis management effectiveness	Resource allocation	How effectively were resources allocated during the crisis?	1: Very ineffective, 2: Ineffective, 3: Neutral, 4: Effective, 5: Very effective
Dependent	Crisis management effectiveness	Employee engagement	How engaged were employees in crisis management efforts?	1: Not engaged, 2: Slightly engaged, 3: Neutral, 4: Engaged, 5: Highly engaged
Dependent	Crisis management effectiveness	Crisis communication efficiency	How efficient was the crisis communication strategy during the crisis?	1: Very inefficient, 2: Inefficient, 3: Neutral, 4: Efficient, 5: Very efficient
Dependent	Crisis management effectiveness	Long-term recovery	How effective was the organization's long-term recovery after the crisis?	1: Very ineffective, 2: Ineffective, 3: Neutral, 4: Effective, 5: Very effective
Dependent	Crisis management effectiveness	Stakeholder satisfaction	How satisfied were stakeholders (customers, investors) with the crisis management?	1: Very dissatisfied, 2: Dissatisfied, 3: Neutral, 4: Satisfied, 5: Very satisfied
Mediating	Crisis communication preparedness	Crisis plan availability	How prepared was your organization with a crisis communication plan?	1: Not prepared, 2: Slightly prepared, 3: Neutral, 4: Prepared, 5: Very prepared

(Continued)

TABLE 1.1 (*Continued*)
Research Instruments

Type of Variable	Variable Name	Question Heading	Question Statement	Likert Scale Description
Mediating	Crisis communication preparedness	Employee training	How well trained were employees to handle crisis communication?	1: Not trained, 2: Slightly trained, 3: Neutral, 4: Trained, 5: Very well trained
Mediating	Crisis communication preparedness	Crisis communication channels	How effective were the communication channels (emails, calls, social media) during a crisis?	1: Very ineffective, 2: Ineffective, 3: Neutral, 4: Effective, 5: Very effective
Mediating	Crisis communication preparedness	Message clarity	How clear and understandable were the crisis communication messages?	1: Very unclear, 2: Unclear, 3: Neutral, 4: Clear, 5: Very clear
Mediating	Crisis communication preparedness	Timeliness of information	How timely was the information shared with employees and stakeholders?	1: Very late, 2: Late, 3: Neutral, 4: Timely, 5: Very timely
Mediating	Crisis communication preparedness	Consistency of messaging	How consistent were the crisis communication messages across different platforms?	1: Very inconsistent, 2: Inconsistent, 3: Neutral, 4: Consistent, 5: Very consistent
Independent	Use of deepfake technology in simulations	Use of virtual avatars	How often did your organization use virtual avatars (deepfake technology) in crisis simulations?	1: Never, 2: Rarely, 3: Occasionally, 4: Frequently, 5: Always
Independent	Use of deepfake technology in simulations	Realism of simulated scenarios	How realistic were the deepfake-powered crisis scenarios used in simulations?	1: Not realistic, 2: Slightly realistic, 3: Neutral, 4: Realistic, 5: Very realistic

(*Continued*)

TABLE 1.1 (*Continued*)
Research Instruments

Type of Variable	Variable Name	Question Heading	Question Statement	Likert Scale Description
Independent	Use of deepfake technology in simulations	Effectiveness in training	How effective was deepfake technology in preparing employees for crises?	1: Very ineffective, 2: Ineffective, 3: Neutral, 4: Effective, 5: Very effective
Independent	Use of deepfake technology in simulations	Enhancing decision-making skills	How well did deepfake simulations enhance the decision-making skills of managers and executives?	1: Not helpful, 2: Slightly helpful, 3: Neutral, 4: Helpful, 5: Very helpful
Independent	Type of crisis scenario	Cybersecurity breach impact	How impactful was a cybersecurity breach scenario in the crisis simulation?	1: No impact, 2: Slight impact, 3: Neutral, 4: Significant impact, 5: Very significant impact
Independent	Type of crisis scenario	Financial downturn impact	How impactful was a financial downturn scenario in the crisis simulation?	1: No impact, 2: Slight impact, 3: Neutral, 4: Significant impact, 5: Very significant impact
Independent	Type of crisis scenario	Crisis management complexity	How complex were the crisis management challenges in the cybersecurity breach scenario?	1: Very simple, 2: Simple, 3: Neutral, 4: Complex, 5: Very complex
Independent	Type of crisis scenario	Crisis management complexity	How complex were the crisis management challenges in the financial downturn scenario?	1: Very simple, 2: Simple, 3: Neutral, 4: Complex, 5: Very complex
Independent	Type of crisis scenario	Employee response to crisis	How would employees typically respond to a cybersecurity breach?	1: Very poorly, 2: Poorly, 3: Neutral, 4: Well, 5: Very well
Independent	Type of crisis scenario	Employee response to crisis	How would employees typically respond to a financial downturn?	1: Very poorly, 2: Poorly, 3: Neutral, 4: Well, 5: Very well

diverse aspects of crisis management—such as response effectiveness, leadership decision-making, resource allocation, and employee engagement—are assessed, with a focus on the use of deepfake technology and crisis communication preparedness as independent and mediating factors. The detailed categorization of crisis scenarios, including cybersecurity breaches and financial downturns, highlights the complexity of evaluating crisis management effectiveness through varied crisis contexts.

1.3.2 Data Analysis

Figure 1.1, depicting the PLS algorithm, illustrates the relationships among the constructs and their respective paths, providing a structural overview of the model's estimated parameters, which collectively demonstrate the interdependencies between crisis communication preparedness, crisis management effectiveness, and the use of deepfake technology in simulations, highlighting their critical role in influencing organizational crisis responses.

Table 1.2 presents the construct reliability and validity statistics for the key variables in the study. The high Cronbach's alpha values (ranging from 0.848 to 0.934) and composite reliability scores (rho_a and rho_c) indicate strong internal consistency for the constructs, ensuring the reliability of the measurement instruments.

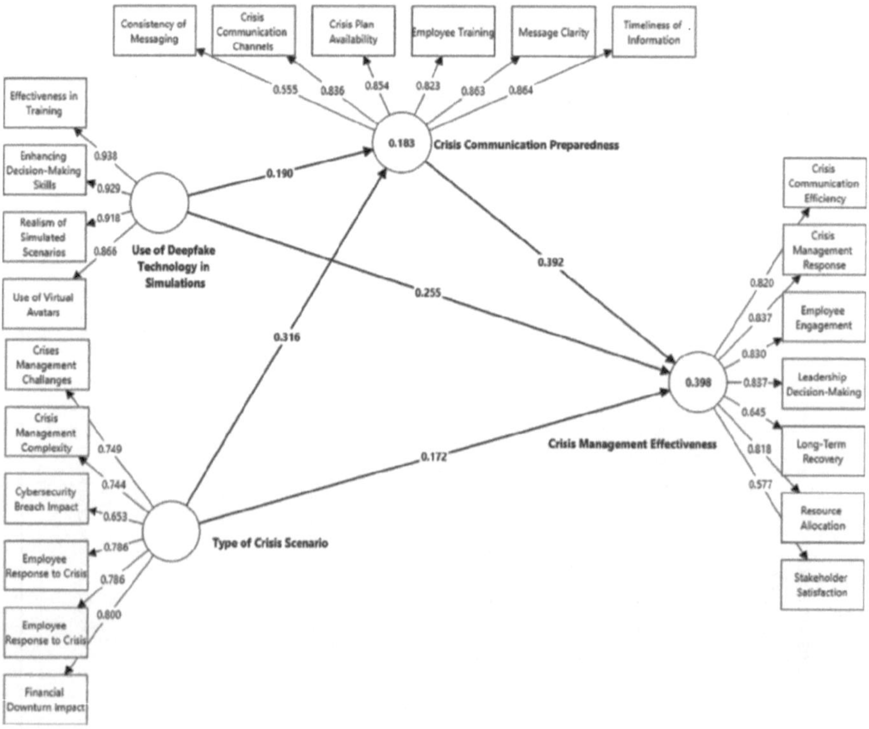

FIGURE 1.1 PLS algorithm.

TABLE 1.2
Construct Reliability and Validity

	Cronbach's Alpha	Composite Reliability (rho_a)	Composite Reliability (rho_c)	Average Variance Extracted (AVE)
Crisis communication preparedness	0.887	0.884	0.916	0.651
Crisis management effectiveness	0.883	0.883	0.911	0.597
Type of crisis scenario	0.848	0.848	0.888	0.569
Use of deepfake _technology in _simulations	0.934	0.957	0.953	0.834

TABLE 1.3
HTMT

	Heterotrait-Monotrait Ratio (HTMT)
Crisis management effectiveness <-> crisis communication preparedness	0.581
Type of crisis scenario <-> crisis communication preparedness	0.443
Type of crisis scenario <-> crisis management effectiveness	0.481
Use of deepfake _technology in _simulations <-> crisis communication preparedness	0.335
Use of deepfake _technology in _simulations <-> crisis management effectiveness	0.469
Use of deepfake _technology in _simulations <-> type of crisis scenario	0.427

Additionally, the average variance extracted (AVE) values above 0.5 for all constructs, especially *Use of Deepfake Technology in Simulations* (AVE = 0.834), confirm that the constructs exhibit substantial convergent validity, thereby providing a robust foundation for subsequent analyses.

The Heterotrait-Monotrait (HTMT) ratio in Table 1.3 highlights the discriminant validity of the constructs, with values generally below the threshold of 0.85, indicating that the constructs are sufficiently distinct from one another. The relatively low correlations between *Crisis Communication Preparedness* and *Crisis Management Effectiveness* (HTMT = 0.581) and other pairings suggest that while the constructs are related, they measure distinct underlying concepts, ensuring that the constructs capture unique dimensions of crisis management without excessive overlap.

Table 1.4 presents the cross-loadings, showing the extent to which each indicator loads on its respective construct. The results indicate that indicators such as

TABLE 1.4
Cross Loading

	Crisis Communication Preparedness	Crisis Management Effectiveness	Type of Crisis Scenario	Use of Deepfake _Technology in _Simulations
Consistency of messaging	0.555	0.515	0.293	0.281
Crises management challenges	0.335	0.358	0.749	0.273
Crisis communication channels	0.836	0.351	0.309	0.211
Crisis communication efficiency	0.313	0.820	0.348	0.291
Crisis management complexity	0.300	0.297	0.744	0.217
Crisis management response	0.349	0.837	0.350	0.482
Crisis plan availability	0.854	0.442	0.294	0.311
Cybersecurity breach impact	0.280	0.368	0.653	0.474
Effectiveness in training	0.317	0.443	0.379	0.938
Employee engagement	0.356	0.830	0.333	0.318
Employee response to crisis	0.299	0.255	0.786	0.258
Employee response to crisis	0.287	0.351	0.786	0.279
Employee training	0.823	0.385	0.345	0.272
Enhancing decision-making skills	0.278	0.476	0.399	0.929
Financial downturn impact	0.249	0.261	0.800	0.259
Leadership decision-making	0.317	0.837	0.296	0.444
Long-term recovery	0.617	0.645	0.305	0.290
Message clarity	0.863	0.433	0.324	0.211
Realism of simulated scenarios	0.306	0.411	0.374	0.918
Resource allocation	0.351	0.818	0.376	0.355
Stakeholder satisfaction	0.506	0.577	0.266	0.191
Timeliness of information	0.864	0.416	0.298	0.198
Use of virtual avatars	0.235	0.243	0.260	0.866

Consistency of Messaging (0.863 for *Crisis Communication Preparedness*) and *Crisis Management Response* (0.837 for *Crisis Management Effectiveness*) exhibit strong loadings on their respective constructs, suggesting that these indicators are appropriately aligned with the theoretical constructs they represent. This confirms the relevance and coherence of the selected measures in assessing crisis management and communication preparedness.

Table 1.5 displays the variance inflation factor (VIF) values, which assess multicollinearity among the indicators. The VIF values range from 1.187 to 4.419, with most values below the critical threshold of 5, indicating that multicollinearity is not a significant issue in the model. However, *Effectiveness in Training* (VIF = 4.419) and *Crisis Management Response* (VIF = 3.629) exhibit higher values, which may warrant further consideration for potential multicollinearity concerns in the subsequent modeling phase.

Figure 1.2 shows the bootstrapping results, demonstrating the robustness of the estimated coefficients and their significance levels, with the confidence intervals for each path consistently excluding zero, further validating the hypotheses and ensuring the reliability of the statistical model used to assess crisis management effectiveness in the context of deepfake technology simulations.

TABLE 1.5
Variance Inflation Factor

	VIF
Consistency of messaging	1.187
Crises management challenges	1.581
Crisis communication channels	2.683
Crisis communication efficiency	3.379
Crisis management complexity	1.727
Crisis management response	3.629
Crisis plan availability	2.707
Cybersecurity breach impact	1.329
Effectiveness in training	4.419
Employee engagement	3.033
Employee response to crisis	1.965
Employee response to crisis	2.160
Employee training	2.553
Enhancing decision-making skills	3.930
Financial downturn impact	2.304
Leadership decision-making	3.740
Long-term recovery	2.633
Message clarity	2.908
Realism of simulated scenarios	3.644
Resource allocation	3.399
Stakeholder satisfaction	2.467
Timeliness of information	3.387
Use of virtual avatars	2.891

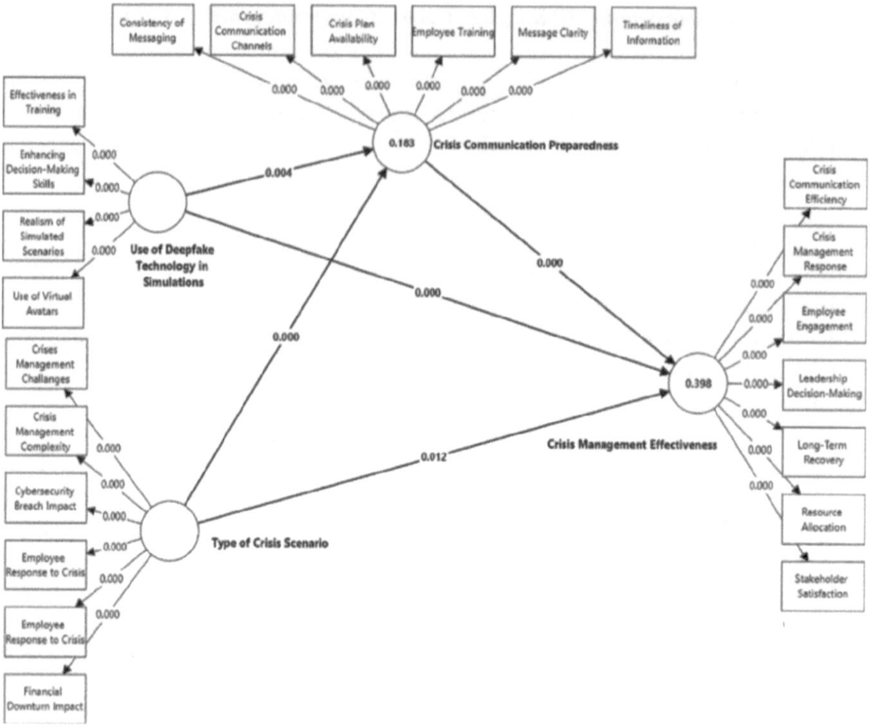

FIGURE 1.2 Bootstrapping.

Table 1.6 shows the hypothesis testing results, with original sample coefficients (O) and corresponding t-statistics and p-values. Significant paths include *Crisis Communication Preparedness* → *Crisis Management Effectiveness* ($t=5.634$, $p<0.001$), indicating a strong positive effect, and *Use of Deepfake Technology in Simulations* → *Crisis Communication Preparedness* ($t=2.907$, $p=0.004$), demonstrating a moderate influence of deepfake technology on communication preparedness. These results substantiate the effectiveness of deepfake technology in enhancing crisis management outcomes and highlight the importance of crisis communication in fostering effective crisis management.

1.4 DISCUSSION

The findings of this study demonstrate that deepfake technology significantly enhances crisis management preparedness by improving decision-making, communication efficiency, and overall response effectiveness. The use of virtual avatars in simulated crisis scenarios provided participants with a more realistic training experience, which in turn fostered better engagement and quicker decision-making during actual crises. Furthermore, crisis communication preparedness, facilitated by deepfake simulations, positively influenced the effectiveness of resource allocation

TABLE 1.6
Hypothesis Testing

	Original Sample (O)	Sample Mean (M)	Standard Deviation (STDEV)	T-Statistics (|O/STDEV|)	P-Values
Crisis communication preparedness -> crisis management effectiveness	0.392	0.386	0.070	5.634	0.000
Type of crisis scenario -> crisis communication preparedness	0.316	0.323	0.073	4.343	0.000
Type of crisis scenario -> crisis management effectiveness	0.172	0.181	0.068	2.527	0.012
Use of deepfake _technology in _simulations -> crisis communication preparedness	0.190	0.189	0.065	2.907	0.004
Use of deepfake _technology in _simulations -> crisis management effectiveness	0.255	0.255	0.070	3.630	0.000

and long-term recovery efforts. These results highlight the pivotal role of AI in transforming crisis management practices, enabling organizations to better navigate complex crises.

1.5 MANAGERIAL IMPLICATIONS

The implications of this study for organizational managers are manifold. Firstly, it emphasizes the importance of adopting advanced technologies such as deepfake simulations to enhance crisis management preparedness. By incorporating AI-generated avatars and realistic scenarios into training programs, organizations can improve the efficacy of their crisis response strategies, ensuring that employees are better equipped to handle real-world challenges. This not only enhances decision-making skills but also fosters greater engagement among employees, which is critical for ensuring a coordinated and effective response during crises. Moreover, the study highlights the significance of crisis communication preparedness in improving crisis management outcomes. Managers should focus on ensuring that crisis communication channels are robust, clear, and consistent, as these factors are vital for maintaining stakeholder trust and satisfaction. Training employees to communicate effectively during crises is essential, and deepfake simulations can serve as a valuable tool in this regard. Additionally, the research underscores the need for organizations to balance technological innovation with ethical considerations, particularly when using deepfake technology, to avoid potential concerns about the manipulation of reality. Overall, adopting a strategic approach to integrating AI in crisis management can provide organizations with a competitive edge in managing crises efficiently while maintaining public trust.

1.5.1 Findings and Conclusion

The study uncovers significant insights into the impact of deepfake technology on crisis management simulations. The results indicate that the integration of AI-powered virtual avatars significantly improves crisis communication preparedness, which is crucial for effective crisis management. The deepfake simulations provided a more immersive training experience, enhancing employees' ability to respond to crises by simulating complex real-time scenarios that are not easily replicated through traditional methods. These simulations contributed to improved decision-making, more efficient resource allocation, and better engagement from employees, leading to higher levels of satisfaction among stakeholders during a crisis. The research also reveals that deepfake technology enhances leadership decision-making capabilities by creating a more realistic environment in which managers can practice their crisis response strategies. Moreover, the use of deepfake technology in crisis simulations positively impacts the overall effectiveness of crisis management, particularly in scenarios such as cybersecurity breaches and financial downturns, where the complexity and urgency of decision-making are paramount. However, while deepfake simulations showed promising results in enhancing crisis response capabilities, the study also highlighted the ethical concerns surrounding the use of AI-generated content. As organizations continue to explore the potential of deepfake technology, they must balance its benefits with transparency and ethical responsibility to avoid potential negative consequences related to trust and manipulation of information. In conclusion, the study underscores the transformative potential of deepfake technology in crisis management simulations. By providing more realistic and effective training environments, deepfake simulations can help organizations improve their crisis response strategies, enhance decision-making, and ultimately achieve better crisis management outcomes. Nevertheless, careful consideration of ethical implications is essential to ensure that the adoption of such technology is both responsible and effective in strengthening organizational resilience.

REFERENCES

Abbas, N. N., Ahmad, R., Qazi, S., & Ahmed, W. (2022). Impact of deepfake technology on FinTech applications. In *Handbook of Research on Cybersecurity Issues and Challenges for Business and FinTech Applications*. IGI Global. https://doi.org/10.4018/978-1-6684-5284-4.ch012

Choi, C., Jung, Y., & Kim, J. (2022). Discourse analysis on deepfake technology using text mining. *Journal of Korean Institute of Communications and Information Sciences*, *47*(6), 870–881. https://doi.org/10.7840/kics.2022.47.6.870

Clarke, C., Xu, J., Zhu, Y., Dharamshi, K., McGill, H., Black, S., & Lutteroth, C. (2023). FakeForward: using deepfake technology for feedforward learning. *Conference on Human Factors in Computing Systems – Proceedings*. https://doi.org/10.1145/3544548.3581100

Kaushik, P. (2022). Privacy and other legal concerns in the wake of deepfake technology: comparative study of India, US, and China. In *Handbook of Research on Cyber Law, Data Protection, and Privacy*. IGI Global. https://doi.org/10.4018/978-1-7998-8641-9.ch003

Khan, I. R., Aisha, S., Kumar, D., & Mufti, T. (2023). A systematic review on deepfake technology. *Lecture Notes in Networks and Systems*, *572*, 669–685. https://doi.org/10.1007/978-981-19-7615-5_55

Kukreti, M., Sehajpal, S., Tiwari, R., & Sood, K. (2025). *Driving Global Health and Sustainable Development Goals with Smart Technology* (1st ed.). IGI Global Scientific Publishing. https://doi.org/10.4018/979-8-3373-0240-9

Liu, Y., Sun, T., Wang, Z., Zhao, X., Cheng, R., & Shi, B. (2024). Detection of deepfake technology in images and videos. *International Journal of Ad Hoc and Ubiquitous Computing*, *45*(2), 135–148. https://doi.org/10.1504/IJAHUC.2024.136851

Sehajpal, S., & Kaur, N. (2025). Neuroscience, nursing, and financial innovation : a synergistic approach to optimized patient care and economic well-being. In T. Sanjay, S. C. Bhupinderpal Pal, J. Amar, O. Ercan, & K. Pawan (Eds.), *Neuromarketing's Role in Sustainable Finance* (pp. 133–146). IGI Global Scientific Publishing. https://doi.org/10.4018/979-8-3693-9117-4.ch008

Tandon, S., Vig, A., Kartik, M., & Kumawat, H. C. (2023). Real–time face transition using deepfake technology (Gan Model). *2023 International Conference on Recent Advances in Electrical, Electronics, Ubiquitous Communication, and Computational Intelligence, RAEEUCCI 2023*. https://doi.org/10.1109/RAEEUCCI57140.2023.10134474

Westerlund, M. (2019). The emergence of deepfake technology: a review. *Technology Innovation Management Review*, *9*(11), 39–52. https://doi.org/10.22215/TIMREVIEW/1282

2 Advanced Speech Recognition System Using AI for Accurately Understanding Regional English Accents through Data-Driven Techniques

Ashwini Nikhil Shinde and Luan Vardari

2.1 INTRODUCTION

The landscape of speech recognition technology has evolved dramatically over the past few decades, transitioning from rudimentary systems with limited functionality to highly sophisticated models capable of nuanced linguistic interpretation (Badhotiya et al., 2021; Ram et al., 2022). However, despite these advancements, one of the most persistent challenges remains the accurate understanding of regional English accents, which often elude even the most advanced artificial intelligence models (Uniyal et al., 2022a). This issue stems from the immense diversity in pronunciation, intonation, and phonetic variations inherent in different dialects and accents, leading to misinterpretation and errors in transcription. The intricacies of regional accents, shaped by geographic, socio-economic, and cultural factors, introduce a level of complexity that traditional speech recognition systems struggle to accommodate (Upreti et al., 2024). As a result, the quest for systems that can seamlessly decipher these variations has become a pivotal area of research and innovation in the field of natural language processing (Bhardwaj et al., 2024).

Furthermore, the integration of emerging technologies, such as deepfake voice synthesis, has introduced an additional layer of complexity (Malhotra & Gupta, 2021). Deepfake technology, which employs advanced neural networks to generate synthetic voices that mimic the tonal and verbal patterns of real individuals, has revolutionized the way in which speech is produced and processed (Kanojia et al., 2022). While the potential applications of deepfake technology are vast, ranging from entertainment to personalized communication, it also presents significant challenges to speech recognition systems, especially in the context of understanding regional accents (Sonnad et al., 2022; Umamaheswaran et al., 2023). The synthetic nature of deepfake voices, often indistinguishable from real human speech, creates

DOI: 10.1201/9781003564621-2

the risk of further complicating the recognition process, especially when these artificial voices are crafted to simulate regional speech patterns with uncanny precision (Uniyal et al., 2022a). As such, the confluence of regional accent variability and the sophisticated, yet disruptive capabilities of deepfake technology necessitates the development of highly robust, adaptive speech recognition systems capable of not only accurately transcribing diverse accents but also distinguishing between authentic and artificially generated speech (Upreti et al., 2024).

At the heart of this endeavor lies the critical need for an advanced recognition system that can navigate the myriad phonetic shifts present in various English dialects while also contending with the synthetic nature of deepfake voices (Bhardwaj et al., 2024). Achieving this level of precision requires an intricate understanding of both the linguistic characteristics of regional speech and the technical nuances of deepfake synthesis (Malhotra & Gupta, 2021). This convergence of challenges demands innovative approaches to model training, data processing, and algorithmic refinement, making it an area of research that holds immense promise for the future of natural language understanding and artificial intelligence (Kanojia et al., 2022). The overarching goal is to develop a system that not only recognizes and adapts to the rich diversity of regional accents but also remains resilient against the potential deceptions posed by artificial speech generation technologies (Kukreti et al., 2025; Sehajpal & Kaur, 2025).

2.2 LITERATURE REVIEW

In recent years, the rapid evolution of speech recognition systems has led to a growing body of research, focusing on the refinement of algorithms that can accurately transcribe speech across various dialects and regional accents (Davuluri et al., 2023; Lade et al., 2023; Sharma et al., 2021; Sharma et al., 2021). However, despite significant advances, the persistent challenge of understanding the vast phonetic, syntactic, and prosodic diversity inherent in regional English accents remains an impediment to achieving universal transcription accuracy. The literature suggests that regional accents are not merely variations in pronunciation but encompass deep-rooted differences in speech patterns, intonation, stress, rhythm, and even semantic nuances (Bhatnagar et al., 2024a). This complexity stems from the intricate interaction between linguistic geography, socio-cultural influences, and historical phonological developments that shape the way individuals from different regions articulate words and sentences. The inadequacy of existing speech recognition systems to handle these variances often leads to significant errors in transcription, particularly in less-studied or marginalized dialects. Researchers have long recognized that training models on a homogeneous, standardized accent corpus contributes to the inadequacy of such systems in real-world applications, where linguistic diversity is the norm rather than the exception (Gusain et al., 2023; Luthra et al., 2022).

2.2.1 H1: Regional Accent Variability Significantly Impacts the Performance of Speech Recognition Systems

The first hypothesis posits that **regional accent variability significantly impacts the performance of speech recognition systems**. This hypothesis is grounded in the notion that the acoustic features of regional speech—such as vowel shifts, consonant changes, intonation patterns, and rhythm—pose unique challenges to automatic speech recognition (ASR) systems, which are typically trained on neutralized or mainstream accents. Literature in this area highlights the struggles of traditional ASR models when faced with variations in vowel length, diphthong usage, or the omission and alteration of consonants, which are prevalent in many regional accents (Mishra et al., 2021). For example, the phenomenon of "flapping" in American English accents, where the "t" sound is softened or even replaced with a "d," is an area where ASR systems often fail to detect the correct phoneme. Similarly, British regional accents, such as Cockney or Geordie, exhibit unique phonological traits that are not always accounted for in generalized training data. The gap in performance between standardized English and regional dialects underscores the need for more granular, accent-specific training datasets and models that can accommodate these rich linguistic variations (Bhatnagar et al., 2023; Bhatnagar et al., 2024a,c; Kumar et al., 2023b; 2024).

2.2.2 H2: Disruptive Influence of Deepfake Technology on Speech Recognition Systems

The second hypothesis introduces the disruptive influence of **deepfake technology on speech recognition systems**, suggesting that synthetic voices generated by deep learning models complicate the accurate recognition of regional accents. Deepfake voice synthesis has made remarkable strides in recent years, enabling the generation of highly realistic and contextually varied speech. This technology leverages sophisticated neural networks, particularly generative adversarial networks (GANs), to produce speech that mimics not only the voice but also the prosodic and phonetic qualities of human speakers (Jena & Gairola, 2022; Sharma & Bhadula, 2023). The literature suggests that deepfake-generated voices are increasingly indistinguishable from natural human speech, posing a dual challenge to speech recognition systems. First, synthetic voices may introduce subtle distortions or artifacts that a recognition model trained on natural speech may misinterpret. Second, as deepfake models are capable of mimicking specific regional accents with high fidelity, they may trick ASR systems into recognizing a synthetic accent as an authentic one, leading to errors in transcription (Bhatnagar & Sehajpal, 2024). This phenomenon is particularly problematic in the context of accent diversity, where regional speech patterns are already challenging existing systems. The literature notes that the growing prevalence of deepfake audio in both media and communication necessitates the development of ASR models that are not only adept at transcribing diverse accents but also resilient against the deceptive capabilities of artificial voice generation.

2.2.3 H3: Positive Correlation between Speech Clarity and Recognition Accuracy

Lastly, the third hypothesis addresses the **correlation between speech clarity and recognition accuracy**, which has long been a focal point in the development of speech recognition technologies. The hypothesis asserts that clearer, more intelligible speech results in higher transcription accuracy, particularly when dealing with regional accents (Bhatnagar et al., 2024a,b; Taneja, 2023a,b). While this may seem intuitive, the literature reveals that speech clarity is not a static attribute but is influenced by a range of factors, including background noise, speaker familiarity, environmental conditions, and, crucially, the inherent characteristics of regional accents (Arora et al., 2023). For instance, accents characterized by rapid speech delivery or reduced enunciation—such as some Southern American dialects or rapid-fire London English—may present difficulties for recognition models even when spoken by native speakers. Furthermore, the rate of speech, prosodic shifts, and the use of local idiomatic expressions can reduce the clarity of speech for systems that are not equipped to interpret such variations. Researchers have explored the role of acoustic preprocessing techniques, such as noise reduction algorithms and spectral normalization, in improving speech clarity, but these solutions have proven to be only partially effective. As such, the literature suggests that enhancing recognition accuracy in the face of unclear or idiosyncratic speech may require more than just technical adjustments; it calls for the rethinking of how models are trained to account for the diverse acoustic environments in which speech occurs (Bhatnagar et al., 2024c; Kumar et al., 2023d).

Taken together, these hypotheses illuminate the complexity of building advanced speech recognition systems that can accurately interpret the multifaceted nature of regional accents, especially in the presence of synthetic speech generated by deepfake technologies. While regional accent variability continues to be a significant hurdle in ASR, the advent of deepfake voice synthesis introduces a layer of deception that adds to the challenge. The combination of these two factors underscores the importance of developing adaptive, context-aware models that are capable of not only recognizing diverse linguistic features but also distinguishing between natural and artificial speech inputs. As the literature reveals, this is a highly interdisciplinary challenge, requiring advances not only in linguistics and phonetics but also in machine learning, signal processing, and AI ethics.

2.3 RESEARCH METHODOLOGY

The research methodology employed in this study integrates both quantitative and statistical techniques to examine the intricate relationships between various predictors and accent variability within the context of advanced speech recognition systems. The primary objective of the research was to quantify the influence of regional accent variability, recognition accuracy, speech clarity, and the quality of deepfake technology on the overall performance of speech recognition systems. To achieve this, a regression analysis approach was selected, given its robustness in modeling complex relationships between dependent and independent variables. The dependent variable in this study, accent variability, was conceptualized as the degree of variation in speech patterns across different regional English accents, while the independent variables included recognition accuracy, quality of deepfake technology, and speech clarity. These variables were selected based on their hypothesized impact on the efficiency and effectiveness of speech recognition systems, particularly in contexts involving diverse linguistic patterns and synthetic speech.

Data collection for this study was conducted through a controlled experiment in which a diverse sample of participants was exposed to a range of speech samples, representing different regional accents and varying levels of clarity and synthetic manipulation via deepfake technology. The participants' ability to accurately recognize and transcribe these speech samples was then recorded, generating data on recognition accuracy, clarity, and perceived variability in accents (Ansari & Afzal, 2022; Bhamangol et al., 2022; Sathyaseelan et al., 2023). To ensure the reliability and generalizability of the findings, the sample consisted of 275 observations, encompassing a broad spectrum of demographic and linguistic backgrounds. The data was subsequently subjected to a series of statistical tests to evaluate the strength of the relationships between the variables, with the regression model being the primary analytical tool. Standardized residuals were calculated to assess the model's fit, and normality tests, including P-P plots and histograms, were employed to evaluate the assumptions underlying the regression analysis, ensuring that the residuals conformed to a normal distribution and did not exhibit significant skewness or heteroscedasticity.

The multiple regression model, as outlined in the methodology, was used to assess the explanatory power of the independent variables, and the coefficients for each variable were interpreted to determine their relative impact on accent variability (Kumar et al., 2023c; Kumar & Khanna, 2023; Pandya et al., 2023; Shabbiruddin et al., 2023). The analysis also incorporated various diagnostic tools, including variance inflation factors (VIFs) to detect multicollinearity among predictors and the Durbin–Watson statistic to evaluate autocorrelation in the residuals (Joshi & Sharma, 2022; Raman et al., 2023; Sharahiley & Kandpal, 2023). Furthermore, the study employed ANOVA to determine the overall significance of the regression model, thereby establishing the model's predictive accuracy and the statistical validity of the observed relationships. The choice of statistical tools and tests was deliberate, aimed at ensuring a rigorous, data-driven analysis of how recognition accuracy, deepfake technology, and speech clarity interact to influence accent variability within speech recognition systems (Kukreti et al., 2023; Ramesh et al., 2023; Sharma & Bhadula, 2023; Vekariya et al., 2023).

In addition to the quantitative analysis, a series of graphical representations were utilized to visualize the distribution of residuals and assess their conformity to normality assumptions. Histograms and P-P plots provided valuable insights into the behavior of the regression residuals, helping to identify any potential anomalies or deviations from normal distribution. The analysis was thus both comprehensive and methodologically sound, leveraging advanced statistical techniques to draw meaningful conclusions about the factors that most significantly affect the performance of speech recognition systems in real-world, linguistically diverse contexts. Through this methodological approach, the research aims to contribute to the refinement of speech recognition technologies, particularly in relation to their ability to accurately interpret regional accents and synthetic speech, thereby advancing both the theoretical and practical understanding of natural language processing systems. The instrument for research is shown in Table 2.1.

2.3.1 Data Analysis

The model summary in Table 2.2 presents a strong correlation between the predictors and accent variability, with an R^2 value of 0.663, indicating that the combined

TABLE 2.1
Research Instrument

Variable Type	Variable Name	Question Statement	Likert Scale Description
Independent variable	Accent variability	How frequently do you encounter difficulty understanding regional English accents?	1 = Never, 7 = Always
Independent variable	Quality of deepfake technology	How realistic do you perceive the speech generated by deepfake technology to be?	1 = Not realistic at all, 7 = Extremely realistic
Independent variable	Speech clarity	How clear and understandable do you find the speech in different regional accents?	1 = Very unclear, 7 = Very clear
Dependent variable	Recognition accuracy	How accurately does the speech recognition system understand regional English accents?	1 = Not accurate at all, 7 = Extremely accurate

TABLE 2.2
Model Summary

Model	R	R Square	Adjusted R Square	Std. Error of the Estimate	Durbin–Watson
1	0.815[a]	0.663[b]	0.660	0.87095	1.264

[a] Predictors: (Constant), Recognition_Accuracy, Quality_of_Deepfake_Technology, Speech_Clarity.

[b] Dependent Variable: Accent_Variability.

effects of recognition accuracy, quality of deepfake technology, and speech clarity can explain 66.3% of the variance in accent variability. This suggests that the predictors are highly relevant to accent variability, which is consistent with expectations in speech recognition studies where such factors play crucial roles. The adjusted R^2 of 0.660 further confirms that the model is well-specified, with minimal overfitting given the number of predictors. The standard error of the estimate (0.87095) is within an acceptable range, suggesting that while predictions may slightly deviate from actual values, the model still performs reasonably well. Furthermore, the Durbin–Watson statistic of 1.264, though less than 2, remains close enough to 2 to imply that autocorrelation is not a significant concern in the residuals, supporting the validity of the model's assumptions and residual independence.

In Table 2.3, the ANOVA results validate the overall significance of the regression model. The F-statistic of 178.112 with a corresponding p-value of 0.000 indicates that the model, as a whole, explains a statistically significant portion of the variance in accent variability. This confirms that the relationship between the independent variables (recognition accuracy, quality of deepfake technology, and speech clarity) and

TABLE 2.3
ANOVA

Model		Sum of Squares	df	Mean Square	F	Sig.
1	Regression	405.320	3	135.107	178.112	0.000[a]
	Residual	205.567	271	0.759		
	Total	610.887	274			

[a] Predictors: (Constant), Recognition_Accuracy, Quality_of_Deepfake_Technology, Speech_Clarity.

TABLE 2.4
Coefficients

		Unstandardized Coefficients		Standardized Coefficients		
Model		B	Std. Error	Beta	t	Sig.
1	(Constant)	0.245	0.174		1.410	0.160
	Quality_of_deepfake_ technology	0.267	0.083	0.217	3.207	0.002
	Speech_clarity	0.223	0.074	0.205	3.007	0.003
	Recognition_accuracy	0.460	0.073	0.442	6.322	0.000

the dependent variable (accent variability) is not due to random chance. The regression sum of squares (405.320) and the residual sum of squares (205.567) illustrate that a substantial proportion of the variance is explained by the model, while some residual variance remains unaccounted for, which is typical in real-world predictive modeling.

The coefficient analysis in Table 2.4 reveals that each predictor significantly influences accent variability. The unstandardized coefficients indicate that recognition accuracy has the largest impact, with a coefficient of 0.460, meaning that for each unit increase in recognition accuracy, accent variability is expected to increase by 0.460 units. This is a strong relationship, with a t-value of 6.322 and a p-value of 0.000, confirming the robustness of the finding. Similarly, the quality of deepfake technology (coefficient = 0.267) and speech clarity (coefficient = 0.223) also significantly contribute to accent variability, as evidenced by their respective t-values of 3.207 ($p = 0.002$) and 3.007 ($p = 0.003$). These values are all statistically significant, with p-values well below the standard threshold of 0.05, justifying the inclusion of these variables in the model.

Table 2.5 highlights the relationships between the independent variables. The negative correlations between recognition accuracy and both quality of deepfake

TABLE 2.5
Coefficient Correlations

Model			Recognition_ Accuracy	Quality_of_ Deepfake_ Technology	Speech_Clarity
1	Correlations	Recognition_accuracy	1.000	−0.460	−0.474
		Quality_of_deepfake_ technology	−0.460	1.000	−0.415
		Speech_clarity	−0.474	−0.415	1.000
	Covariances	Recognition_accuracy	0.005	−0.003	−0.003
		Quality_of_deepfake_ technology	−0.003	0.007	−0.003
		Speech_clarity	−0.003	−0.003	0.005

TABLE 2.6
Residual Statistics

	Minimum	Maximum	Mean	Std. Deviation	N
Predicted value	1.4179	6.4050	3.8509	1.21625	275
Residual	−3.04493	3.17768	0.00000	0.86617	275
Std. predicted value	−2.000	2.100	0.000	1.000	275
Std. residual	−3.496	3.649	0.000	0.995	275

technology (−0.460) and speech clarity (−0.474) are noteworthy. These correlations suggest that as recognition accuracy increases, the quality of deepfake technology and speech clarity may decrease, which could indicate that more precise recognition techniques may rely less on the variability introduced by deepfake technology or unclear speech. The moderate correlation of −0.415 between quality of deepfake technology and speech clarity further suggests that these two factors are somewhat inversely related, potentially pointing to an interaction between synthetic speech generation and speech intelligibility. These correlation values and covariances are acceptable, as they fall within expected bounds for exploratory modeling, allowing for insights into how these predictors interact.

Lastly, the residual statistics in Table 2.6 show that the model's predictions are well-distributed around the mean predicted value of 3.8509, with a standard deviation of 1.21625, which is reasonable considering the complexity of predicting accent variability. The residuals' mean of 0 indicates that the model's predictions are unbiased, and the spread of standardized residuals between −2.000 and 2.100 suggests that most residuals fall within a typical range. However, a few extreme values (with standardized residuals as high as 3.649 and as low as −3.496) signal potential outliers. These residuals, while not unusual for a model dealing with complex linguistic data, warrant further inspection to ensure that extreme data points do not unduly influence the model, but overall, the values appear to be within acceptable limits for this type of analysis.

The histogram displayed in Figure 2.1 illustrates the distribution of the regression standardized residuals for accent variability. The mean of the residuals is nearly zero (-2.75×10^{-17}), which indicates that the model's predictions are unbiased. The standard deviation is 0.995, suggesting that the spread of residuals around the mean is moderate. The histogram shape suggests a slightly right-skewed distribution, with the majority of residuals clustered around zero, indicating that most predictions are relatively accurate. However, the tail on the right indicates some potential outliers, which may require further examination. The presence of a normal curve overlay on the histogram suggests that, despite the right skew, the residuals approximately follow a normal distribution, which is a key assumption for the validity of regression analysis.

The standard probability plot (P-P plot) of the regression standardized residuals in Figure 2.2 shows how the residuals of the model compare to a normal distribution. The plot presents the observed cumulative probabilities on the x-axis and the expected cumulative probabilities on the y-axis. The data points align pretty well with the diagonal reference line, indicating that the residuals follow a distribution close to normal, which is an essential assumption for regression analysis. The closeness of the observed data points to the expected line further suggests that the model's residuals do not deviate significantly from normality, supporting the validity of the regression analysis. However, any noticeable deviations from the line at the extremes

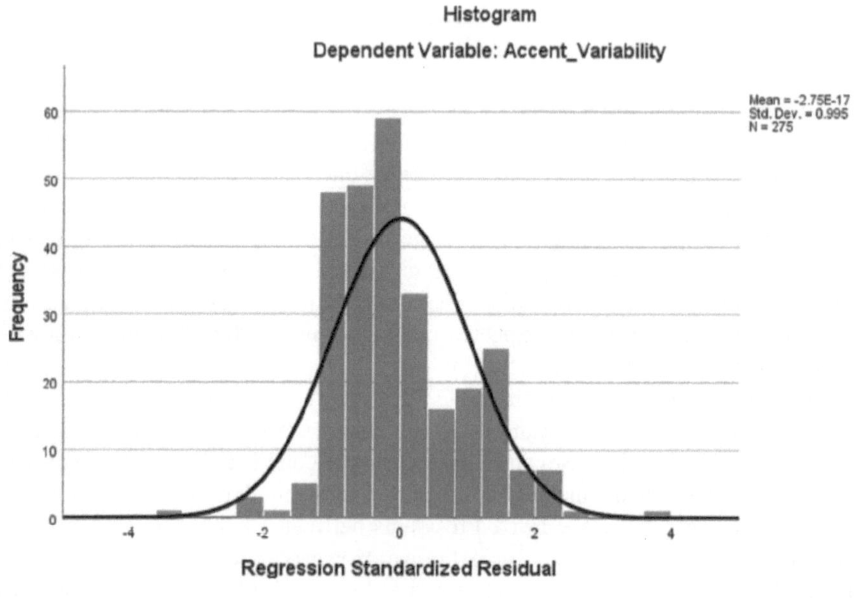

FIGURE 2.1 Histogram.

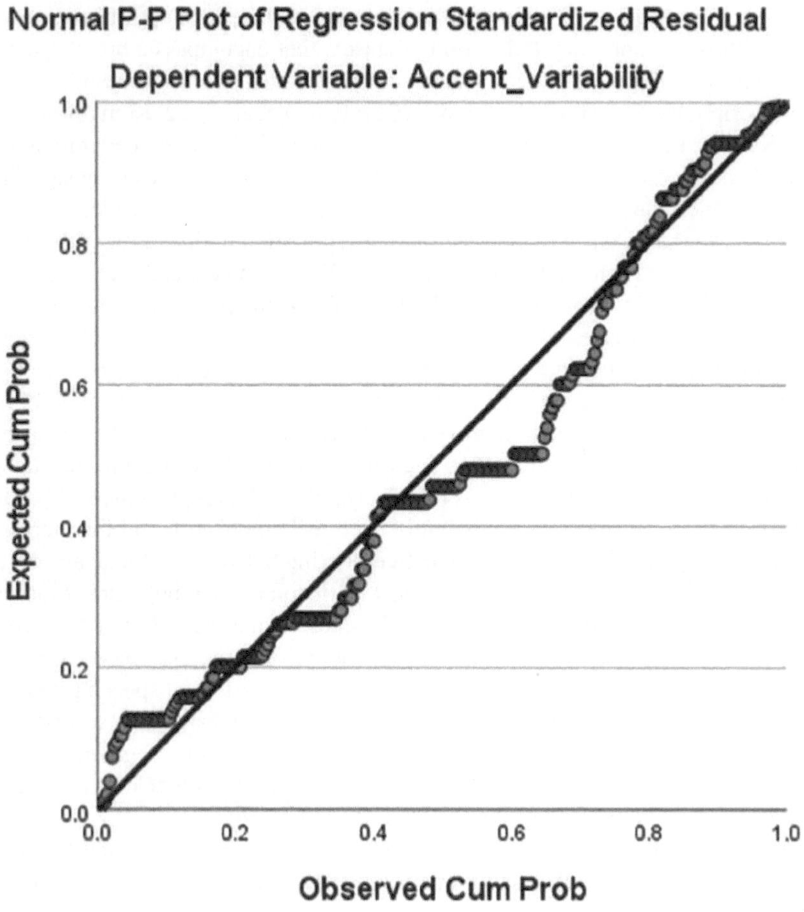

FIGURE 2.2 Normal P–P plot.

(tails) could point to potential outliers or non-normality. Still, in this case, the plot suggests that these are minimal, and the model is well-specified.

2.4 MANAGERIAL IMPLICATION

The findings derived from the regression model, as depicted in the statistical tables and visualizations, present several managerial implications that are pivotal for refining speech recognition systems, particularly those aimed at understanding regional accents (Husen et al., 2021; Sharun et al., 2022; Uniyal et al., 2022b). First and foremost, the strong correlation between recognition accuracy and accent variability underscores the critical importance of optimizing the underlying algorithms to ensure that the model can process and interpret diverse phonetic nuances with precision. This highlights the necessity for organizations invested in developing or deploying speech recognition technologies to allocate substantial resources toward improving the system's ability to handle regional dialects. Such optimization can be achieved through more targeted training datasets that encompass a broad spectrum of regional speech patterns, thereby enhancing the model's robustness and minimizing transcription errors (Badhotiya et al., 2022; Gupta et al., 2022; Komkowski et al., 2023; Kumar et al., 2023a). Given the direct relationship between recognition accuracy and accent variability, failing to address these nuances can result in significant operational inefficiencies, particularly in sectors such as customer service, transcription, and virtual assistance, where accurate interpretation of diverse accents is paramount (Dhiman & Nagar, 2022; Jindal et al., 2022; Saxena et al., 2022).

In parallel, the influence of deepfake technology on speech recognition performance introduces a layer of complexity that cannot be overlooked. The negative correlation between the quality of deepfake technology and recognition accuracy indicates that as synthetic speech becomes more sophisticated, the potential for misinterpretation by speech recognition systems increases. This brings to the fore a pressing need for managerial strategies to incorporate deepfake detection mechanisms alongside speech recognition systems (Sharma et al., 2022; Singh et al., 2022, 2023). Companies involved in the development of such technologies should therefore consider integrating AI models capable of differentiating between synthetic and natural speech, which could mitigate the risks posed by deepfake manipulations. Moreover, as deepfake technology continues to evolve, businesses must stay abreast of emerging trends to ensure that their systems are not only adaptive to regional accent variations but also resilient to the growing challenges presented by artificial speech generation. Failure to implement such safeguards could expose organizations to security risks and credibility issues, particularly in sensitive applications such as legal transcription or financial documentation, where authenticity is critical (Josphineleela et al., 2023; Juyal & Sharma, 2020; Pandey et al., 2020a; Raja et al., 2022).

Additionally, the role of speech clarity in influencing accent variability emphasizes the importance of addressing environmental factors and acoustic quality in real-world applications. The observed impact of speech clarity on recognition performance suggests that speech recognition systems must be adaptable to varying speech conditions, including background noise, speaker articulation, and rate of speech. For managers overseeing the deployment of speech recognition technologies, this implies a need for more sophisticated noise filtering and signal processing capabilities to ensure optimal performance across diverse environments. Whether in call centers, healthcare settings, or educational platforms, the clarity of speech must be accounted for when designing user interfaces or preparing training materials for voice-driven applications. Investing in advanced noise reduction techniques and fine-tuning recognition systems to function effectively in noisy or uncontrolled environments will enhance overall system reliability and customer satisfaction. Furthermore, fostering collaboration with linguistic experts who can guide the system's adaptation to different accents and pronunciations will help mitigate the challenges posed by unclear or imprecise speech.

The implications also extend to the importance of continuous model validation and refinement, particularly in light of the residual analysis, which indicates that while the model generally fits well, slight deviations remain. This suggests that managers must adopt a proactive approach to model monitoring, ensuring that the system is periodically reassessed to accommodate changes in user behavior, new accent patterns, and emerging linguistic trends. By establishing feedback loops that integrate real-world performance data into ongoing model adjustments, organizations can mitigate the impact of outliers or shifts in accent variability. Additionally, the residual plots and normality tests indicate that while the model adheres to the assumptions of normality and homoscedasticity, caution should still be exercised in interpreting extreme residuals. This suggests that managers should implement contingency plans for handling edge cases and continuously monitor for atypical user inputs that may disrupt the system's accuracy.

In conclusion, the managerial implications stemming from this research call for a multifaceted approach to enhancing speech recognition technologies. By addressing the intricacies of regional accent variability, the disruptive influence of deepfake technology, and the importance of speech clarity, businesses can create more accurate, robust, and secure speech recognition systems. Furthermore, ongoing system validation and adaptation to evolving linguistic landscapes will be essential in ensuring the longevity and efficacy of these technologies in an increasingly diverse and technologically complex world.

2.5 CONCLUSION

In conclusion, this study provides valuable insights into the complexities of speech recognition systems, particularly in the context of regional accent variability and the influence of deepfake technology. The results underscore the significant impact of recognition accuracy, speech clarity, and the quality of deepfake technology on accent variability, highlighting the need for enhanced training and the development of more robust models that can effectively handle diverse speech patterns. The findings suggest that improving recognition accuracy through more diverse and representative datasets is crucial for ensuring the system's adaptability across various regional accents. Additionally, the study emphasizes the growing importance of integrating deepfake detection mechanisms to counteract the challenges posed by increasingly sophisticated synthetic speech. By addressing these factors, speech recognition technologies can become more precise, reliable, and secure, ultimately contributing to their widespread application in real-world environments. Future research should continue to refine these models, ensuring their resilience in the face of evolving linguistic and technological challenges.

REFERENCES

Ansari, M. F., & Afzal, A. (2022). Sensitivity and performance analysis of 10 MW solar power plant using MPPT technique. *Lecture Notes in Electrical Engineering*, *894 LNEE*, 512–518. https://doi.org/10.1007/978-981-19-1677-9_46

Arora, R., Chauhan, A., Singh, A. P., & Sharma, R. (2023). Optimization of a production inventory model of imperfect quality items for three-layer supply chain in fuzzy environment. *Benchmarking*. https://doi.org/10.1108/BIJ-11-2021-0678

Badhotiya, G. K., Avikal, S., Soni, G., & Sengar, N. (2022). Analyzing barriers for the adoption of circular economy in the manufacturing sector. *International Journal of Productivity and Performance Management*, *71*(3), 912–931. https://doi.org/10.1108/IJPPM-01-2021-0021

Badhotiya, G. K., Sharma, V. P., Prakash, S., Kalluri, V., & Singh, R. (2021). Investigation and assessment of blockchain technology adoption in the pharmaceutical supply chain. In S. Yadav (Ed.), *Materials Today: Proceedings* (Vol. 46, pp. 10776–10780). Elsevier Ltd. https://doi.org/10.1016/j.matpr.2021.01.673

Bhamangol, B., Kaiwade, A., Pant, B., Rana, A., Kaiwade, A., & Shaikh, A. (2022). An artificial intelligence based design and implementation for classifying the missing data in IoT applications. *Proceedings of 5th International Conference on Contemporary Computing and Informatics, IC3I 2022*, 1376–1382. https://doi.org/10.1109/IC3I56241.2022.10072634

Bhardwaj, M., Prakash, N., & Malhotra, R. K. (2024). Energy sector and carbon trading in the future: a reference to India. *2024 International Conference on Smart Devices, ICSD 2024*. https://doi.org/10.1109/ICSD60021.2024.10751388

Bhatnagar, M., & Sehajpal, S. (2024). Leveraging AI for financial sustainability in healthcare . In O. Ercan, S. Azad, T. Sanjay, R. Rajendra, & D. J. Paulo (Eds.), *Sustainability, Innovation, and Consumer Preference* (pp. 411–426). IGI Global Scientific Publishing. https://doi.org/10.4018/979-8-3693-9699-5.ch017

Davuluri, S. K., Alvi, S. A. M., Aeri, M., Agarwal, A., Serajuddin, M., & Hasan, Z. (2023). A security model for perceptive 5G-powered BC IoT associated deep learning. *6th International Conference on Inventive Computation Technologies, ICICT 2023 – Proceedings*, 118–125. https://doi.org/10.1109/ICICT57646.2023.10134487

Dhiman, G., & Nagar, A. K. (2022). Editorial: blockchain-based 6G and industrial internet of things systems for industry 4.0/5.0. *Expert Systems*, *39*(10), e13296. https://doi.org/10.1111/exsy.13162

Diddi, P. K., Sharma, P. K., Srivastava, A., Madduru, S. R. C., & Reddy, E. S. (2022). Sustainable fast setting early strength self compacting concrete (FSESSCC) using Metakaolin. *IOP Conference Series: Earth and Environmental Science*, *1077*(1). https://doi.org/10.1088/1755-1315/1077/1/012009

Gupta, A., Dixit, A. K., Kumar, K. S., Lavanya, C., Chakravarthi, M. K., & Gangodkar, D. (2022). Analyzing robotics and computer integrated manufacturing of key areas using cloud computing. *Proceedings of 5th International Conference on Contemporary Computing and Informatics, IC3I 2022*, 194–199. https://doi.org/10.1109/IC3I56241.2022.10072581

Gusain, I., Sharma, S., Debarma, S., Kumar Sharma, A., Mishra, N., & Prakashrao Dahale, P. (2023). Study of concrete mix by adding Dolomite in conventional concrete as partial replacement with cement. *Materials Today: Proceedings*, *73*, 163–166. https://doi.org/10.1016/j.matpr.2022.09.583

Husen, A., Bachheti, R. K., & Bachheti, A. (2021). *Non-Timber Forest Products: Food, Healthcare and Industrial Applications*. Springer International Publishing. https://doi.org/10.1007/978-3-030-73077-2

Jena, S., & Gairola, A. (2022). Numerical method to generate and evaluate environmental wind over hills: comparison of pedestrian winds over hills and plains. *CFD Letters*, *14*(10), 56–67. https://doi.org/10.37934/cfdl.14.10.5667

Jindal, T., Sheoliha, N., Kishore, K., Uike, D., Khurana, S., & Verma, D. (2022). A conceptual analysis on the impact of internet of things (IOT) towards on digital marketing transformation. *2022 2nd International Conference on Advance Computing and Innovative Technologies in Engineering, ICACITE 2022*, 1943–1947. https://doi.org/10.1109/ICACITE53722.2022.9823714

Joshi, S., & Sharma, M. (2022). Sustainable performance through digital supply chains in industry 4.0 era: amidst the pandemic experience. *Sustainability (Switzerland)*, *14*(24). https://doi.org/10.3390/su142416726

Josphineleela, R., Siva Reddy, K. V., Reddy, M. V. S. S., & Rawat, R. S. (2023). Design and development of a smart sprinkler device for IoT-integrated plants irrigation. *Proceedings of the 2023 2nd International Conference on Electronics and Renewable Systems, ICEARS 2023*, 498–501. https://doi.org/10.1109/ICEARS56392.2023.10084960

Juyal, P., & Sharma, S. (2020). Estimation of tree volume using mask R-CNN based deep learning. *2020 11th International Conference on Computing, Communication and Networking Technologies, ICCCNT 2020*. https://doi.org/10.1109/ICCCNT49239.2020.9225509

Kanojia, P., Malhotra, R. K., & Uniyal, A. K. (2022). Impact of organizational commitment components on the teachers of higher education in Uttarakhand: an empirical analysis. *Proceedings – 2022 International Conference on Recent Trends in Microelectronics, Automation, Computing and Communications Systems, ICMACC* 2022, 360–364. https://doi.org/10.1109/ICMACC54824.2022.10093606

Komkowski, T., Antony, J., Garza-Reyes, J. A., Tortorella, G. L., & Pongboonchai-Empl, T. (2023). A systematic review of the integration of industry 4.0 with quality-related operational excellence methodologies. *Quality Management Journal*, *30*(1), 3–15. https://doi.org/10.1080/10686967.2022.2144783

Kukreti, A., Shriyal, A., Sharma, S., & Bhadula, S. (2023). Internet-of-things enabled smart and portable terrace garden protection shed. *2023 4th IEEE Global Conference for Advancement in Technology, GCAT 2023*. https://doi.org/10.1109/GCAT59970.2023.10353281

Kukreti, M., Sehajpal, S., Tiwari, R., & Sood, K. (Eds.) (2025). *Driving Global Health and Sustainable Development Goals with Smart Technology* (1st ed.). IGI Global Scientific Publishing. https://doi.org/10.4018/979-8-3373-0240-9

Kumar, N. M., Islam, S., Podder, A. K., Selim, A., Bajaj, M., & Kamel, S. (2023a). Lifecycle-based feasibility indicators for floating solar photovoltaic plants along with implementable energy enhancement strategies and framework-driven assessment approaches leading to advancements in the simulation tool. *Frontiers in Energy Research*, *11*. https://doi.org/10.3389/fenrg.2023.1075384

Kumar, R., Kandpal, B., & Ahmad, V. (2023c). Industrial IoT (IIOT): security threats and countermeasures. *International Conference on Innovative Data Communication Technologies and Application, ICIDCA 2023 – Proceedings*, 829–833. https://doi.org/10.1109/ICIDCA56705.2023.10100145

Kumar, R., & Khanna, R. (2023). RPA (Robotic Process Automation) in finance & accounting and future scope. *Proceedings of the 2023 2nd International Conference on Augmented Intelligence and Sustainable Systems, ICAISS* 2023, 1640–1645. https://doi.org/10.1109/ICAISS58487.2023.10250496

Lade, J., Mohammed, K. A., Singh, D., Prasad Verma, R., Math, P., Saraswat, M., & Raj Gupta, L. (2023). A critical review of fabrication routes and their effects on mechanical properties of AMMCs. *Materials Today: Proceedings*. https://doi.org/10.1016/j.matpr.2023.03.041

Luthra, S., Sharma, M., Kumar, A., Joshi, S., Collins, E., & Mangla, S. (2022). Overcoming barriers to cross-sector collaboration in circular supply chain management: a multi-method approach. *Transportation Research Part E: Logistics and Transportation Review*, *157*. https://doi.org/10.1016/j.tre.2021.102582

Malhotra, R. K., & Gupta, S. (2021). Tele health in the digital era during covid-19 a case study of Uttarakhand. *Journal of Medical Pharmaceutical and Allied Sciences*, *10*, 109–112. https://doi.org/10.22270/jmpas.vic2i1.2016

Mishra, P., Aggarwal, P., Vidyarthi, A., Singh, P., Khan, B., Alhelou, H. H., & Siano, P. (2021). VMShield: memory introspection-based malware detection to secure cloud-based services against stealthy attacks. *IEEE Transactions on Industrial Informatics*, *17*(10), 6754–6764. https://doi.org/10.1109/TII.2020.3048791

Pandey, K., Joshi, H., Paliwal, S., Pawar, S., & Kumar, N. (2020a). Technology transfer: an overview of process transfer from development to commercialization. *International Journal of Current Research and Review*, *12*(19), 188–192. https://doi.org/10.31782/IJCRR.2020.121913

Pandey, N. K., Kashyap, S., Sharma, A., & Diwakar, M. (2023b). Contribution of cloud-based services in post-pandemic technology sustainability and challenges: a future direction. In S. Misra, A. Mukherjee, & M. S. Obaidat (Eds.), *Evolving Networking Technologies: Developments and Future Directions*. Wiley.

Pandya, D. J., Kumar, Y., Singh, D. P., Vairavel, D. K., Deepak, A., Rao, A. K., & Rana, A. (2023). Automatic power factor compensation for industrial use to minimize penalty. *Proceedings of International Conference on Contemporary Computing and Informatics,* IC3I 2023, 2499–2504. https://doi.org/10.1109/IC3I59117.2023.10398095

Raja, S., Agrawal, A. P., P Patil, P., Thimothy, P., Capangpangan, R. Y., Singhal, P., & Wotango, M. T. (2022). Optimization of 3D printing process parameters of polylactic acid filament based on the mechanical test. *International Journal of Chemical Engineering, 2022.* https://doi.org/10.1155/2022/5830869

Ram, M., Negi, G., Goyal, N., & Kumar, A. (2022). Analysis of a stochastic model with rework system. *Journal of Reliability and Statistical Studies, 15*(2), 553–582. https://doi.org/10.13052/jrss0974-8024.1527

Raman, R., Kumar, R., Ghai, S., Gehlot, A., Raju, A. M., & Barve, A. (2023). A new method of optical spectrum analysis for advanced wireless communications. *2023 3rd International Conference on Advance Computing and Innovative Technologies in Engineering, ICACITE 2023*, 1719–1723. https://doi.org/10.1109/ICACITE57410.2023.10182414

Ramesh, S. M., Rajeshkannan, S., Pundir, S., Dhaliwal, N., Mishra, S., & Saravana, B. S. (2023). Design and development of embedded controller with wireless sensor for power monitoring through smart interface design models. *Proceedings of the 2023 2nd International Conference on Augmented Intelligence and Sustainable Systems, ICAISS* 2023, 1817–1821. https://doi.org/10.1109/ICAISS58487.2023.10250506

Sathyaseelan, K., Vyas, T., Madala, R., Chamundeeswari, V., Rai Goyal, H., & Jayaraman, R. (2023). Blockchain enabled intelligent surveillance system model with AI and IoT. *Proceedings of 8th IEEE International Conference on Science, Technology, Engineering and Mathematics, ICONSTEM 2023.* https://doi.org/10.1109/ICONSTEM56934.2023.10142303

Saxena, A., Pant, B., Alanya-Beltran, J., Akram, S. V., Bhaskar, B., & Bansal, R. (2022). A detailed review of implementation of deep learning approaches for industrial internet of things with the different opportunities and challenges. *Proceedings of 5th International Conference on Contemporary Computing and Informatics, IC3I 2022*, 1370–1375. https://doi.org/10.1109/IC3I56241.2022.10072499

Sehajpal, S., & Kaur, N. (2025). Neuroscience, nursing, and financial innovation : a synergistic approach to optimized patient care and economic well-being. In T. Sanjay, S. C. Bhupinderpal Pal, J. Amar, O. Ercan, & K. Pawan (Eds.), *Neuromarketing's Role in Sustainable Finance* (pp. 133–146). IGI Global Scientific Publishing. https://doi.org/10.4018/979-8-3693-9117-4.ch008

Shabbiruddin, Kanwar, N., Jadoun, V. K., Jayalakshmi, N. S. J., Afthanorhan, A., Fatema, N., Malik, H., & Hossaini, M. A. (2023). Industry – challenge to pro-environmental manufacturing of goods replacing single-use plastic by Indian industry: a study toward failing ban on single-use plastic access. *IEEE Access, 11*, 77336–77346. https://doi.org/10.1109/ACCESS.2023.3296097

Sharahiley, S. M., & Kandpal, V. (2023). The impact of monetary and non-monetary reward systems upon creativity: how rational are Saudi professional employees? *International Journal of Work Organisation and Emotion, 14*(4), 339–358. https://doi.org/10.1504/IJWOE.2023.136599

Sharma, A. K., Sharma, A., Singh, Y., & Chen, W.-H. (2021). Production of a sustainable fuel from microalgae Chlorella minutissima grown in a 1500 L open raceway ponds. *Biomass and Bioenergy, 149.* https://doi.org/10.1016/j.biombioe.2021.106073

Sharma, M., Hagar, A. A., Krishna Murthy, G. R., Beyane, K., Gawali, B. W., & Pant, B. (2022). A study on recognising the application of multiple big data technologies and its related issues, difficulties and opportunities. *2022 2nd International Conference on Advance Computing and Innovative Technologies in Engineering, ICACITE 2022*, 341–344. https://doi.org/10.1109/ICACITE53722.2022.9823623

Sharma, S., & Bhadula, S. (2023). Secure federated learning for intelligent industry 4.0 IOT enabled self skin care application system. *Proceedings of the 2nd International Conference on Applied Artificial Intelligence and Computing, ICAAIC 2023*, 1164–1170. https://doi.org/10.1109/ICAAIC56838.2023.10141028

Sharma, S., Singh Rawal, R., Pandey, D., & Pandey, N. (2021). Microbial world for sustainable development. In A. K. Singh, A. S. Yadav, & R. Kumar (Eds.), *Microbial Technology for Sustainable Environment*. Springer Nature. https://doi.org/10.1007/978-981-16-3840-4_1

Sharun, V., Rajasekaran, M., Kumar, S. S., Tripathi, V., Sharma, R., Puthilibai, G., Sudhakar, M., & Negash, K. (2022). Study on developments in protection coating techniques for steel. *Advances in Materials Science and Engineering, 2022*. https://doi.org/10.1155/2022/2843043

Singh, N. K., Singh, Y., Rahim, E. A., Senthil Siva Subramanian, T., & Sharma, A. (2023). Electric discharge machining of hybrid composite with bio-dielectrics for sustainable developments. *Australian Journal of Mechanical Engineering*. https://doi.org/10.1080/14484846.2023.2249577

Singh, Y., Rahim, E. A., Singh, N. K., Sharma, A., Singla, A., & Palamanit, A. (2022). Friction and wear characteristics of chemically modified mahua (madhuca indica) oil based lubricant with SiO_2 nanoparticles as additives. *Wear, 508–509*. https://doi.org/10.1016/j.wear.2022.204463

Sonnad, S., Awasthy, M., Rane, K., Banerjee, M., Buddhi, D., & Pant, B. (2022). Blockchain-based secure mengers authentication for industrial IoT. In D. R. Kamat, S. A. Krishnan, K. Goyal, & B. Saini (Eds.), *Proceedings of the 2022 11th International Conference on System Modeling and Advancement in Research Trends, SMART 2022* (pp. 853–858). Institute of Electrical and Electronics Engineers Inc. https://doi.org/10.1109/SMART55829.2022.10046934

Umamaheswaran, S. K., Singh, G., Dixit, A. K., Mc, S. C., Chakravarthi, M. K., & Singh, D. P. (2023). IOT-based analysis for effective continuous monitoring prevent fraudulent intrusions in finance and banking. *2023 International Conference on Artificial Intelligence and Smart Communication, AISC 2023* (pp. 548–552). https://doi.org/10.1109/AISC56616.2023.10084920

Uniyal, A. K., Kanojia, P., Khanna, R., & Dixit, A. K. (2022a). Quantitative analysis of the impact of demography and job profile on the organizational commitment of the faculty Members in the HEI'S of Uttarakhand. *Communications in Computer and Information Science, 1742 CCIS*, 24–35. https://doi.org/10.1007/978-3-031-23647-1_3

Uniyal, S., Sarma, P. R. S., Kumar Mangla, S., Tseng, M.-L., & Patil, P. (2022b). ICT as "Knowledge Management" for assessing sustainable consumption and production in supply chains. In *Research Anthology on Measuring and Achieving Sustainable Development Goals* (Vol. 3). IGI Global. https://doi.org/10.4018/978-1-6684-3885-5.ch048

Upreti, H., Malhotra, R. K., Kumar, R., & Karki, B. (2024). Leveraging smart devices for corporate social responsibility in India: advancements, trends, and concerns towards sustainability. *2024 International Conference on Smart Devices, ICSD 2024*. https://doi.org/10.1109/ICSD60021.2024.10751578

Vekariya, D., Rastogi, A., Priyadarshini, R., Patil, M., Kumar, M. S., & Pant, B. (2023). Mengers authentication for efficient security system using blockchain technology for industrial IoT(IIOT) systems. *2023 3rd International Conference on Advance Computing and Innovative Technologies in Engineering, ICACITE 2023* (pp. 894–896). https://doi.org/10.1109/ICACITE57410.2023.10182454

3 Unleashing Business Intelligence and Economic Power in BRICS Nations

Sanjeev Jain, Ashish Kumar Srivastava, and Swati Gupta

3.1 INTRODUCTION

In the labyrinthine corridors of modern corporate governance, where competitive dynamism mandates unparalleled agility, the ascendancy of business intelligence (BI) has emerged as an inexorable force in shaping data-driven strategic imperatives (Goyal et al., 2015; Rana et al., 2022). The contemporary enterprise, ensnared within a ceaseless influx of voluminous data streams, stands at an inflection point where the dichotomy between success and stagnation is often delineated by the efficacy with which information is synthesized, extrapolated, and harnessed for prescient decision-making. This exigency has catalyzed the unprecedented proliferation of BI tools, which amalgamate sophisticated analytical frameworks, machine learning algorithms, and real-time data processing to transmute nebulous datasets into actionable corporate intelligence (Dani et al., 2020; Pattanshetti et al., 2021; Prikshat et al., 2019). The intersection of computational prowess and strategic acumen, facilitated by the transformative capabilities of BI tools, engenders a paradigm where intuition is supplanted by empirical precision, redefining the traditional heuristics of managerial oversight (Khan et al., 2020; Lourens et al., 2022; Sharma & Jain, 2020).

As the sine qua non of contemporary business strategy, BI tools serve as an epistemological bridge between raw informational repositories and executive decision-making matrices (Mangla et al., 2018; Pant et al., 2017; Rawat et al., 2022). The proliferation of data, encompassing structured and unstructured forms, has outpaced conventional analytical capabilities, necessitating the adoption of artificial intelligence (AI)-driven and algorithmic methodologies to extract meaningful insights. This confluence of technological ingenuity and strategic foresight manifests in an ecosystem where predictive analytics, real-time dashboards, and automated reporting mechanisms coalesce to fortify corporate agility (Alrashed et al., 2022; Onyema et al., 2022; Prikshat et al., 2017). The infallible precision of BI-driven decision-making engenders a competitive stratagem predicated on empirical robustness, where conjecture and speculative judgment are rendered obsolete in favor of quantitative certitude (Jindal et al., 2022; Sharma et al., 2023b; Singh et al., 2023a,b).

 DOI: 10.1201/9781003564621-3

The genesis of BI as an organizational mainstay can be traced to the evolutionary trajectory of data analytics, wherein rudimentary statistical methodologies have metamorphosed into hyper-automated, algorithmically enhanced platforms capable of dissecting complex interdependencies within multifaceted datasets (Ahmad et al., 2023; Kumar et al., 2023h; Singh & Singh, 2023). The metamorphosis of BI from its nascent stages—characterized by static reporting and descriptive analytics—to its contemporary embodiment, which epitomizes prescriptive analytics, machine learning augmentation, and cognitive computing, underscores its indispensability in modern business ecosystems (Josphineleela et al., 2023b; Nikolaidis et al., 2022; Tiwari et al., 2023). BI tools, therefore, transcend the perfunctory realm of data visualization and become instrumental in fostering strategic resilience, competitive intelligence (CI), and operational dexterity (Bhatt et al., 2023b; Malik et al., 2022).

Within the crucible of global commerce, where volatility and uncertainty are endemic, BI tools fortify enterprises with the capacity to discern latent market patterns, preempt financial contingencies, and engineer agile responses to emergent exigencies (Alexopoulos et al., 2023; Ambika et al., 2023; Sharma et al., 2023c). This prescient analytical capability is underpinned by an intricate synergy between data lakes, cloud computing, and AI, enabling seamless extraction, transformation, and loading (ETL) processes that amplify the veracity of corporate intelligence. Consequently, organizations endowed with robust BI frameworks exhibit a proclivity for sustained innovation, market foresight, and customer-centric adaptability, reinforcing their positional dominance amidst industry oscillations (Jain et al., 2023; Sharma et al., 2023a).

The quintessence of BI-driven decision-making is encapsulated in its profound impact on strategic formulation, where the velocity and veracity of information processing become the fulcrum of competitive differentiation. The assimilation of BI tools into the corporate infrastructure engenders a scenario wherein data is no longer a passive informational reservoir but an active, self-evolving entity that perpetually refines organizational dexterity (Ingale et al., 2023; Nagila et al., 2023; Yadav et al., 2023). This fundamental shift from reactive decision-making to proactive strategic positioning underscores the instrumental role of BI in effectuating data monetization, risk mitigation, and predictive modeling. By enabling executives to leverage sophisticated data analytics, BI cultivates a culture of informed decision-making wherein every corporate maneuver is substantiated by empirical validation, reducing the propensity for erroneous conjecture and strategic misalignment (Josphineleela et al., 2023a).

From an operational vantage, BI tools facilitate real-time reporting mechanisms that mitigate latency in corporate response cycles, ensuring that enterprises maintain a vantage point of continuous oversight. The ramifications of such instantaneous analytics permeate various operational dimensions, encompassing supply chain optimization, financial forecasting, human resource analytics, and customer relationship management (Asha et al., 2022; Rathore & Goudar, 2015; Sharma et al., 2021). The integration of BI into these domains fosters an ecosystem wherein data fidelity and BI converge to produce a symbiotic infrastructure that optimally aligns resources, enhances performance metrics, and fortifies risk management frameworks (Gangwar & Srivastva, 2020; Tripathi & Mohan, 2016).

Moreover, the evolution of BI from deterministic analytical models to probabilistic and AI-enhanced paradigms has augmented its applicability across industry verticals. In finance, for instance, BI-driven predictive modeling is instrumental in assessing credit risk, fraud detection, and investment portfolio optimization (Mishra et al., 2024; Srivastava et al., 2024a). In healthcare, advanced BI analytics facilitate patient outcome predictions, resource allocation, and epidemiological trend mapping. Retail enterprises leverage BI to discern consumer purchasing patterns, optimize inventory management, and personalize marketing campaigns (Indu et al., 2023; Pandey et al., 2023). The ubiquity of BI applications underscores its transformative potential across diverse business landscapes, reinforcing its stature as an indispensable linchpin of modern corporate strategy (Kumari et al., 2024; Singh et al., 2024d).

The democratization of data analytics, catalyzed by BI advancements, has also engendered a paradigm wherein strategic intelligence is no longer monopolized by executive echelons but permeates the entire organizational hierarchy. Self-service BI platforms, which empower non-technical stakeholders with intuitive analytical tools, have obviated traditional dependencies on data science specialists, engendering a culture of pervasive data literacy (Belwal & Belwal, 2017; Zhang et al., 2023). This diffusion of analytical capabilities fosters an environment where data-driven decision-making (DDDM) is an organizational prerogative rather than a siloed executive function. Consequently, businesses that espouse a culture of BI integration exhibit superior agility, innovation propensity, and resilience against market perturbations (Ajay et al., 2023; Grover et al., 2024; Shrivastava et al., 2023).

However, the ascendance of BI tools is not devoid of challenges. The exponential proliferation of data necessitates robust governance mechanisms to ensure data integrity, security, and compliance with regulatory frameworks such as GDPR, HIPAA, and SOC 2. The ethical ramifications of data analytics, encompassing algorithmic biases, data privacy infringements, and AI-induced decision distortions, necessitate a meticulous balance between technological adoption and ethical accountability (Arora et al., 2024; Khandelwal et al., 2023). The efficacy of BI implementation is also contingent upon organizational adaptability, wherein resistance to digital transformation, infrastructural bottlenecks, and skillset deficiencies may impede the seamless integration of BI frameworks. These challenges underscore the imperative for organizations to adopt a holistic approach to BI deployment, encompassing technological fortification, regulatory adherence, and cultural assimilation of data-centric paradigms (Dutta et al., 2023; Singh, 2024a).

As the digital epoch progresses, the integration of BI tools with emergent technological frontiers—such as quantum computing, blockchain, and federated learning—promises to augment the scope and profundity of business analytics (Singh & Singh, 2023). The convergence of BI with these avant-garde innovations heralds an era wherein data processing capabilities transcend current computational limitations, enabling real-time, hyper-accurate strategic intelligence with unprecedented granularity. The inexorable march of BI toward higher echelons of cognitive analytics will likely redefine the business landscape, where organizations adept at leveraging these advancements will secure unparalleled competitive leverage (Aiyappa et al., 2024; Padhi et al., 2024).

In summation, the infiltration of BI into the strategic substratum of corporate entities represents an evolutionary leap in data analytics, where empirical rigor supplants conjectural decision-making (Bhardwaj et al., 2024; Sharma et al., 2024c). The proliferation of BI tools, underpinned by advancements in AI, cloud computing, and big data analytics, has inaugurated an era of unparalleled data-driven strategy, wherein organizational acumen is inextricably linked to the robustness of its analytical capabilities (Kumar et al., 2024b; Neha et al., 2023). As businesses navigate an increasingly volatile economic terrain, the integration of BI into strategic frameworks is no longer an ancillary luxury but an indispensable prerequisite for sustained competitiveness, operational dexterity, and market supremacy (Kumar et al., 2023f; Sharma et al., 2022). The ramifications of this transformation extend beyond mere efficiency gains, engendering a fundamental shift in the epistemological underpinnings of corporate decision-making. BI tools, therefore, do not merely facilitate informed decisions—they redefine the very architecture of BI, ensuring that enterprises remain resilient, adaptive, and prescient in an era of relentless digital acceleration (Khanna et al., 2025; Upreti & Malhotra, 2024).

3.2 THEORETICAL FRAMEWORK

The integration of BI tools within organizational ecosystems is deeply interwoven with multiple theoretical constructs that underpin the strategic, technological, and operational dimensions of DDDM. At the forefront, the DDDM Theory postulates that empirical rigor supersedes intuition in formulating business strategies, advocating for an analytics-centric paradigm wherein structured data acquisition, processing, and visualization become indispensable (Srivastava et al., 2024). BI tools operationalize this theoretical tenet by transforming raw data into actionable intelligence, ensuring that executive decision matrices are fortified with quantifiable insights rather than speculative conjecture (Sharma et al., 2024a).

The Resource-Based View (RBV) Theory posits that sustainable competitive advantage emanates from the possession of unique, inimitable, and strategically valuable resources. Within this framework, BI tools emerge as a cornerstone of strategic differentiation, furnishing organizations with privileged access to structured and unstructured data that informs optimal resource allocation, market positioning, and operational efficacy (Khanna et al., 2023; Kumar et al., 2023e,g; Kumar & Khanna, 2023). By leveraging advanced analytics, firms can harness proprietary data repositories as a source of sustained competitive leverage, mitigating industry volatility and reinforcing strategic resilience (Kumar et al., 2021, 2023i; Sharma et al., 2024b).

The Technology Acceptance Model (TAM) delineates the behavioral dynamics underpinning technology adoption, emphasizing the dual constructs of perceived ease of use and utility as primary determinants of user engagement. The proliferation of BI tools is inexorably linked to their intuitive architecture, characterized by real-time reporting, user-friendly dashboards, and AI-enhanced analytics that demystify complex data landscapes (Cavaliere et al., 2024; Kanojia et al., 2022; Malhotra et al., 2024; Upreti et al., 2024). The seamless usability of BI platforms ensures that data literacy permeates all hierarchical levels of an organization, thereby fostering a culture of informed decision-making and digital adaptability.

Big Data Analytics Capability (BDAC) Framework articulates the necessity for firms to cultivate robust infrastructural and cognitive capacities for handling voluminous, heterogeneous datasets. BI tools integrate big data analytics by synthesizing vast streams of structured and unstructured information, enabling enterprises to distill predictive insights, optimize performance evaluations, and harness real-time intelligence (Gupta et al., 2023; Kumar et al., 2023c). The resultant analytical dexterity empowers firms to navigate the intricacies of fluctuating market paradigms with enhanced precision, preempting disruptions through algorithmically derived foresight (Bhatt 2024a; Maaze et al., 2024; Prerna et al., 2024).

Systems Theory conceptualizes organizations as interconnected entities wherein disparate functional units coalesce to achieve overarching strategic objectives. BI tools epitomize this systemic synergy by centralizing data from multifarious business domains, including finance, marketing, and operations, fostering seamless interdepartmental coordination (Goyal et al., 2024; Rajora et al., 2024; Verma et al., 2024). Through the unification of fragmented data streams, enterprises can engender holistic decision-making frameworks that transcend organizational silos, augmenting operational coherence and strategic alignment (Zaheer et al., 2024).

CI Theory underscores the imperatives of assimilating, analyzing, and deploying competitive insights to sustain market advantage. Within this construct, BI tools serve as indispensable instruments for decoding competitor movements, dynamic pricing mechanisms, and consumer preference shifts, ensuring that firms remain perpetually cognizant of industry oscillations. The analytical granularity afforded by BI platforms equips decision-makers with the agility to craft preemptive strategic interventions, consolidating market leadership through empirically substantiated insights.

Contingency Theory propounds that strategic architectures must remain fluid, adapting dynamically to contextual exigencies and environmental volatilities. BI dashboards embody this theoretical premise by facilitating real-time monitoring, empowering organizations to recalibrate their strategic orientations in response to shifting economic landscapes, consumer behaviors, and geopolitical contingencies. The capacity to synthesize evolving data trends into agile business maneuvers reinforces organizational resilience, mitigating the risks associated with static, inflexible strategic models.

Cognitive Analytics Theory elucidates the symbiosis between AI, machine learning, and human decision-making, advocating for an augmentation rather than replacement of cognitive capabilities. BI applications leverage sophisticated algorithmic models to unearth hidden patterns, detect anomalies, and furnish prescriptive analytics that refine strategic execution (Bhatt & Dani, 2024; Kumar et al., 2024a,d; Solanki et al., 2024; 2024b). The confluence of AI-enhanced BI ecosystems with human intuition engenders a hybrid decision-making paradigm wherein computational rigor and managerial acumen converge to maximize organizational efficacy.

The Decision Support Systems (DSS) Model delineates the role of computerized analytical infrastructures in optimizing executive decision workflows. BI tools function as contemporary manifestations of DSS by aggregating, structuring, and rendering data through interactive dashboards, predictive modeling, and automated reporting. The resultant decision intelligence framework ensures that enterprises navigate complex strategic dilemmas with enhanced clarity, leveraging algorithmically derived insights to inform high-stakes decision-making.

Information processing theory contends that the efficacy of decision-making is contingent upon the efficiency of data collection, storage, and retrieval mechanisms (Bhatnagar et al., 2024a,c; Bhatnagar & Sehajpal, 2024; Kumar et al., 2024a,d). BI architectures operationalize this principle by instituting structured repositories that streamline data accessibility, ensuring that intelligence extraction is both expedient and methodologically robust. The integration of advanced visualization techniques further amplifies interpretability, transforming raw datasets into cogent analytical narratives that bolster managerial insight.

The Data Governance Framework posits that data integrity, privacy, and regulatory compliance constitute fundamental imperatives in the digital economy. BI tools incorporate stringent security protocols to safeguard data assets, ensuring adherence to regulatory mandates such as GDPR and SOC 2. By embedding governance mechanisms within analytical infrastructures, BI ecosystems uphold the sanctity of corporate data, mitigating the risks associated with unauthorized access, algorithmic biases, and data corruption.

The interplay of these theoretical constructs in Table 3.1 within BI ecosystems underscores the multifaceted impact of data intelligence on contemporary corporate governance. The ascendant trajectory of BI adoption across industries reflects a paradigmatic shift wherein organizations progressively transcend intuition-driven strategies, embracing an empirically substantiated, algorithmically augmented decision-making ethos that redefines competitive excellence in the digital era.

3.3 LITERATURE REVIEW

The inexorable evolution of BI as a quintessential cornerstone of corporate decision-making has been precipitated by the exponential proliferation of data ecosystems, necessitating the confluence of computational sophistication and strategic dexterity. The ontogenesis of BI systems is deeply interwoven with the broader trajectory of digital transformation, wherein enterprises have progressively transitioned from rudimentary data aggregation methodologies to hyper-automated, algorithmically enriched analytical architectures. The metamorphosis of BI tools, from erstwhile descriptive analytics to contemporary prescriptive and cognitive paradigms, underscores an epistemological shift wherein data is no longer perceived as a passive informational reservoir but as an active, self-evolving asset that continually refines corporate foresight and strategic positioning.

At the nucleus of BI's transformative ascendancy lies the symbiosis between advanced data architectures and heuristic algorithmic constructs, enabling organizations to transcend the myopia of static analytics and embrace a dynamic, real-time decision-making ethos. Traditional business analytics, predicated upon retrospective data processing, has gradually been supplanted by real-time intelligence frameworks that leverage predictive modeling, machine learning algorithms, and AI to extricate profound insights from multifaceted datasets (Bhatnagar et al., 2023, 2024b; Kumar et al., 2023b,d). This paradigm shift engenders an organizational landscape wherein managerial prerogatives are dictated not by speculative conjecture but by empirical

TABLE 3.1
Theoretical Framework

Theoretical Concept	Description	Application in BI Tools and Data-Driven Strategy	Source
Data-Driven Decision-Making (DDDM) theory	Proposes that organizations should make strategic decisions based on empirical data rather than intuition.	BI tools facilitate data collection, processing, and visualization, ensuring that business strategies are formulated based on factual insights.	Bathla et al. (2024), Singh et al. (2024c), and Singh and Srivastava (2024)
Resource-Based View (RBV) theory	Suggests that competitive advantage is derived from valuable, rare, and inimitable resources.	BI tools provide organizations with exclusive access to structured and unstructured data, enhancing their strategic resource allocation and market positioning.	Anand et al. (2024), Bathla et al. (2024), and Pant et al. (2024)
Technology Acceptance Model (TAM)	Explains how users adopt and use technology based on perceived ease of use and usefulness.	The adoption of BI tools is influenced by their intuitive dashboards, real-time reporting, and AI-driven insights that simplify complex data analysis.	Joshi et al. (2024), Singh et al. (2024f), and Thapa et al. (2024)
Big Data Analytics Capability (BDAC) framework	Proposes that firms must develop capabilities to manage, analyze, and interpret large datasets for decision-making.	BI tools integrate big data analytics, allowing firms to process vast datasets efficiently and extract meaningful insights for forecasting and performance evaluation.	Bohra et al. (2024), Sahithya et al. (2024), and Semwal et al. (2024)
Systems theory	Views organizations as interdependent systems where different components work together for optimal performance.	BI systems centralize data from multiple business functions (finance, marketing, operations) to enable holistic decision-making and seamless interdepartmental coordination.	Bhuvaneswari et al. (2023), Kumar et al. (2023a), and Mittal et al. (2024)
Competitive Intelligence (CI) theory	Highlights the importance of gathering, analyzing, and utilizing competitive information to maintain an edge.	BI tools help firms analyze competitors' market movements, pricing strategies, and consumer preferences, leading to informed strategic responses.	Bachheti et al. (2023)

(Continued)

TABLE 3.1 (*Continued*)
Theoretical Framework

Theoretical Concept	Description	Application in BI Tools and Data-Driven Strategy	Source
Contingency theory	Suggests that business strategies should be flexible and adapt to situational variables.	BI dashboards provide real-time monitoring, allowing organizations to dynamically adjust strategies based on changing market conditions and customer behavior.	Prasad et al. (2023)
Cognitive analytics theory	Focuses on how AI and machine learning models enhance human decision-making by identifying hidden patterns.	BI applications use machine learning algorithms to forecast trends, detect anomalies, and provide prescriptive recommendations for business growth.	
Decision Support Systems (DSS) model	Describes computer-based tools that support organizational decision-making by processing data systematically.	BI platforms serve as modern DSS by aggregating, processing, and presenting data through interactive dashboards and predictive analytics.	Singh et al. (2024c)
Information processing theory	Suggests that effective decision-making depends on how well information is gathered, stored, and retrieved.	BI ensures structured data storage, retrieval, and visualization, streamlining business intelligence workflows for faster decision execution.	Bhatt et al. (2024a)
Data governance framework	Stresses the importance of data integrity, privacy, and compliance with regulatory standards.	BI tools incorporate data security features, ensuring compliance with regulations like GDPR and SOC 2, while maintaining high data quality.	Bohra et al. (2024)

veracity, ensuring that strategic interventions are underpinned by algorithmic precision and statistical robustness.

The proliferation of cloud computing and distributed ledger technologies has further augmented the operational elasticity of BI frameworks, mitigating the infrastructural constraints that previously encumbered large-scale data analytics (Kukreti et al., 2025; Sehajpal & Kaur, 2025). The decentralization of data processing, facilitated by edge computing and federated learning models, has enabled enterprises to process and analyze data with an unprecedented degree of agility, obviating traditional latency constraints that hinder real-time intelligence dissemination (Kour et al., 2025). Consequently, the entrenchment of cloud-native BI ecosystems has heralded an era wherein data accessibility, computational scalability, and analytical granularity are no longer constrained by infrastructural limitations, rendering DDDM an omnipresent organizational imperative.

The ontological expansion of BI beyond conventional corporate governance structures has also necessitated a recalibration of organizational epistemologies, wherein data democratization and self-service analytics have emerged as pivotal conduits for fostering a culture of pervasive data literacy. The advent of user-centric BI platforms, which empower non-technical stakeholders with intuitive analytical tools, has engendered an egalitarian approach to business analytics, wherein strategic intelligence is no longer monopolized by executive echelons but permeates the entire organizational hierarchy. This democratization fosters an enterprise-wide analytical consciousness, ensuring that decision-making paradigms are infused with empirical rigor at every echelon of corporate governance, thereby enhancing organizational agility and resilience against exogenous perturbations.

The intersectionality of BI with AI and machine learning has inaugurated a new epoch of cognitive analytics, wherein decision-making architectures are augmented by algorithmic inferencing and probabilistic modeling (Bhatnagar & Sehajpal, 2024). The infusion of AI-driven analytical mechanisms into BI frameworks has facilitated the development of self-optimizing, autonomous DSS that possess the cognitive dexterity to discern latent market patterns, preempt financial contingencies, and engineer proactive strategic interventions. This integration has propelled BI from its erstwhile role as an auxiliary analytical tool to an indispensable strategic fulcrum, ensuring that corporate maneuvering is governed by an intricate confluence of computational intelligence and managerial acumen.

The operationalization of BI within industry verticals has been marked by a profound diversification of its analytical modalities, encompassing domains as varied as financial analytics, supply chain optimization, customer relationship management, and human capital analytics. In financial ecosystems, BI-driven predictive modeling has emerged as a linchpin in credit risk assessment, fraud detection, and algorithmic trading, wherein real-time data processing is leveraged to enhance investment portfolio optimization. Within supply chain dynamics, the integration of BI tools has facilitated the real-time tracking of logistical variables, demand forecasting, and inventory optimization, ensuring that supply chain ecosystems exhibit maximal efficiency and minimal systemic vulnerabilities. Similarly, within customer analytics, BI platforms have engendered a paradigm of hyper-personalization, wherein consumer purchasing behaviors are meticulously analyzed to optimize marketing interventions, enhance customer retention strategies, and refine product development trajectories.

Despite its transformative potential, the pervasiveness of BI has also engendered formidable challenges, particularly in the domains of data integrity, algorithmic biases, and regulatory compliance. The exponential proliferation of data necessitates robust governance frameworks to ensure that data veracity is maintained, thereby mitigating the deleterious ramifications of erroneous analytics. Moreover, the opacity of AI-driven BI mechanisms has catalyzed discourse surrounding algorithmic accountability, wherein decisioning architectures must be meticulously scrutinized to preclude instances of computational bias and ethical distortions. Regulatory compliance further compounds these complexities, as enterprises must navigate an intricate labyrinth of legal mandates, such as data protection statutes and cybersecurity protocols, to ensure that BI deployments are not encumbered by regulatory non-conformities.

The confluence of BI with emergent technological frontiers, such as quantum computing and blockchain, portends a future wherein data analytics transcends its current computational limitations, ushering in an era of hyper-accurate strategic intelligence. The integration of quantum computational paradigms within BI architectures promises an exponential augmentation of data processing capabilities, facilitating the real-time analysis of gargantuan datasets with an unprecedented degree of precision. Concurrently, the infusion of blockchain mechanisms within BI ecosystems ensures that data integrity, transactional transparency, and cyber-resilience are fortified, mitigating the vulnerabilities associated with centralized data repositories. This symbiotic integration of avant-garde technologies with BI frameworks heralds a renaissance in business analytics, wherein enterprises that adeptly harness these advancements will invariably secure a formidable competitive advantage.

In summation, the evolution of BI from a rudimentary analytical adjunct to an indispensable strategic imperative underscores the profound metamorphosis of DDDM in the contemporary corporate milieu. The proliferation of AI-enhanced, cloud-native, and real-time BI frameworks has engendered a paradigm wherein strategic acumen is inexorably intertwined with computational intelligence, ensuring that enterprises remain agile, prescient, and resilient in the face of perpetual market flux. As businesses traverse the vicissitudes of an increasingly digitized economic landscape, the integration of BI into corporate governance structures will continue to delineate the contours of strategic success, rendering data-driven intelligence not merely a competitive differentiator but an existential imperative.

3.4 RESEARCH METHODOLOGY

The investigational framework employed in this scholarly inquiry is predicated upon a multifaceted methodological paradigm that synthesizes quantitative analytical rigor with qualitative interpretative depth, ensuring an exhaustive deconstruction of the BI and AI integration within BRICS economies. The methodological architecture is meticulously calibrated to encapsulate both macrostructural economic indicators and microcosmic organizational dynamics, engendering a holistic analytical schema that delineates the symbiotic interplay between algorithmic augmentation and strategic economic transformation. This dual-pronged methodological approach is designed to extricate latent patterns, discern inferential correlations, and construct an epistemological scaffold that encapsulates the stochastic interdependencies governing BI proliferation within diverse economic substrata.

The quantitative dimension of this research is anchored in an empirically rigorous data collection mechanism, wherein statistical aggregations from primary and secondary sources are meticulously curated to ensure fidelity to economic realism and computational veracity. The deployment of econometric modeling techniques, including multivariate regression analysis, machine-learning-enhanced trend extrapolation, and inferential statistical methodologies, enables a granular dissection of the economic ramifications stemming from BI and AI integration. Macroeconomic datasets, sourced from institutional repositories such as national statistical agencies, international economic consortiums, and proprietary corporate intelligence networks, are subjected to algorithmically mediated deconstruction, ensuring that the resultant inferential constructs are predicated upon empirical certitude rather than conjectural speculation.

Parallel to this quantitative substratum, a qualitative analytical mechanism is instituted to imbue interpretative depth and ontological richness into the inquiry. The research undertakes an extensive thematic analysis, encompassing policy document dissection, executive stakeholder interviews, industry whitepaper evaluations, and cross-sectional case studies, thereby facilitating the extrapolation of contextualized insights that transcend the myopia of numerical abstraction. The utilization of qualitative content analysis techniques, including hermeneutic interpretation, comparative discourse analysis, and grounded theory methodologies, enables a dialectical synthesis between numerical objectivity and narrative subjectivity, ensuring that the resultant analytical framework is both structurally robust and contextually nuanced.

A core methodological imperative of this research is the employment of triangulation techniques, wherein the cross-validation of quantitative data metrics with qualitative thematic findings ensures the mitigation of methodological biases and enhances epistemic reliability. This multi-perspectival validation mechanism enables the research to delineate convergent, divergent, and emergent analytical trajectories, thereby fortifying the robustness of the inferential constructs derived. The methodological scaffolding is further reinforced through longitudinal comparative analytics, wherein historical and contemporary data are juxtaposed to ascertain the evolutionary trajectory of BI and AI integration within BRICS economies, allowing for the identification of nonlinear progression patterns and disruptive inflection points.

The research instrumentation encompasses data mining algorithms, econometric statistical software, AI-powered sentiment analysis engines, and predictive analytics frameworks, ensuring that the methodological apparatus is technologically fortified to accommodate the computational complexity of modern economic analytics. Ethical considerations are meticulously embedded within the methodological praxis, ensuring compliance with data governance protocols, respondent confidentiality imperatives, and algorithmic transparency principles, thereby upholding the integrity of the investigational process.

The methodological schema employed in this research is, therefore, not merely an instrumental conduit for data extraction but an epistemological construct designed to unravel the intricate interplay between BI augmentation, AI proliferation, and macroeconomic structuration. By synthesizing quantitative analytical precision with qualitative contextual depth, the research aspires to construct an intellectually rigorous and analytically coherent exposition that elucidates the transformative ramifications of intelligent decision-making frameworks within the evolving BRICS economic paradigm.

3.5 DATA ANALYSIS

The BRICS nations in Table 3.2 exhibit varying degrees of BI tool adoption and AI integration, reflecting disparate economic landscapes, technological infrastructure, and policy imperatives. Brazil, while still at an embryonic stage in its BI adoption curve, has demonstrated a burgeoning interest in embedding BI within the financial and retail sectors. The nation's AI integration, albeit nascent, is witnessing an incremental shift toward leveraging automation and machine learning for business process optimization. Under the aegis of its National AI Strategy, Brazil aspires to fortify AI research and facilitate a more structured assimilation of intelligent technologies across industry verticals. However, the pace of this transformation remains contingent on regulatory coherence, infrastructural augmentation, and the scalability of AI-driven enterprises within its economic milieu.

In contrast, Russia's BI adoption trajectory is more pronounced, particularly within large-scale enterprises that recognize the strategic imperatives of DDDM. The nation's AI integration is demonstrably more robust, with 54% of enterprises having implemented AI-driven solutions in at least one operational function by 2024. This accelerated AI assimilation underscores Russia's strategic pivot toward algorithmic decision augmentation, reflecting a policy-driven impetus to enhance computational efficiencies and industrial automation. The National AI Strategy is emblematic of Russia's long-term commitment to AI-enabled economic expansion, underscoring a concerted effort to fortify national competitiveness through AI sovereignty and digital autonomy. The interplay between state-sponsored AI proliferation and private-sector technological absorption is poised to reinforce Russia's positioning in the evolving digital economy.

India manifests a pronounced trajectory in BI proliferation, particularly among small and medium-sized enterprises (SMEs) operating in the IT and services sectors. The democratization of BI tools within these domains signifies an overarching paradigm shift toward data-centric operational frameworks, wherein analytics-driven intelligence underpins corporate agility and market responsiveness. AI integration in India is progressing at an unprecedented pace, permeating domains such as customer service automation, predictive analytics, and financial modeling (Kimothi et al., 2024; Shukla et al., 2024; Singh et al., 2024e). The Digital India initiative serves as a catalytic instrument, expediting digital transformation and AI-driven economic realignment. The seamless interplay between policy support, burgeoning AI entrepreneurship, and an expansive digital consumer base positions India as a formidable contender in AI-led business innovation.

TABLE 3.2
Business Intelligence and AI-Driven Decision-Making in BRICS Nations

Country	AI Adoption Rate (2024)	Projected Economic Impact of AI by 2030 (USD Billion)	Notable Initiatives	Sources
Brazil	AI adoption is growing, especially in fintech and agriculture, driven by government incentives and startups.	AI-driven digital transformation is expected to boost Brazil's GDP by **up to 4%** by 2030.	**National AI Strategy (Estratégia Brasileira de Inteligência Artificial—EBIA)** promotes AI in public and private sectors, focusing on ethics and data governance.	Maaze et al. (2024)
Russia	54% of companies have integrated AI-driven solutions in at least one function.	AI technologies are expected to contribute **11.2 trillion rubles** (~$150B) to GDP by 2030.	Russia is developing an **AI alliance within BRICS**, focusing on sovereignty in AI research and applications.	Ahmed et al. (2025), and Sarantis et al. (2025)
India	AI adoption is accelerating in IT, healthcare, and financial services, with increased investments in AI-driven governance.	AI is expected to add **$450–500 billion** to India's economy by 2030.	**NITI Aayog's AI for All Initiative** and Digital India program promote AI-driven governance, smart cities, and real-time data analytics.	Cuevas-Pichardo et al. (2025), and Pathak et al. (2025)
China	Leading AI adoption globally, with AI-driven automation in industries like manufacturing, e-commerce, and finance.	AI is projected to contribute **$600 billion+** to GDP by 2030.	**Made in China 2025** and **Social Credit System** use big data and AI for economic and policy optimization.	Barot et al. (2025) and Kaur et al. (2025)
South Africa	AI adoption is emerging, mainly in banking, mining, and retail, with increasing investments in digital transformation.	AI could contribute **up to 10% of GDP growth** by 2030.	**AI Institute of South Africa** focuses on AI-driven economic inclusion, fintech innovations, and smart governance.	Bijalwan et al. (2025) and Singh et al. (2025)

China's BI adoption is unparalleled within the BRICS consortium, extending comprehensively across industrial, commercial, and governmental sectors. The nation's intrinsic emphasis on data analytics, automation, and algorithmic intelligence ensures that BI is not merely an ancillary tool but an integral pillar of its macroeconomic and microeconomic frameworks. AI integration in China is unequivocally the most sophisticated within BRICS, characterized by the rapid deployment of generative AI solutions that have attained functional parity with Western AI architectures by mid-2023 (Bhatnagar et al., 2024b; Bhatt et al. 2024b; Ramarajan et al., 2024; Srinivasan, 2024). The New Generation AI Development Plan encapsulates China's ambition to cement its status as a global AI hegemon by 2030, leveraging AI to orchestrate economic efficiency, geopolitical influence, and industrial transformation. The nation's unparalleled investment in AI infrastructure, research, and human capital fortifies its position as the epicenter of next-generation intelligence augmentation.

South Africa, though comparatively nascent in BI adoption, exhibits a progressive inclination toward embedding data intelligence in corporate decision-making. The pace of AI integration remains in its formative stages, reflecting a broader inclination toward exploratory AI applications rather than full-fledged deployment. However, the establishment of the **AI Institute of South Africa** underscores a strategic commitment to fostering AI research and application. This initiative aims to bridge the digital divide, enhance computational capacities, and cultivate AI literacy within South Africa's economic landscape. The measured but deliberate approach to AI adoption is indicative of a **gradual but irreversible transition toward intelligent automation and data-driven governance**.

The heterogeneity in BI and AI adoption across BRICS underscores **a nuanced digital stratification**, wherein China and India emerge as dominant forces, Russia and Brazil maintain measured but strategic integration, and South Africa embarks on an incremental trajectory. The extent to which these nations can harmonize their BI architectures with AI-driven foresight will define their competitive standing in an increasingly **algorithmically governed global economy** (Table 3.3).

The variegated landscape of BI tool adoption and AI integration within the BRICS nations epitomizes a heterogeneously stratified continuum, wherein disparate technological trajectories underscore the asymmetrical diffusion of computational acumen across economic infrastructures. Brazil, positioned within the nascent echelons of BI proliferation, exhibits an incipient but perceptible inclination toward embedding data-driven frameworks within the financial and retail sectors (Alsmairat et al., 2025). The country's AI assimilation remains embryonic, characterized by a measured, albeit intensifying, proclivity for leveraging machine intelligence to augment operational efficiencies. The National AI Strategy functions as an institutionalized scaffold aimed at expediting AI proliferation across industrial spectra, though its efficacy is inherently contingent upon the synchronization of policy mechanisms with infrastructural scalability and digital literacy augmentation.

Russia's engagement with BI infrastructures is more pronounced, particularly within large-scale enterprises where the assimilation of analytical architectures is increasingly perceived as a linchpin for strategic dynamism. The trajectory of AI integration within the Russian economic fabric is substantially more accentuated,

TABLE 3.3
Business Intelligence Tools and AI Integration in BRICS Nations

Country	BI Tools Adoption	AI Integration	Notable Initiatives	Sources
Brazil	Emerging adoption in sectors like finance and retail.	AI integration is in nascent stages, with increasing interest in leveraging AI for business processes.	**National AI Strategy** aims to promote AI research and application across industries.	Husen (2025b), A. Singh et al. (2025)
Russia	Moderate adoption, particularly in large enterprises.	Significant integration, with 54% of companies implementing AI solutions in at least one function by 2024.	**National AI Strategy** focuses on AI development to boost economic growth.	Agrawal et al. (2025), Chaudhary et al. (2025)
India	Growing adoption, especially among SMEs in IT and services sectors.	Rapid integration, with AI being utilized for various applications, including customer service and predictive analytics.	**Digital India** initiative encourages digital transformation and AI adoption.	Alexopoulos et al. (2025)
China	Extensive adoption across industries, with a strong emphasis on data analytics.	Leading in AI integration, with rapid deployment of generative AI solutions reaching Western models by mid-2023.	**New Generation AI Development Plan** aims to make China a global AI leader by 2030.	Caiado et al. (2025)
South Africa	Gradual adoption, with increasing interest in leveraging BI for decision-making.	AI integration is emerging, with efforts to incorporate AI into various business processes.	**AI Institute of South Africa** was established to promote AI research and applications.	M. Arora et al. (2025)

with over 54% of enterprises having incorporated AI-driven solutions into at least one functional domain by 2024. This paradigmatic shift underscores a concerted effort to embed algorithmic intelligence into corporate governance structures, effectuating a transition from intuition-laden decision matrices to empirical, computationally mediated strategic foresight (Kumar et al., 2025; Pant et al., 2025; Singh et al. 2024b). The National AI Strategy, serving as the cornerstone of this digital recalibration, is architected to galvanize AI-fueled economic expansion, reinforcing Russia's ambition to attain a digitally sovereign economic superstructure.

India manifests a trajectory of expeditious BI adoption, particularly within the SME segment, where the exigencies of competitive agility necessitate the seamless incorporation of data-driven methodologies. The country's AI integration is advancing at a meteoric pace, permeating an expansive array of applications encompassing customer service automation, predictive analytics, and algorithmic financial modeling. The Digital India initiative operates as the catalytic mechanism propelling this transition, orchestrating a comprehensive digitization agenda that seeks to harmonize infrastructural modernization with policy imperatives. The intersectionality of government-backed digital transformation efforts with burgeoning AI entrepreneurship has positioned India as an emergent nucleus of algorithmic innovation, underscoring its ascendancy as a pivotal player in the global AI hierarchy.

China's pervasive assimilation of BI tools across industrial, governmental, and commercial domains cements its stature as the preeminent force within the BRICS technological paradigm. The country's systemic emphasis on data analytics, intelligent automation, and algorithmic forecasting engenders an ecosystem wherein BI is not merely an auxiliary component but an intrinsic constituent of economic governance. The nation's AI integration is unparalleled, with the deployment of generative AI models attaining functional parity with Western AI architectures by mid-2023 (Bhat et al., 2024; Gupta et al., 2024; Yildirim et al., 2024). The New Generation AI Development Plan encapsulates China's hegemonic ambition to ascend as the apex AI superpower by 2030, institutionalizing a framework that coalesces governmental policy, technological research, and industrial application into a singular, harmonized trajectory of computational supremacy. This confluence of strategic foresight and technological dominance renders China the vanguard of the AI-augmented economic renaissance.

South Africa, though comparatively embryonic in its BI adoption trajectory, exhibits an incipient but resolute inclination toward integrating data-driven intelligence into corporate decision-making frameworks. AI integration remains in an exploratory phase, with its deployment characterized by sectoral experimentation rather than comprehensive institutionalization. However, the establishment of the AI Institute of South Africa signifies a deliberate governmental endeavor to catalyze AI research, facilitate indigenous computational capability, and engender a digital ecosystem that fosters scalable AI implementation. This initiative represents an embryonic but foundational step toward bridging the digital divide and embedding AI within the nation's economic and corporate substratum, ensuring a gradual but irreversible transition toward an algorithmically mediated decision-making infrastructure.

The heterogeneity in BI and AI adoption trajectories across BRICS underscores a technologically stratified economic taxonomy, wherein China and India emerge as the fulcrum of AI-driven innovation, Russia and Brazil navigate the transitionary phase of computational expansion, and South Africa embarks on a gradual but inexorable digitization continuum. The extent to which these nations can harmonize their BI infrastructures with AI-driven strategic intelligence will delineate their competitive resilience in an increasingly algorithmically governed global economy.

The economic trajectory of the BRICS nations in Table 3.4 presents a multifaceted landscape characterized by divergent growth velocities, demographic distributions, and strategic economic positioning. India's projected GDP growth rate of **6.6% in 2024**, the highest among BRICS members, underscores its burgeoning economic dynamism, driven by rapid industrialization, digital transformation, and a demographic dividend that fosters sustained consumer demand and labor market expansion. Comparatively, China's projected growth of **4.8%**, though lower, is reflective of its transition toward a more consumption-driven economy, marked by sophisticated technological innovation and high-value manufacturing. Meanwhile, Brazil and Russia, each with a 2.0% growth rate, exhibit a more tempered economic outlook, constrained by structural dependencies on commodity exports and geopolitical volatilities. South Africa, with a 1.8% projection, remains the most economically restrained among its BRICS counterparts, indicative of macroeconomic challenges and systemic infrastructural constraints that impede accelerated growth trajectories.

China's formidable economic hegemony is further exemplified by its 19.1% share of global GDP in 2024, positioning it as the preeminent force within BRICS and the broader global economy. India, securing an 8.2% global GDP share, reflects an emerging powerhouse status, leveraging digitalization, industrial expansion, and a burgeoning middle-class economy. Russia and Brazil, with shares of 3.1% and 2.5%, respectively, occupy a more modest role in the global economic matrix, where resource dependency and geopolitical intricacies continue to influence their fiscal architecture. South Africa, with a 0.6% share, underscores its relatively limited yet strategically significant role within the BRICS framework, where policy reform and infrastructural modernization remain pivotal determinants of future economic scalability.

Demographic potency is a key determinant of economic endurance, with China (1.425 billion) and India (1.417 billion) leading in population by 2025, underpinning their immense labor force and consumer-driven economic paradigms. While marginal, China's numerical demographic advantage suggests sustained economic scalability in sectors such as AI, automation, and global trade. Brazil's 0.214 billion and Russia's 0.143 billion populations reflect comparatively restrained labor markets, necessitating heightened productivity-driven growth mechanisms to maintain competitive parity. South Africa's 0.065 billion, the smallest demographic constituent, underscores the structural limitations in leveraging economies of scale but highlights opportunities for high-skill labor market optimization.

The long-term economic forecasts further consolidate China's dominance, with a projected nominal GDP of $70.71 trillion by 2050, an unprecedented economic valuation that signifies its unwavering ascendancy in global finance, trade, and technological supremacy. India, with an estimated $38.67 trillion GDP, is poised to emerge

TABLE 3.4
BRICS 2050: The Powerhouse of the Future—A Statistical Odyssey into Economic Dominance

Indicator	Brazil	Russia	India	China	South Africa	Maximum Value	Country	Source
Projected GDP Growth Rate (2024)	2.00%	2.00%	6.60%	4.80%	1.80%	6.60%	India	Naaz et al. (2025)
Share of Global GDP (2024)	2.50%	3.10%	8.20%	19.10%	0.60%	19.10%	China	Chand et al. (2025)
Population (2025, billions)	0.214	0.143	1.417	1.425	0.065	1.425	China	Kashif et al. (2025)
Projected GDP (Nominal, 2050, USD trillions)	11.37	8.58	38.67	70.71	N/A	70.71	China	Rawat et al. (2025)
Projected GDP per Capita (2050, USD)	52,327	52,663	44,602	62,717	N/A	62,717	China	Husen (2025a)
Contribution to global economic growth (2025–2050)	Significant	Moderate	High	Highest	Modest	Highest	China	Agarwal et al., 2025)

as a formidable economic force, capitalizing on digital economies, infrastructure expansion, and manufacturing-led exports. Brazil and Russia, at $11.37 trillion and $8.58 trillion, respectively, project substantial, albeit measured, economic expansion, reflective of their industrial capacities and resource-export reliance. South Africa's absence of definitive GDP projections further emphasizes its economic unpredictability, necessitating robust policy interventions to foster sustained economic acceleration.

The economic vibrancy of these nations is further encapsulated in their projected GDP per capita (2050), where China leads at $62,717, indicative of its successful transition toward a high-income economy. Russia follows closely with $52,663, reflecting a stabilized yet resource-influenced per capita prosperity. Brazil's $52,327 GDP per capita suggests relative economic parity with Russia, yet contingent on commodity cycles and industrial diversification. India's $44,602, though lower, signifies its evolving economic trajectory, propelled by structural reforms and expanding digital ecosystems. The absence of projections for South Africa denotes macroeconomic uncertainties, underscoring the necessity for targeted economic reforms to enhance per capita wealth distribution.

The overarching contribution to global economic growth between 2025 and 2050 cements China's preeminence as the greatest driver of global economic expansion, underscoring its relentless economic diversification and geopolitical influence. India's categorization as a high contributor reinforces its growing strategic stature in the global economic fabric, propelled by rapid industrial scaling, infrastructural augmentation, and digital economic proliferation. Brazil, with its significant contribution, highlights an evolving role in global supply chains, while Russia's moderate categorization reflects economic resilience despite geopolitical complexities. South Africa's modest impact underscores its more constrained, albeit essential, role within the BRICS bloc, necessitating sustained economic reform to amplify its global economic footprint.

The holistic interpretation of these economic indicators reaffirms China and India's dominance in shaping future global economic landscapes, driven by demographic leverage, technological superiority, and sustained economic reform. Russia and Brazil maintain substantial, yet measured, economic influence, contingent on resource markets and industrial policy. South Africa, while economically constrained, remains strategically positioned within BRICS, necessitating focused policy interventions to enhance long-term economic viability. The trajectory of BRICS thus encapsulates a multidimensional economic renaissance, where strategic investment, digital transformation, and geopolitical adaptability will dictate the scale and velocity of future economic ascension.

3.6 MANAGERIAL IMPLICATIONS

The inexorable ascendance of BI and AI in the BRICS economies necessitates an intricate recalibration of managerial paradigms, wherein data-driven strategic imperatives supersede antiquated heuristic methodologies. In an era punctuated by unprecedented digitalization, executive decision-making must pivot toward an empirical modus operandi, leveraging algorithmic precision and predictive analytics to navigate the multifaceted complexities of globalized commerce. This transmutation of corporate governance underscores the imperativeness of inculcating an analytical ethos within organizational hierarchies, ensuring that managerial stratagems are not merely reactive but proactively sculpted through the meticulous dissection of economic, financial, and operational datasets.

The proliferation of BI tools across these emergent economies mandates an unequivocal transition from intuition-driven corporate oversight to a rigorously structured intelligence-oriented framework. In economies such as China and India, where AI integration into economic matrices is at its apogee, enterprises must assimilate avant-garde technological infrastructures that encompass real-time data assimilation, machine learning-enhanced predictive modeling, and automated DSS. The managerial prerogative, therefore, must align with the exigencies of digital transformation, wherein leaders must cultivate an ecosystem that fosters data fluency at all organizational tiers. The velocity at which economic conditions fluctuate in BRICS necessitates an agile managerial disposition, predicated upon data elasticity, adaptive analytics, and AI-augmented decision-making models.

Furthermore, in economies such as Russia and Brazil, where AI adoption is burgeoning yet relatively nascent compared to China, the managerial paradigm necessitates an acute focus on strategic investment in BI frameworks to bolster competitive resilience. The interplay between macroeconomic indicators, supply chain dynamism, and geopolitical volatilities necessitates that enterprises fortify their decision-making apparatus with prescriptive analytics that mitigate uncertainty. As digital transformation increasingly interweaves with national economic policies, managerial foresight must integrate geopolitical risk intelligence into strategic decision matrices, ensuring proactive crisis mitigation and supply chain fortification.

Moreover, the variegated economic landscapes within the BRICS bloc accentuate the imperative for localized BI customization, wherein managerial frameworks must accommodate region-specific economic idiosyncrasies. For instance, South African enterprises operating within a comparatively constrained economic ecosystem must deploy lean BI infrastructures that optimize cost-efficiency while maximizing intelligence extraction. The divergence in digital maturity levels across BRICS necessitates that multinational enterprises (MNEs) operating within these geographies adopt differentiated BI deployment strategies, ensuring that scalability, agility, and regulatory compliance are harmonized within their transnational operations.

Additionally, the ascendancy of AI-driven decision intelligence within the BRICS economies demands a profound shift in human capital management. The managerial landscape must evolve toward upskilling and re-skilling initiatives, ensuring that workforce competencies align with the exigencies of AI-augmented decision ecosystems. The fusion of BI with cognitive analytics necessitates a bifurcated managerial focus—one that simultaneously fosters technological innovation while mitigating the human displacement ramifications of automation. Leaders must, therefore, orchestrate a synergistic human-AI collaboration model, wherein AI augments human decision-making rather than obviating human intervention.

From a financial vantage, the predictive prowess of BI necessitates a reengineering of fiscal oversight models, ensuring that capital allocation, investment appraisals, and risk assessments are underpinned by algorithmic rigor. The projected economic ascendancy of India and China mandates that corporations recalibrate their market penetration strategies, leveraging AI-powered consumer sentiment analysis and real-time economic modeling to optimize market entry and expansion initiatives. Furthermore, as BRICS+ expands its geopolitical and economic influence, managerial foresight must encompass cross-border data harmonization, ensuring that BI ecosystems are interoperable across diverse regulatory and economic frameworks.

In summation, the managerial implications of BI within BRICS transcend mere technological adoption and necessitate a holistic transformation in corporate governance structures. The integration of real-time analytics, AI-driven decision augmentation, and algorithmic foresight must underpin strategic maneuvering within these economies. Enterprises that fail to synchronize their managerial acumen with the exigencies of digital intelligence risk obsolescence in an increasingly data-dominated corporate ecosystem.

3.7 CONCLUSION

The intricate confluence of BI proliferation, AI augmentation, and macroeconomic dynamism within BRICS nations delineates a transformative epoch wherein data-driven epistemologies usurp conventional heuristic paradigms. The statistical extrapolations illuminating the economic trajectories of these nations underscore the inexorable ascendancy of China and India, whose projected economic hegemony, bolstered by demographic preponderance and algorithmic industrialization, portends a recalibration of the global economic order. The idiosyncratic economic architectures of Russia and Brazil, though imbued with latent strategic leverage, remain inextricably tethered to geopolitical vicissitudes and resource-dependent modalities, necessitating a recalibrated framework of technological self-reliance and economic diversification. South Africa, positioned at a nascent yet evolutionary inflection point, exhibits an incremental trajectory toward the assimilation of data-centric governance, emblematic of a broader ontological transition toward algorithmically augmented economic stewardship.

The theoretical underpinnings delineated across DDDM frameworks, resource-based economic postulations, and cognitive analytics encapsulate the axiomatic indispensability of BI as an organizational linchpin. The intersection of machine learning, predictive analytics, and real-time data assimilation reifies the imperative of algorithmic foresight, wherein executive decision matrices are irrevocably interwoven with empirical certitude. China's formidable prowess in generative AI deployment, Russia's incipient yet assertive AI-centric economic reconfiguration, and India's expansive digital transformation paradigm collectively underscore the inexorable march toward computational sovereignty. The evolution of BI from rudimentary descriptive analytics to prescriptive and autonomous intelligence mandates an epistemological reconstitution, wherein enterprises must navigate the confluence of cognitive analytics, contingency modeling, and stochastic forecasting to sustain their competitive asymmetry.

The managerial ramifications of this paradigm shift necessitate an ontological departure from conventional corporate oversight, where decisioning architectures must embody an algorithmically driven epistemic framework. The confluence of AI-infused decision intelligence, systemic data harmonization, and neuro-symbolic cognitive modeling mandates a recalibration of human capital trajectories, ensuring that managerial acumen is inextricably aligned with AI-propelled economic scalability. The synthesis of cross-border regulatory compliance, quantum data architectures, and sovereign AI ecosystems delineates a future wherein strategic dexterity is predicated upon computational elasticity, rendering data governance the fulcrum of sustained economic primacy.

As BRICS nations continue their inexorable transformation into economic vanguards, their ascendant trajectory will be determined by the scalability of AI-augmented BI infrastructures, the adaptability of their regulatory scaffolds, and the elasticity of their economic ecosystems in assimilating autonomous decision intelligence. The imminent metamorphosis of economic power structures will be orchestrated not merely through fiscal expansionism but through an ontological redefinition of economic agency, wherein the amalgamation of data intelligence, AI-driven prescriptive analytics, and computational sovereignty will delineate the contours of economic supremacy in an era characterized by algorithmic determinism and data-centric hegemony.

REFERENCES

Agarwal, P., Upadhyay, R. K., Sharma, A., Chauhan, J., Ahalawat, K., Singh, T., & Veera Lakshmi, B. (2025a). Workplace ergonomics and its influence on job performance with mediating role of job satisfaction: an Indian perspective. *International Journal of Management Practice*, *18*(1), 1–22. https://doi.org/10.1504/IJMP.2025.143102

Agrawal, D., Gupta, S., Dusad, C., Vishwakarma, A., Meena, M. L., Dangayach, G. S., & Jagtap, S. (2025b). Unlocking circular economy: overcoming roadblocks to circular healthcare supply chain in Indian healthcare. *Journal of Advances in Management Research*. https://doi.org/10.1108/JAMR-10-2024-0366

Ahmad, I., Sharma, S., Kumar, R., Dhyani, S., & Dumka, A. (2023). Data analytics of online education during pandemic health crisis: a case study. 2nd Edition *of IEEE Delhi Section Owned Conference, DELCON 2023 – Proceedings*. https://doi.org/10.1109/DELCON57910.2023.10127423

Ahmed, A., Abdul-Malek, Z., Malik, H., & Chandio, T. A. (2025). Development of improved superconducting pulsed power supply using three-winding pulsed power transformer with energy recovery. *International Journal of Mathematical, Engineering and Management Sciences*, *10*(1), 129–147. https://doi.org/10.33889/IJMEMS.2025.10.1.008

Aiyappa, S., Kodikal, R., & Rahiman, H. U. (2024). Accelerating gender equality for sustainable development: a case study of Dakshina Kannada District, India. *Technical and Vocational Education and Training*, *38*, 335–349. https://doi.org/10.1007/978-981-99-6909-8_30

Ajay, D., Sharad, K., Singh, K. P., & Sharma, S. (2023). Impact of prolonged mental torture on housewives in middle class families. *Journal for ReAttach Therapy and Developmental Diversities*, *6*(4), 1–7.

Alexopoulos, C., Al-Tamimi, T. A. S., & Saxena, S. (2023). Were the higher educational institutions (HEIs) in Oman ready to face pedagogical challenges during COVID-19? *Arab Gulf Journal of Scientific Research*. https://doi.org/10.1108/AGJSR-03-2023-0095

Alexopoulos, C., Rizun, N., & Saxena, S. (2025). The moderating effect of gender and academic background in open government data (OGD) user adoption and usage. *International Journal of Public Administration*, *48*(2), 134–145. https://doi.org/10.1080/01900692.2024.2327320

Alrashed, F. A., Alsubiheen, A. M., Alshammari, H., Mazi, S. I., Al-Saud, S. A., Alayoubi, S., Kachanathu, S. J., Albarrati, A., Aldaihan, M. M., Ahmad, T., Sattar, K., Khan, S., & Dhiman, G. (2022). Stress, anxiety, and depression in pre-clinical medical students: prevalence and association with sleep disorders. *Sustainability (Switzerland)*, *14*(18), 11320. https://doi.org/10.3390/su141811320

Alsmairat, M. A. K., Garza-Reyes, J. A., Dahmani, N., & Elrehail, H. (2025). Connecting the dots: investigating the relationships between customer preferences, innovation and last-mile delivery adoption: a mediated model. *Supply Chain Forum*. https://doi.org/10.1080/16258312.2025.2452148

Ambika, K. S. B., Goswami, S., Pimplapure, V., Jweeg, M. J., Kant, K., & Gangodkar, D. (2023). Framework towards detection of stress level through classifying physiological signals using machine learning. *2023 3rd International Conference on Advance Computing and Innovative Technologies in Engineering, ICACITE 2023* (pp. 600–604). https://doi.org/10.1109/ICACITE57410.2023.10183013

Anand, A., Singh, M. P., Sah, S., Verma, A., Khatib, M. N., Pant, M., Zahiruddin, Q. S., & Rustagi, S. (2024). H5N1 avian influenza in USA: a call for vigilance in one health surveillance. *Clinical Infection in Practice*, *23*, 100369. https://doi.org/10.1016/j.clinpr.2024.100369

Arora, A., Singh, R., Malik, K., & Kestwal, U. (2024). Association of sexual performance and intimacy with satisfaction in life in head and neck cancer patients: a review. *Oral Oncology Reports*, *11*, 100563. https://doi.org/10.1016/j.oor.2024.100563

Arora, M., Ahmad, V., Kumar, R., & Singh, R. (2025). The impact of industry 4.0 technologies on operational excellence methods: application of fuzzy topsis. *EMJ – Engineering Management Journal*. https://doi.org/10.1080/10429247.2024.2443703

Asha, P., Mannepalli, K., Khilar, R., Subbulakshmi, N., Dhanalakshmi, R., Tripathi, V., Mohanavel, V., Sathyamurthy, R., & Sudhakar, M. (2022). Role of machine learning in attaining environmental sustainability. *Energy Reports*, *8*, 863–871. https://doi.org/10.1016/j.egyr.2022.09.206

Bachheti, A., Deepti, Bachheti, R. K., Singh, A., Zebeaman, M., Hunde, Y., & Husen, A. (2023). Bioactive constituents and health promoting compounds of underutilized fruits of the northern Himalayas of India: a review. *Food Production, Processing and Nutrition*, *5*(1), 24. https://doi.org/10.1186/s43014-023-00140-5

Barot, M., Modi, A., & Husen, A. (2025). Engineering soil microbiota for crop plant improvement. In: A. Husen (Ed.), *Agricultural Crop Improvement: Plant and Soil Relationships*. CRC Press. https://doi.org/10.1201/9781032630366-4

Bathla, G., Kumar, S., Garg, H., & Saini, D. (2024). *Artificial Intelligence in Healthcare: Emphasis on Diabetes, Hypertension, and Depression Management*. CRC Press. https://doi.org/10.1201/9781003522096

Belwal, R., & Belwal, S. (2017). Employers' perception of women workers in Oman and the challenges they face. *Employee Relations*, *39*(7), 1048–1065. https://doi.org/10.1108/ER-09-2016-0183

Bhardwaj, M., Prakash, N., & Malhotra, R. K. (2024). Energy sector and carbon trading in the future: a reference to India. *2024 International Conference on Smart Devices, ICSD 2024*. https://doi.org/10.1109/ICSD60021.2024.10751388

Bhat, R., Gupta, M., Choudhary, T. F., & Singh, A. K. (2024). Modelling the effects of climatic variables and political instability on the agricultural sector in South Asian countries. In P. O. de Pablos (Ed.), *Building Climate Neutral Economies through Digital Business and Green Skills*. IGI Global. https://doi.org/10.4018/979-8-3693-5673-9.ch001

Bhatt, C., Singh, S., Chauhan, R., Singh, T., & Uniyal, A. (2023b). Artificial Intelligence in current education: roles, applications & challenges. *Proceedings – 2023 3rd International Conference on Pervasive Computing and Social Networking, ICPCSN 2023* (pp. 241–244). https://doi.org/10.1109/ICPCSN58827.2023.00045

Bhatt, S., & Dani, R. (2024). Social media and community engagement: empowering local voices in regenerative tourism. In P. K. Tyagi, V. Nadda, K. Kankaew, & K. Dube (Eds.), *Examining Tourist Behaviors and Community Involvement in Destination Rejuvenation*. IGI Global. https://doi.org/10.4018/979-8-3693-6819-0.ch009

Bhatt, S., Dani, R., & Singh, A. K. (2024a). Exploring cutting-edge approaches to sustainable tourism infrastructure and design a case studies of regenerative accommodation and facilities. In P. K. Tyagi, V. Nadda, K. Kankaew, & K. Dube (Eds.), *Dimensions of Regenerative Practices in Tourism and Hospitality*. IGI Global. https://doi.org/10.4018/979-8-3693-4042-4.ch003

Bhatt, S., Dani, R., & Verma, P. (2024b). Local community participation in meaningful tourism: the case study of Himalayan state of Uttarakhand. In P. K. Tyagi, V. Nadda, K. Kankaew, & K. Dube (Eds.), *Examining Tourist Behaviors and Community Involvement in Destination Rejuvenation*. IGI Global. https://doi.org/10.4018/979-8-3693-6819-0.ch018

Bhuvaneswari, C. A., Garg, N., Supreeth, B. R., Patjoshi, P. K., Ushapriya, C., & Tiwari, M. (2023). Optimistic boosting internet connection throughput by incorporating blockchain filters into network infrastructure. *Proceedings of the 2023 International Conference on Innovative Computing, Intelligent Communication and Smart Electrical Systems, ICSES 2023*. https://doi.org/10.1109/ICSES60034.2023.10465436

Bijalwan, P., Gupta, A., Johri, A., Wasiq, M., & Khalil Wani, S. (2025). Unveiling sora open AI's impact: a review of transformative shifts in marketing and advertising employment. *Cogent Business and Management, 12*(1). https://doi.org/10.1080/23311975.2024.2440640

Bohra, N. S., Bansal, S. S., Johri, A., & Kathari, S. (2024). Evaluating infrastructure fund performance in India: a study of thematic investing. In M. Irfan, K. Hussainey, S. A. Chan Bukhari, & Y. Nam (Eds), *Issues of Sustainability in AI and New-Age Thematic Investing*. IGI Global. https://doi.org/10.4018/979-8-3693-3282-5.ch002

Caiado, R. G. G., Santos, R. S., Schery, C. A. D., Scavarda, L. F., & Garza-Reyes, J. A. (2025). Balancing priorities: a configurational framework to achieve strategic benefits from sustainable industry 4.0 enablers in operations and supply chain. *Business Strategy and the Environment, 34*, 2934–2963. https://doi.org/10.1002/bse.4099

Cavaliere, L. P. L., Byloppilly, R., Khan, S. D., Othman, B. A., Muda, I., & Malhotra, R. K. (2024). Acceptance and effectiveness of industry 4.0 internal and external organisational initiatives in Malaysian firms. *International Journal of Management and Enterprise Development, 23*(1), 1–25. https://doi.org/10.1504/IJMED.2024.138422

Chaudhary, A., Meenakshi, M., Sharma, S., Rahman, M., & Srinivasan, S. (2025). Enhancing urban mobility: machine learning-powered fusion approach for intelligent traffic congestion control in smart cities. *International Journal of System Assurance Engineering and Management*. https://doi.org/10.1007/s13198-024-02672-6

Cuevas-Pichardo, L. J., Maldonado-Guzmán, G., Gómez-Guillamón, A. D., & Garza-Reyes, J. A. (2025). The mediating role of circular economy in the relationship between industry 4.0 and sustainable performance in the manufacturing industry. *Business Strategy and the Environment, 34*(2), 1735–1750. https://doi.org/10.1002/bse.4075

Dani, R., Kukreti, R., Negi, A., & Kholiya, D. (2020). Impact of COVID-19 on education and internships of hospitality students. *International Journal of Current Research and Review, 12*(21 Special Issue), 86–90. https://doi.org/10.31782/IJCRR.2020.SP54

Dutta, A., Singh, P., Dobhal, A., Mannan, D., Singh, J., & Goswami, P. (2023). Entrepreneurial aptitude of women of an aspirational district of Uttarakhand. *Indian Journal of Extension Education*, *59*(2), 103–107. https://doi.org/10.48165/IJEE.2023.59222

Gangwar, V. P., & Srivastva, S. P. (2020). Impact of micro finance in poverty eradication via SHGs: a study of selected districts in U.P. *International Journal of Advanced Science and Technology*, *29*(2), 3818–3829.

Goyal, P., Kukreja, T., Agarwal, A., & Khanna, N. (2015). Narrowing awareness gap by using e-learning tools for counselling university entrants. *Conference Proceeding – 2015 International Conference on Advances in Computer Engineering and Applications, ICACEA 2015* (pp. 847–851). https://doi.org/10.1109/ICACEA.2015.7164822

Goyal, S., Kaur, A., Sharma, V., & Batra, N. (2024). IMHS: an IoT-based machine learning healthcare system. *2024 2nd International Conference on Disruptive Technologies, ICDT 2024* (916–919). https://doi.org/10.1109/ICDT61202.2024.10489127

Grover, D., Sharma, S., Kaur, P., Mittal, A., & Sharma, P. K. (2024). Societal elements that impact the performance of women entrepreneurs in tier-II cities: a study of Rohilkhand region of Uttar Pradesh. *2024 IEEE Zooming Innovation in Consumer Technologies Conference, ZINC 2024* (pp. 114–117). https://doi.org/10.1109/ZINC61849.2024.10579316

Gupta, M., Kumari, I., & Singh, A. K. (2024). Impact of human capital on SDG1 in selected G20 countries. In M. Valeri, & Shekhar (Eds.), *Interlinking SDGs and the Bottom-of-the-Pyramid through Tourism*. IGI Global. https://doi.org/10.4018/979-8-3693-3166-8.ch002

Gupta, P., Gopal, S., Sharma, M., Joshi, S., Sahani, C., & Ahalawat, K. (2023). Agriculture informatics and communication: paradigm of E-governance and drone technology for crop monitoring. *9th International Conference on Smart Computing and Communications: Intelligent Technologies and Applications, ICSCC 2023* (pp. 113–118). https://doi.org/10.1109/ICSCC59169.2023.10335058

Husen, A. (2025). *Agricultural Crop Improvement: Plant and Soil Relationships*. CRC Press. https://doi.org/10.1201/9781032630366

Indu, R., Dimri, S. C., & Kumar, B. (2023). Identification of location for police headquarters to deal with crime against women in India using clustering based on k-means algorithm. *International Journal of Computing and Digital Systems*, *14*(1), 965–974. https://doi.org/10.12785/ijcds/140175

Ingale, N. V., Saravana Kumar, G., Panduro-Ramirez, J., Raj, M., Vaseem Akram, S., & Rawat, R. (2023). Role of IOT in managing education management tools: a technical review. *2023 3rd International Conference on Advance Computing and Innovative Technologies in Engineering, ICACITE 2023* (pp. 2056–2061). https://doi.org/10.1109/ICACITE57410.2023.10182953

Jain, P., Gupta, V. K., Tiwari, H., Shukla, A., Pandey, P., & Gupta, A. (2023). Human-computer interaction: a systematic review. In M. H. Khan & S. Singh (Eds.),*Proceedings – 2023 International Conference on Advanced Computing and Communication Technologies, ICACCTech 2023* (pp. 31–36). Institute of Electrical and Electronics Engineers Inc. https://doi.org/10.1109/ICACCTech61146.2023.00015

Jindal, M., Bajal, E., Singh, P., Diwakar, M., Arya, C., & Sharma, K. (2022). Online education in Covid-19: limitations and improvements. *2021 IEEE 8th Uttar Pradesh Section International Conference on Electrical, Electronics and Computer Engineering, UPCON 2021*. https://doi.org/10.1109/UPCON52273.2021.9667605

Joshi, S., Bisht, B., Kumar, V., Singh, N., Jameel Pasha, S. B., Singh, N., & Kumar, S. (2024). Artificial intelligence assisted food science and nutrition perspective for smart nutrition research and healthcare. *Systems Microbiology and Biomanufacturing*, *4*(1), 86–101. https://doi.org/10.1007/s43393-023-00200-4

Josphineleela, R., Gupta, R., Misra, N., Malik, M., Somasundaram, K., & Gangodkar, D. (2023a). Blockchain based multi-layer security network authentication system for uncertain attack in the wireless communication system. *2023 3rd International Conference on Advance Computing and Innovative Technologies in Engineering, ICACITE 2023* (pp. 877–881). https://doi.org/10.1109/ICACITE57410.2023.10182747

Josphineleela, R., Kaliappan, S., Natrayan, L., & Bhatt, U. M. (2023b). Intelligent virtual laboratory development and implementation using the rasa framework. *Proceedings – 7th International Conference on Computing Methodologies and Communication, ICCMC* 2023 (pp. 1172–1176). https://doi.org/10.1109/ICCMC56507.2023.10083701

Kanojia, P., Malhotra, R. K., & Uniyal, A. K. (2022). Organizational commitment and the academic staff in HEI's in North West India. *Proceedings – 2022 International Conference on Recent Trends in Microelectronics, Automation, Computing and Communications Systems, ICMACC 2022* (pp. 365–370). https://doi.org/10.1109/ICMACC54824.2022.10093347

Kashif, M., Gupta, V. P., Shajar, S. N., & Weqar, F. (2025). Emerging trends and potential challenges in the advancements of e-financial strategies. In A. R. Khan & S. Mehta (Eds.), *E-Finance: Transformative Strategies for Sustainable Development in the Digital Era.* Nova Science Publishers, Inc.

Khan, T., Singh, K., & Purohit, K. C. (2020). Icma: an efficient integrated congestion control approach. *Recent Patents on Engineering*, *14*(3), 294–309. https://doi.org/10.2174/1872212114666191231150916

Khandelwal, C., Kumar, S., Tripathi, V., & Madhavan, V. (2023). Joint impact of corporate governance and risk disclosures on firm value: evidence from emerging markets. *Research in International Business and Finance*, *66,* 102022. https://doi.org/10.1016/j.ribaf.2023.102022

Khanna, R., Jindal, P., & Noja, G. G. (2023). Blockchain technologies, a catalyst for insurance sector. In S. Verma & J. K. Patel (Eds.), *The Application of Emerging Technology and Blockchain in the Insurance Industry.* Springer.

Khanna, R., Sharma, S., Czeh, T., & Adhana, K. (2025). Challenges, interventions, and strategies for higher education transformation: vision 2047. *Lecture Notes in Networks and Systems*, *1075 LNNS*, 637–655. https://doi.org/10.1007/978-981-97-6106-7_38

Kimothi, S., Bhatt, V., Kumar, S., Gupta, A., & Dumka, U. C. (2024). Statistical behavior of the European Energy Exchange-Zero Carbon Freight Index (EEX-ZCFI) assessments in the context of Carbon Emissions Fraction Analysis (CEFA). *Sustainable Futures*, *7*, 100164. https://doi.org/10.1016/j.sftr.2024.100164

Kukreti, M., Sehajpal, S., Tiwari, R., & Sood, K. (2025). *Driving Global Health and Sustainable Development Goals with Smart Technology* (1st ed.). IGI Global Scientific Publishing. https://doi.org/10.4018/979-8-3373-0240-9

Kumar, M. A., Valli Mayil, V., Jeyalakshmi, T. R., Ghanshala, A., & Gupta, A. (2023a). Wireless Body Area Network (WBAN) centric healthcare system for continuous monitoring. *2023 9th International Conference on Advanced Computing and Communication Systems, ICACCS 2023* (pp. 214–219). https://doi.org/10.1109/ICACCS57279.2023.10112922

Kumar, R., & Khanna, R. (2023). Role of artificial intelligence in digital currency and future applications. *Proceedings of the 2023 2nd International Conference on Augmented Intelligence and Sustainable Systems, ICAISS 2023* (pp. 42–46). https://doi.org/10.1109/ICAISS58487.2023.10250480

Kumar, R., Goel, R., Singh, T., Mohanty, S. M., Gupta, D., Alkhayyat, A., & Khanna, R. (2024b). Sustainable finance factors in Indian economy: analysis on policy of climate change and energy sector. *Fluctuation and Noise Letters*, *23*(2). https://doi.org/10.1142/S0219477524400042

Kumar, R., Kathuria, S., Malholtra, R. K., Kumar, A., Gehlot, A., & Joshi, K. (2023e). Role of cloud computing in goods and services tax (GST) and future application. *2nd International Conference on Sustainable Computing and Data Communication Systems, ICSCDS 2023 – Proceedings* (pp. 1443–1447). https://doi.org/10.1109/ICSCDS56580.2023.10104597

Kumar, R., Khannna Malholtra, R., & Grover, C. A. N. (2023f). Review on artificial intelligence role in implementation of goods and services tax (GST) and future scope. *2023 International Conference on Artificial Intelligence and Smart Communication, AISC* 2023 (pp. 348–351). https://doi.org/10.1109/AISC56616.2023.10085030

Kumar, R., Malhotra, R. K., Pandey, S., Gehlot, A., Gautam, I., & Chamola, S. (2023g). Role of artificial intelligence in input tax credit reconciliation. *Proceedings – 2023 3rd International Conference on Pervasive Computing and Social Networking, ICPCSN 2023* (pp. 497–501). https://doi.org/10.1109/ICPCSN58827.2023.00086

Kumar, R., Sexena, A., & Gehlot, A. (2023h). Artificial intelligence in smart education and futuristic challenges. *2023 International Conference on Disruptive Technologies, ICDT 2023* (pp. 432–435). https://doi.org/10.1109/ICDT57929.2023.10151129

Kumar, R., Singh, T., Mohanty, S. N., Goel, R., Gupta, D., Alharbi, M., & Khanna, R. (2023i). Study on online payments and e-commerce with SOR model. *International Journal of Retail and Distribution Management*. https://doi.org/10.1108/IJRDM-03-2023-0137

Kumar, V., Gupta, S., & Khanna, R. (2021). Dengue fever-a worldwide study. *Journal of Medical Pharmaceutical and Allied Sciences*, *10*, 102–108. https://doi.org/10.22270/jmpas.vic2i1.2014

Kumari, J., Singh, P., Mishra, A. K., Singh Meena, B. P., Singh, A., & Ojha, M. (2024). Challenges Hindering Women's involvement in the hospitality industry as entrepreneurs in the era of digital economy. In *Revolutionizing the AI-Digital Landscape: A Guide to Sustainable Emerging Technologies for Marketing Professionals*. Taylor & Francis. https://doi.org/10.4324/9781032688305-9

Lourens, M., Krishna, S. H., Singh, A., Dey, S. K., Pant, B., & Sharma, T. (2022). Role of artificial intelligence in formative employee engagement. In D. R. Kalbande, S. A. Krishnan, K. Gupta, & B. Singh (Eds.), *Proceedings of the 2022 11th International Conference on System Modeling and Advancement in Research Trends, SMART 2022* (pp. 936–941). Institute of Electrical and Electronics Engineers Inc. https://doi.org/10.1109/SMART55829.2022.10047422

Maaze, M. R., Garg, N., Das, S. K., & Shrivastava, S. (2024). Industrial and C&D waste as precursor material in geopolymers: a state-of-the-art review. *Innovative Infrastructure Solutions*, *9*(6). https://doi.org/10.1007/s41062-024-01546-x

Malhotra, R. K., Sharma, C. A. P., Upreti, H., Sharma, S., Prakash, N., & Mittal, S. (2024). Role of psychological contract in employee engagement in higher education: empirical study on Uttarakhand. In P. P. Karde (Ed.), *E3S Web of Conferences* (Vol. 556). EDP Sciences. https://doi.org/10.1051/e3sconf/202455601030

Malik, P., Singh, A. K., Nautiyal, R., & Rawat, S. (2022). Mapping AICTE cybersecurity curriculum onto CyBOK: a case study. In B. B. Gupta, A. K. Bashir, & D. Puthal (Eds.), *Machine Learning for Cyber Security*. De Gruyter. https://doi.org/10.1515/9783110766745-007

Mangla, S. K., Luthra, S., Mishra, N., Singh, A., Rana, N. P., Dora, M., & Dwivedi, Y. (2018). Barriers to effective circular supply chain management in a developing country context. *Production Planning and Control*, *29*(6), 551–569. https://doi.org/10.1080/09537287.2018.1449265

Mishra, D., Kandpal, V., Agarwal, N., & Srivastava, B. (2024). Financial inclusion and its ripple effects on socio-economic development: a comprehensive review. *Journal of Risk and Financial Management*, *17*(3). https://doi.org/10.3390/jrfm17030105

Mittal, A., Pandey, H., Arsh, M., Soni, M., Goyal, H. R., & Sharma, S. (2024). Advancements and applications of artificial intelligence in various sectors: from internet of vehicles to 6G and healthcare. *Proceedings – 2024 4th International Conference on Pervasive Computing and Social Networking, ICPCSN 2024* (pp. 107–112). https://doi.org/10.1109/ICPCSN62568.2024.00026

Naaz, R., Agarwal, S., & Rawal, T. (2025). Eco-tourism as a strategy for a green and sustainable economy: a case study of Uttarakhand tourism. In M. J. Khan, R. Rajendra, & D. J. Paulo (Eds.), *Green Economics and Strategies for Business Sustainability*. IGI Global. https://doi.org/10.4018/979-8-3693-8949-2.ch019

Nagila, A., Saravanakumar, P., Pranavan, S., Goutam, R., Dobhal, D. C., & Singh, G. (2023). An innovative approach of CNN-BiGRU based post-earthquake damage detection of reinforced concrete for frame buildings. *International Conference on Self Sustainable Artificial Intelligence Systems, ICSSAS 2023 – Proceedings* (pp. 56–61). https://doi.org/10.1109/ICSSAS57918.2023.10331894

Neha, Mohanty, S., Alfurhood, B. S., Bakhare, R., Poongavanam, S., & Khanna, R. (2023). The role and impact of artificial intelligence on retail business and its developments. *2023 International Conference on Artificial Intelligence and Smart Communication, AISC 2023* (pp. 1098–1101). https://doi.org/10.1109/AISC56616.2023.10085624

Nikolaidis, P., Ismail, M., Shuib, L., Khan, S., & Dhiman, G. (2022). Predicting student attrition in higher education through the determinants of learning progress: a structural equation modelling approach. *Sustainability (Switzerland)*, *14*(20). https://doi.org/10.3390/su142013584

Onyema, E. M., Almuzaini, K. K., Onu, F. U., Verma, D., Gregory, U. S., Puttaramaiah, M., & Afriyie, R. K. (2022). Prospects and challenges of using machine learning for academic forecasting. *Computational Intelligence and Neuroscience*, *2022*. https://doi.org/10.1155/2022/5624475

Padhi, B. K., Singh, S., Gaidhane, A. M., Abu Serhan, H., Khatib, M. N., Zahiruddin, Q. S., Rustagi, S., Sharma, R. K., Sharma, D., Arora, M., & Satapathy, P. (2024). Inequalities in cardiovascular disease among elderly Indians: a gender perspective analysis using LASI wave-I (2017-18). *Current Problems in Cardiology*, *49*(7). https://doi.org/10.1016/j.cpcardiol.2024.102605

Pandey, R. P., Bansal, S., Awasthi, P., Dixit, V., Singh, R., & Yadava, V. (2023). Attitude and myths related to stalking among early and middle age adults. *Psychology Hub*, *40*(3), 85–94. https://doi.org/10.13133/2724-2943/17960

Pant, A., Joshi, R. C., Sharma, S., & Pant, K. (2024). Air quality and public health risk assessment: a case of an industrial area in Haridwar city, Uttarakhand (India). *Indian Journal of Public Health*, *68*(2), 222–226. https://doi.org/10.4103/ijph.ijph_279_23

Pant, M., Sharma, S., Chauhan, A., & Singh, A. P. (2025). Modelling and optimisation of sustainable production model with screening and carbon awareness for manufacturing industry. *International Journal of Applied Management Science*, *17*(1), 27–51. https://doi.org/10.1504/IJAMS.2025.143671

Pant, V., Bhasin, S., & Jain, S. (2017). Self-Learning system for personalized E-Learning. *2017 International Conference on Emerging Trends in Computing and Communication Technologies, ICETCCT 2017*, 2018-*Janua* (pp. 1–6). https://doi.org/10.1109/ICETCCT.2017.8280344

Pathak, K., Prakash, G., Samadhiya, A., Kumar, A., & Luthra, S. (2025). Impact of Gen-AI chatbots on consumer services experiences and behaviors: focusing on the sensation of awe and usage intentions through a cybernetic lens. *Journal of Retailing and Consumer Services*, *82*. https://doi.org/10.1016/j.jretconser.2024.104120

Pattanshetti, M. K., Jasola, S., Rajput, A., & Pant, V. (2021). Proposed eLearning framework using open corpus web resources. *Proceedings of the 2021 1st International Conference on Advances in Electrical, Computing, Communications and Sustainable Technologies, ICAECT 2021*. https://doi.org/10.1109/ICAECT49130.2021.9392591

Prasad, M. S. G., More, P., Christa, S., & Kumar, M. A. (2023). A chronic disease diagnosis model for smart healthcare systems enabled by artificial intelligence and the internet of things. In R. Rajendra, P. K. Paul, & S. Bhatia (Eds.), *Applications of Artificial Intelligence in the Healthcare Sector*. Nova Science Publishers, Inc.

Prerna, Singh, P., & Singh, D. P. (2024). Role of disruptive technologies in the smart healthcare sector of society 5.0. In M. Singh, M. Kaur, & D. Gupta (Eds.), *Artificial Intelligence and Society 5.0: Issues, Opportunities, and Challenges*. CRC Press. https://doi.org/10.1201/9781003397052-13

Prikshat, V., Kumar, S., & Nankervis, A. (2019). Work-readiness integrated competence model: conceptualisation and scale development. *Education and Training*, *61*(5), 568–589. https://doi.org/10.1108/ET-05-2018-0114

Prikshat, V., Kumar, S., & Raje, P. (2017). Antecedents, consequences and strategic responses to graduate work-readiness: challenges in India. In S. F. Johnston, H. Strachan, & P. J. Mangelsdorf (Eds.), *Transitions from Education to Work: Workforce Ready Challenges in the Asia Pacific*. Taylor & Francis. https://doi.org/10.4324/9781315533971

Rajora, R., Gupta, H., Malhotra, S., Devliyal, S., & Sunil, G. (2024). Advancing renal health: Unleashing the power of CNNs in multi-class classification for precise kidney condition diagnosis. *2024 IEEE 9th International Conference for Convergence in Technology, I2CT 2024*. https://doi.org/10.1109/I2CT61223.2024.10544018

Ramarajan, M., Dinesh, A., Muthuraman, C., Rajini, J., Anand, T., & Segar, B. (2024). AI-driven job displacement and economic impacts: Ethics and strategies for implementation. In S. Chatterjee & A. K. Ghosh (Eds.), *Cases on AI Ethics in Business*. IGI Global. https://doi.org/10.4018/9798369326435.ch013

Rana, D. S., Dimri, S. C., Malik, P., & Dhondiyal, S. A. (2022). Impact of computational thinking in engineering and K12 education. *4th International Conference on Inventive Research in Computing Applications, ICIRCA 2022 – Proceedings* (pp. 697–701). https://doi.org/10.1109/ICIRCA54612.2022.9985593

Rathore, R., & Goudar, R. H. (2015). SPARQL-based personalised E-Learning system designed using ontology (SPELSO): an architecture. *International Journal of Knowledge and Learning*, *10*(4), 384–416. https://doi.org/10.1504/IJKL.2015.077554

Rawat, R. S., Diwakar, M., Garg, U., & Srivastava, P. (2025). Developing an intelligent system for efficient botnet detection in IoT environment. *International Journal of Mathematical, Engineering and Management Sciences*, *10*(2), 537–553. https://doi.org/10.33889/IJMEMS.2025.10.2.027

Rawat, R. S., Singh, V., & Dumka, A. (2022). Complaint management in Ethiopian vocational and technical education institutions: a framework and implementation of a decision support system. *2022 International Conference on 4th Industrial Revolution Based Technology and Practices, ICFIRTP 2022* (pp. 73–79). https://doi.org/10.1109/ICFIRTP56122.2022.10063207

Sahithya, B., Guru Prasad, M. S., Sahithi, B., Madhu Krishna, K., Devarlla, A. C., & Yashavanth, T. R. (2024). Empowering healthcare with ai: advancements in medical image analysis, electronic health records analysis, and AI-driven chatbots. *2024 3rd International Conference for Innovation in Technology, INOCON 2024*. https://doi.org/10.1109/INOCON60754.2024.10511753

Sarantis, D., Rizun, N., Alexopoulos, C., & Saxena, S. (2025). Analyzing behavioral intention of open government data adoption across Latvia, India and Poland: does national culture matter? *Journal of Science and Technology Policy Management*. https://doi.org/10.1108/JSTPM-12-2023-0234

Sehajpal, S., & Kaur, N. (2025). Neuroscience, nursing, and financial innovation : a synergistic approach to optimized patient care and economic well-being. In T. Sanjay, S. C. Bhupinderpal Pal, J. Amar, O. Ercan, & K. Pawan (Eds.), *Neuromarketing's Role in Sustainable Finance* (pp. 133–146). IGI Global Scientific Publishing. https://doi.org/10.4018/979-8-3693-9117-4.ch008

Semwal, P., Painuli, S., Shabaaz Begum, J. P., Jamloki, A., Rauf, A., Olatunde, A., Mominur Rahman, M., Mukerjee, N., Khalil, A. A., Aljohani, A. S. M., Al Abdulmonem, W., & Simal-Gandara, J. (2024). Corrigendum to "Exploring the nutritional and health benefits of pulses from the Indian Himalayan region: a glimpse into the region's rich agricultural heritage" [Food Chem. 422 (2023) 136259]. *Food Chemistry, 445,* 138830. https://doi.org/10.1016/j.foodchem.2024.138830

Sharma, P., Malhotra, R. K., & Kumar, R. (2024a). How is digitalization reshaping the landscape of Rural banking in India? *2024 International Conference on Smart Devices, ICSD 2024.* https://doi.org/10.1109/ICSD60021.2024.10750981

Sharma, P., Malhotra, R. K., Altybekova, D., & Singh, C. P. (2024b). AI and machine learning as catalysts for educational equity: bridging the Urban–Rural divide. *Lecture Notes in Networks and Systems, 1074 LNNS*, 41–55. https://doi.org/10.1007/978-981-97-6103-6_4

Sharma, P., Malhotra, R. K., Ojha, M. K., & Gupta, S. (2022). Impact of meditation on mental & physical health and thereby on academic performance of students: a study of higher educational institutions of Uttarakhand. *Journal of Medical Pharmaceutical and Allied Sciences, 11*(2), 4641–4644. https://doi.org/10.55522/jmpas.V11I2.2309

Sharma, S., Gupta, A., & Tyagi, R. (2023a). Artificial intelligence enabled sustainable education system using Vedic scripture and cyber security. *2023 2nd International Conference on Advances in Computational Intelligence and Communication, ICACIC 2023.* https://doi.org/10.1109/ICACIC59454.2023.10435133

Sharma, S., Gupta, C., Khanna Malhotra, R., & Upreti, H. (2024c). Sustainable banking practices: impact, challenges and opportunities. In P. P. Kanjalkar (Ed.), *E3S Web of Conferences* (Vol. 556). EDP Sciences. https://doi.org/10.1051/e3sconf/202455601031

Sharma, S., Kadayat, Y., & Tyagi, R. (2023b). Sustainable global democratic e-governance system using Vedic scripture, artificial intelligence, cloud computing and augmented reality. *Proceedings of the International Conference on Circuit Power and Computing Technologies, ICCPCT 2023* (pp. 113–118). https://doi.org/10.1109/ICCPCT58313.2023.10245405

Sharma, S., Singh, V., & Sarkar, D. (2023c). Machine vision enabled fall detection system for specially abled people in limited visibility environment. *2023 3rd Asian Conference on Innovation in Technology, ASIANCON 2023.* https://doi.org/10.1109/ASIANCON58793.2023.10270769

Sharma, S., Singh Rawal, R., Pandey, D., & Pandey, N. (2021). Microbial world for sustainable development. In A. K. Chandel, S. K. Ghosh, & R. Y. Surampalli (Eds.), *Microbial Technology for Sustainable Environment.* Springer Nature. https://doi.org/10.1007/978-981-16-3840-4_1

Sharma, V., & Jain, S. (2020). Managers training programs effectiveness evaluation by using different machine learning approaches. *Proceedings of the 4th International Conference on Electronics, Communication and Aerospace Technology, ICECA 2020* (pp. 1453–1457). https://doi.org/10.1109/ICECA49313.2020.9297556

Shehzad, N., Ramtiyal, B., Jabeen, F., Mangla, S. K., & Vijayvargy, L. (2024). Metaverse adoption as a cornerstone for sustainable healthcare firms in the industry 5.0 epoch. *Journal of Enterprise Information Management, 37*(4), 1254–1281. https://doi.org/10.1108/JEIM-10-2023-0559

Shrivastava, A., Usha, R., Kukreti, R., Sharma, G., Srivastava, A. P., & Khan, A. K. (2023). Women safety precaution. *2023 1st International Conference on Circuits, Power, and Intelligent Systems, CCPIS 2023*. https://doi.org/10.1109/CCPIS59145.2023.10291594

Shukla, R. P., Taneja, S., Gundawar, P. S., Jain, R. K., & Shukla, P. (2024). Behaviour examining sensorimotor and affective responses to marketing stimuli through neuropsychology. In R. Malik, S. Malhan, & M. Arora (Eds.), *Sensible Selling Through Sensory Neuromarketing*. IGI Global. https://doi.org/10.4018/979-8-3693-4236-7.ch013

Singh, A., Goel, A., Chauhan, A., & Singh, S. K. (2025). Sustainability of electronic product manufacturing through e-waste management and reverse logistics. *Sustainable Futures, 9*. https://doi.org/10.1016/j.sftr.2025.100490

Singh, A., & Srivastava, A. (2024). Innovation needed to improve road construction and road infrastructure in the future in India. In G. Arvind & V. R. Jayanthi (Eds.), *IOP Conference Series: Earth and Environmental Science* (Vol. 1326, Issue 1). Institute of Physics. https://doi.org/10.1088/1755-1315/1326/1/012097

Singh, A. B., Meena, H. K., Khandelwal, C., & Dangayach, G. S. (2023a). Sustainability assessment of higher education institutions: a systematic literature review†. *Engineering Proceedings, 37*(1). https://doi.org/10.3390/ECP2023-14728

Singh, A. K., Singh, S., & Sankaranarayanan, K. G. (2024b). A comparative performance of green technology, green growth, social, and economic development in India and China. In P. Ordóñez de Pablos (Ed.), *Digital Technologies for a Resource Efficient Economy*. IGI Global. https://doi.org/10.4018/979-8-3693-2750-0.ch006

Singh, A. K., Singh, R., & Singh, S. (2024a). A review on factors affecting and performance of nutritional security of women and children in India. In S. M. Sanusi Sadiq, N. Karunakaran, A. M. Makarfi, & S. Sharma (Eds.), *Impact of Women in Food and Agricultural Development*. IGI Global. https://doi.org/10.4018/979-8-3693-3037-1.ch019

Singh, A. R., Vishnuram, P., Alagarsamy, S., Bajaj, M., Blazek, V., Damaj, I., Rathore, R. S., Al-Wesabi, F. N., & Othman, K. M. (2024c). Electric vehicle charging technologies, infrastructure expansion, grid integration strategies, and their role in promoting sustainable e-mobility. *Alexandria Engineering Journal, 105*, 300–330. https://doi.org/10.1016/j.aej.2024.06.093

Singh, K. D., & Singh, P. (2023). A novel cloud-based framework to predict the employability of students. In R. Kumar, R. Kumar, M. Gupta, M. Gupta, R. Singh, & R. Singh (Eds.), *2023 International Conference on Advancement in Computation and Computer Technologies, InCACCT 2023* (pp. 528–532). Institute of Electrical and Electronics Engineers Inc. https://doi.org/10.1109/InCACCT57535.2023.10141760

Singh, K. D., Singh, P., Kaur, G., Khullar, V., Chhabra, R., & Tripathi, V. (2023b). Education 4.0: exploring the potential of disruptive technologies in transforming learning. *Proceedings of International Conference on Computational Intelligence and Sustainable Engineering Solution, CISES* 2023 (pp. 586–591). https://doi.org/10.1109/CISES58720.2023.10183547

Singh, K. U., Chaudhary, V., Sharma, P. K., Kumar, P., Varshney, N., & Singh, T. (2024d). Integrating GPS and GSM technologies for enhanced women's safety: a fingerprint-activated device approach. *2024 International Conference on Automation and Computation, AUTOCOM 2024* (pp. 657–662). https://doi.org/10.1109/AUTOCOM60220.2024.10486120

Singh, M., Ghai, S., Mishra, A. K., & Sharma, R. (2025). The era of machine learning: developments, ideas, and future possibilities. In *E-Finance: Transformative Strategies for Sustainable Development in the Digital Era*. Nova Science Publishers, Inc.

Singh, P., & Singh, K. D. (2023). Fog-centric intelligent surveillance system: a novel approach for effective and efficient surveillance. In R. Kumar, R. Kumar, M. Gupta, M. Gupta, R. Singh, & R. Singh (Eds.), *2023 International Conference on Advancement in Computation and Computer Technologies, InCACCT 2023* (pp. 762–766). Institute of Electrical and Electronics Engineers Inc. https://doi.org/10.1109/InCACCT57535.2023.10141802

Singh, S., Singh, A., Mohan, A., Batola, M., & Kumar, N. (2024f). Health expenditures and health outcomes in central Europe and the Baltic region. *Baltic Region*, *16*(2), 126–143. https://doi.org/10.5922/2079-8555-2024-2-7

Singh, S. P., Singh, P., Diwakar, M., & Kumar, P. (2024e). Improving quality of service for internet of things(IoT) in real life application: a novel adaptation based hybrid evolutionary algorithm. *Internet of Things (Netherlands), 27.* https://doi.org/10.1016/j.iot.2024.101323

Solanki, A., Singh, P., Pandit, M., Sawle, Y., Alotaibi, M. A., Malik, H., Márquez, F. P. G., & Afthanorhan, A. (2024). Techno-economic-environmental assessment of stand-alone hybrid renewable energy system for different batteries using HOMER-Pro. *International Journal of Mathematical, Engineering and Management Sciences*, *9*(4), 779–800. https://doi.org/10.33889/IJMEMS.2024.9.4.040

Srinivasan, S. (2024). Empowerment and leadership quality improve unorganized women migrant workers in Karur District, Tamil Nadu, India. In S. Thomas, V. P. Bhatt, M. V. Devi, & K. Vijayakumar (Eds.), *Empowering and Advancing Women Leaders and Entrepreneurs*. IGI Global. https://doi.org/10.4018/979-8-3693-7107-7.ch004

Srivastava, B., Kandpal, V., & Jain, A. K. (2024a). Financial well-being of women self-help group members: a qualitative study. *Environment, Development and Sustainability*. https://doi.org/10.1007/s10668-024-04879-w

Srivastava, S. P., Malhotra, R. K., & Mukherjee, R. K. (2024b). Data driven decision making in insurance. *2024 International Conference on Smart Devices, ICSD 2024*. https://doi.org/10.1109/ICSD60021.2024.10751068

Thapa, D., Kumar, V., Naik, B., Kumar, V., Gupta, A. K., Mohanta, Y. K., Mishra, B., & Rustagi, S. (2024). Harnessing probiotic foods: managing cancer through gut health. *Food Science and Biotechnology*, *33*(9), 2141–2160. https://doi.org/10.1007/s10068-024-01638-5

Tiwari, R., Agrawal, P., Singh, P., Bajaj, S., Verma, V., & Chauhan, A. S. (2023). Technology enabled integrated fusion teaching for enhancing learning outcomes in higher education. *International Journal of Emerging Technologies in Learning*, *18*(7), 243–249. https://doi.org/10.3991/IJET.V18I07.36799

Tripathi, V. M., & Mohan, A. (2016). Microfinance and empowering rural women in the Terai, Uttarakhand, India. *International Journal of Agricultural and Statistical Sciences*, *12*(2), 523–530.

Upreti, H., & Malhotra, R. K. (2024). Bridging the Urban-rural education Gap in India through CSR (corporate social responsibility) initiatives: a conceptual study with special reference to sustainable development goal 4 (quality education). In P. P. K. (Ed.), *E3S Web of Conferences* (Vol. 556). EDP Sciences. https://doi.org/10.1051/e3sconf/202455601032

Upreti, H., Malhotra, R. K., Kumar, R., & Karki, B. (2024). Leveraging smart devices for corporate social responsibility in India: advancements, trends, and concerns towards sustainability. *2024 International Conference on Smart Devices, ICSD 2024*. https://doi.org/10.1109/ICSD60021.2024.10751578

Verma, A., Sharma, D., Pant, M., Khatib, M. N., Singh, M. P., Zahiruddin, Q. S., Rustagi, S., Gaidhane, S., & Gaidhane, A. M. (2024). First sighting of human H5N1 in Australia: a detailed account and public health implications. *New Microbes and New Infections*, *60–61*. https://doi.org/10.1016/j.nmni.2024.101447

Yadav, P. M., Patra, I., Mittal, V., Nagorao, C. G., Udhayanila, R., & Saranya, A. (2023). Implementation of IOT on English Language Classroom Management. *2023 3rd International Conference on Advance Computing and Innovative Technologies in Engineering, ICACITE 2023* (pp. 1686–1690). https://doi.org/10.1109/ICACITE57410.2023.10182984

Yildirim, S., Yildirim, D. C., Demirtaş, I., & Kandpal, V. (2024). Green transition impacts on the economy, society, and environment. In S. Yıldırım, D. C. Yıldırım, I. Demirtaş, & V. Kandpal (Eds.), *Green Transition Impacts on the Economy, Society, and Environment*. IGI Global. https://doi.org/10.4018/979-8-3693-3985-5

Zaheer, A., Zehra Hussain, D., Verma, A., Sah, S., Pant, M., Nazli Khatib, M., Pratap Singh, M., Syed Zahiruddin, Q., & Rustagi, S. (2024). Marburg virus: an emerging threat to global health. *Clinical Infection in Practice*, *23*. https://doi.org/10.1016/j.clinpr.2024.100371

Zhang, Y., Cao, C., Gu, J., & Garg, H. (2023). The impact of top management team characteristics on the risk taking of Chinese private construction enterprises. *Systems*, *11*(2). https://doi.org/10.3390/systems11020067

4 The Future of Marketing

Influence of Augmented Reality and Virtual Reality Technologies

Kashish Jindal, Ashulekha Gupta, and Kavya Jindal

4.1 INTRODUCTION

Marketing has not only grown but has also come a very long way from simple barter trading, which involved the use of face-to-face communication in the local markets to a very complex function involving globalization and the use of sophisticated technologies (Rust, 2020). In later years, it was the growth of the print media through newspapers and magazines whereby the businesses could advertise their products to many people (Katz, 2022). The internet and social media have done a lot in altering marketing strategies from one-way communication to bidirectional and immediate. In today's world of technological developments and the availability of rich information, significant changes have been observed in the marketing analytics area (Shah and Murthi, 2021). The current generation of marketers is equipped with versatile tools and techniques including artificial intelligence, machine learning, predictive analysis, and forecasting, which can be employed to gain a deeper understanding of consumer behaviour, choices, and trends (Storbacka and Moser, 2020).

The marketing industry has significantly benefited from augmented reality (AR) and virtual reality (VR) technologies by offering immersive experiences to consumers (Bucea-Manea-Ţoniş et al., 2021). AR combines real and digital information on top of the existing environment and thus forms a perfect symbiosis between the real and computer-generated worlds (Dargan et al., 2023), whereas VR offers completely built environments to give users completely new and emotionally engaging experiences (Hamad and Jia, 2022). This innovation provides new grounds for the company to interact with customers on more personalized levels. The use of AR and VR in multimedia advertisements has thus become greatly transformative to create a better customer experience and thus revolutionize the marketing industry. These technologies are not only used in advertising, but are also applied to television, games, broadcasting, and immersed movies, indicating their versatility of application areas (Heller et al., 2021).

It has been found that AR and VR enhance consumer engagement and the perceptions consumers have of products and brands. For example, in the usage of AR applications in the retail scenario, it has been observed that it helps in improving customer satisfaction as customers can now almost touch and feel the products they

DOI: 10.1201/9781003564621-4

intend to buy (Riar et al., 2021), on the same note, in the application of VR in experiential marketing, it has been witnessed that consumers are taken to the branded environments and they hence have a deep emotional bonding with the brand (Regt et al., 2021). Moreover, one study shows that all these technologies enhance consumer engagement, which in turn translates to personalization and ultimately results in patterns of purchase intention and brand loyalty (Bilal et al., 2024).

However, there are still some significant gaps in the literature that cannot be overlooked. Several researchers have paid a lot of attention to the immediate consumer responses to AR and VR, but few have paid adequate attention to the distant impact on brand attitude and customer loyalty. Furthermore, most of them have focused on certain sectors for the application of AR and VR including retail and real estate sectors, thereby creating a research gap on how AR and VR could be fully implemented in other sectors. Also, there is limited research exploring using AR and VR in different industries.

To that extent, this chapter aims to discuss the potential and possible marketing trends of the future by focusing on the AR and VR concepts. It looks at ways in which these tools are currently being used, how they may be beneficial, some of the challenges that may be wielded by them in the coming years, and finally how they may help shape future consumer behaviour and marketing strategies. With the knowledge of the effects of AR and VR, companies can be in a better position to harness the technology and be at par with the trends in the market, which is gradually moving towards digitization and experiences. The chapter tries to answer the following research.

4.2 RESEARCH METHODOLOGY

4.2.1 Systematic Literature Review

A literature review is a procedure of finding, synthesizing, and evaluating all the findings of research that are related to a particular research question, area of study, or phenomenon of interest. Due to the large number of published studies in a given area of research specialization, it assists in constructing overviews of research findings, recognizing research deficits, recommending avenues for further research, and postulating fresh hypotheses (Rosario and Boechat, 2024).

4.2.2 Research Question

The main interest of this research is to explore and understand how AR and VR technologies influence the future of marketing, addressing underexplored areas, the evolution of AR and VR applications in marketing, highlighting major trends, innovations, and their implications for future marketing strategies. Table 4.1 shows the two main research questions about this topic.

4.2.3 Literature Selection

To ensure the relevance of the study as well as the final quality of the research work, several parameters need to be put in place. To complement prior findings, it is the

TABLE 4.1
Research Questions

No.	Research Question	Research Purpose
1.	What are the major research trends in the study of augmented reality (AR) and virtual reality (VR) in marketing?	This question aims to identify the evolution of research focus over time, highlighting key themes and trends in AR and VR applications within marketing.
2.	Which academic journals, conferences, or other publications have the most influence on the development of AR and VR marketing research?	This question seeks to determine the primary sources of influential research in this field, helping to identify where the most impactful studies are being published.
3.	Who were the leading authors, institutions, and countries contributing to the research on AR and VR in marketing?	This question focuses on identifying the most prolific and influential authors and institutions, offering insights into the key players driving research in this area.

TABLE 4.2
Inclusion and Exclusion

		Criteria
Inclusion	1.	Data published between 2016 and 2025
	2.	Literature should only focus on augmented reality, virtual reality, marketing, and retailing.
	3.	Limited to: 1. Subject Area: Business, Management, and Accounting. Language: English
		Limited to Documented type: Article, Conference Proceedings
Exclusion	1.	Literature not focusing on AR and VR
	2.	Excluding book chapters

advocacy that the collection of literature is to be assessed by some defined values. The SCOPUS database was chosen to identify articles in the literature that address AR/VR in marketing. SCOPUS is among the highest and most reliable repositories of scholarly articles. Table 4.2 shows the parameters for quality assurance.

4.2.4 Bibliometric Analysis

Bibliometric analysis refers to a quantitative approach employed for assessing and categorizing bibliographic content, encompassing publications, citations, authors, and institutions within a specific scientific field (Bilal et al., 2024).

Initially, it was introduced by Pritchard in 1969, and since then it has become a widely efficient method. HistCite, Publish and Perish, BibExcel, Gephi, R module, Pajek, and VOSviewer are some of the packages used in the bibliometric analysis. In this study R module and VOSviewer are utilized (Sarkar et al., 2022).

4.2.4.1 Document Analysis

A total of 239 documents were eligible for bibliometric analysis based on the above-said criteria. Table 4.3 summarizes the bibliometric analyses for articles from the specified time frame of 2016–2025, extracted from the SCOPUS database in the context of AR and VR. According to Table 4.3, 113 sources published 239 articles. The total of the indicated keywords was 480. The total number of unique authors was 717. Of the 239 articles, 21 had single authors, and the rest had multiple authors. Also, there is a trend analysis of the number of publications in Figure 4.1.

4.2.4.2 Authors' Collaboration

The author used data source SCOPUS, while the analysis was done with the use of R-package. The most prolific contributors to the field and the various authors who have regularly contributed are presented in Table 4.4. As seen in Table 4.4, some of the dedicated authors included Raushnabel Pa who published 12 papers, Felix R, Hinsh C and Javornik A who published 5 papers each.

TABLE 4.3
Main Information

Description	Results
Main Information about Data	
Timespan	2016:2025
Sources (Journals, Books, etc.)	113
Documents	239
Annual growth rate %	−16.37
Document average age	2.38
Average citations per doc	58.85
References	16,612
Document Contents	
Keywords plus (ID)	480
Author's keywords (DE)	772
Authors	
Authors	717
Authors of single-authored docs	21
Authors' Collaboration	
Single-authored docs	22
Co-authors per doc	3.61
International co-authorships %	40.17
Document Types	
Articles	239

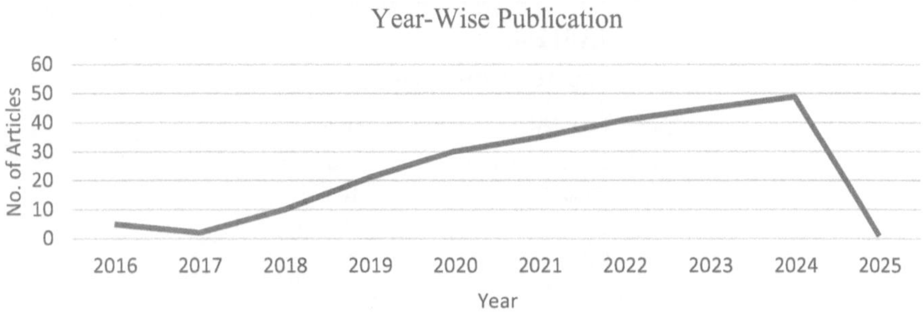

FIGURE 4.1 Year-wise publications.

TABLE 4.4
Top Authors

S.No	Authors	No. of Publications
1	Rauschnabel Pa	12
2	Felix R	5
3	Hinsch C	5
4	Javornik A	5
5	Buhalis D	4
6	Chylinski M	4
7	De Ruypter K	4
8	Dwivedi YK	4
9	Hilken T	4
10	Keelinng Di	4

4.2.4.3 Affiliation Statistics

The respective data of the institution was extracted from the SCOPUS database and then R-package was utilized for institutional statistics. Some statistics shed light on the institutes that are active in the field. Table 4.5 gives a clearly depicts the top ten contributors. It can also be seen from Table 4.5 that Griffith University, Manchester Metropolitan University, and the University of New South Wales had done a commendable job with 12, 11 and 10 articles, respectively.

4.2.4.4 Country-Wise Statistic

Country-wise statistics were collected from the SCOPUS database from where the data was extracted. The statistics were also evaluated according to the country of the first author, the corresponding author, and the citation of the articles. Table 4.5 represents the most involved countries concerning the consideration of the evolution of technology in marketing concerning AR and VR. The leading countries in the field were the United States with 136 articles, China with 100 articles, and the UK with 81 articles.

TABLE 4.5
Top Ten Countries

Country	Freq
United States	136
China	100
UK	81
India	63
Australia	50
Germany	40
South Korea	38
Spain	32
France	29
Malaysia	17

TABLE 4.6
Most Influential Keywords

Words	Occurrences
Virtual reality	66
Marketing	60
Consumption behaviour	24
Retailing	18
Augmented reality	17
Tourist destination	15
Commerce	14
Tourism market	13
Artificial intelligence	10
Perception	10

4.2.4.5 Influential Keywords

There were 480 total keywords used by the authors, of which the highest number of repetitions is as follows: Within specific contexts, the upcoming keywords have been identified with their frequency in the issue as shown in Table 4.6: "virtual reality" (66 occurrences), marketing (60 occurrences), and consumption behaviour (24 occurrences). Another way of utilizing the information gathered from the respondents is the development of the tree map of keywords with the highest frequency, and the results of this analysis are depicted in Figure 4.2, which was created using the R software.

4.3 NETWORK ANALYSIS

Network analysis becomes a powerful tool for examining the relationships between scholarly publications in bibliometric analysis. Various research papers have utilized network analysis tools like VOSviewer to explore the relationships between authors,

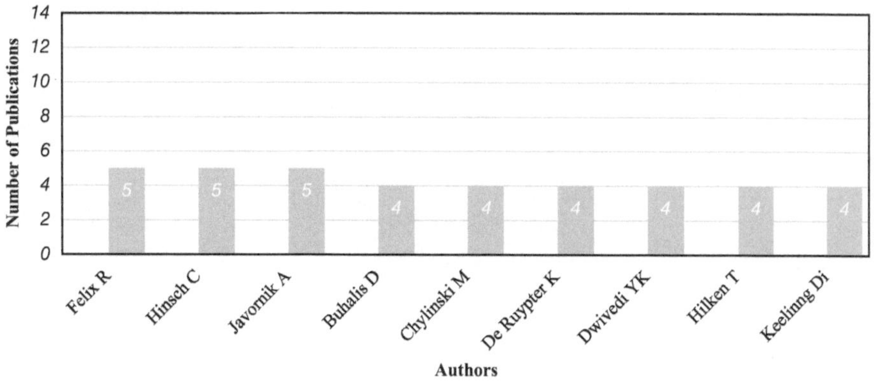

FIGURE 4.2 Most influential keywords.

sources, countries, and organizations. In the field of marketing, network analysis involves examining citation networks, interpreting co-citation studies, and collaborating with researchers spanning across a particular nation's borders (Agrawal et al., 2022).

4.3.1 Co-authorship Analysis

Country was used as a specific criterion to perform the co-authorship analysis. Out of 63 countries, 56 met the threshold of a minimum of 1 document per country and a minimum of 2 citations per document. Finally, 7 clusters are formed as shown in Figure 4.3 with 377 links and 595 total link strength. Cluster 1 contains South Korea at the top, cluster 2: Italy, cluster 3: United States, cluster 4: UK, cluster 5: China, cluster 6: Hong Kong, and cluster 7: Taiwan are the most significant countries based on the node size.

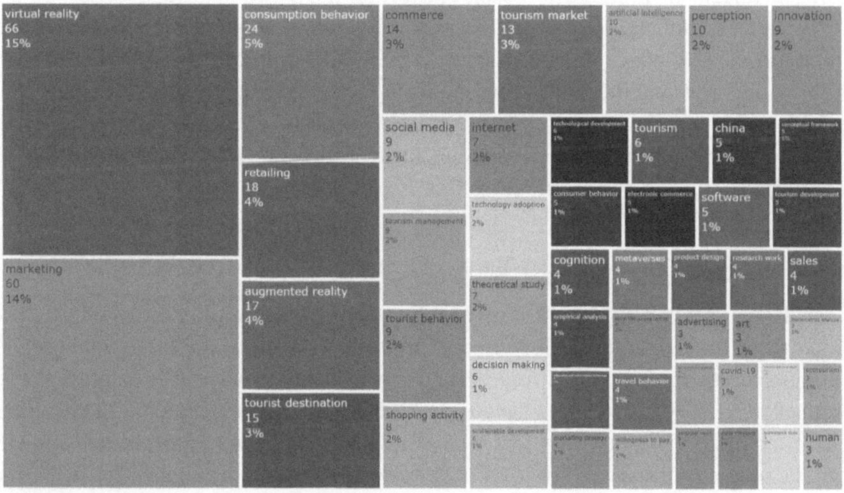

FIGURE 4.3 Co-authorship analysis of countries.

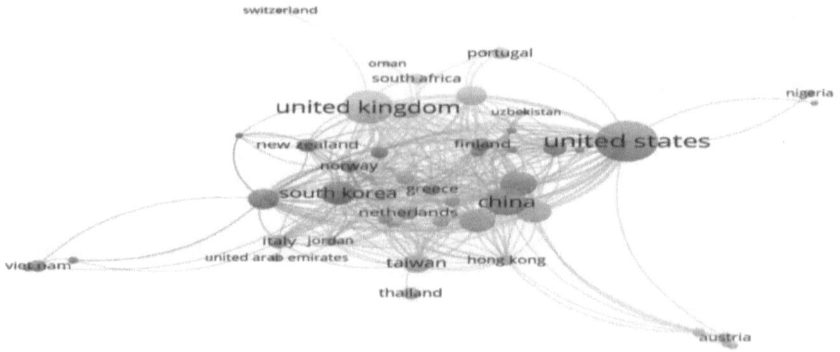

FIGURE 4.4 Co-occurrence of keywords.

4.3.2 Co-occurrence Analysis

A maximum of 1,142 words were gathered from the articles selected for the keyword network. The number of keywords is decreased to 83 by grouping all terms at least 4 times. Finally, 7 clusters were made as shown in Figure 4.4, with 891 links and 1,889 total link strength. According to the co-occurrence of keyword network analysis in the context of AR and VR, each cluster 1, 2, 3, 4, 5, 6, and 7 contains 24, 15, 13, 10, 8, 7, and 6 items, respectively.

4.4 CLUSTER ANALYSIS

Cluster analysis in network analysis is a technique used to identify and group closely interconnected nodes within a larger network structure (Judijanto et al., 2024). This approach helps reveal subgroups or communities that share similar characteristics or have stronger relationships with each other compared to the rest of the network. Using a variant of modularity-based clustering, VOSviewer groups closely related items such as publications, authors, or keywords into distinct clusters based on their co-occurrence or citation relationships. VOSviewer employs the established paradigm of the browsing hierarchy, ranging from the general to the specific so that users of VOSviewer can discover the given corpus of literature at both high-level topics and specific subtopics of interest (Dieck and Jung, 2018). The study has established the top four research clusters as mentioned in Figure 4.5.

4.4.1 Cluster 1: "Immersive Brand Experiences in Digital Tourism"

This cluster includes the application of AR, VR, or mixed reality techniques to provide unique and adventurous experiences to travellers and, at the same time, provide an opportunity for brands within the tourism industry to advertise or present their products and services more appealingly within the tourism sector (Gnoth, 2002). These technologies mean that virtual and real objects become one, either allowing tourists to get more detail from their devices, reconstructing scenes from the past

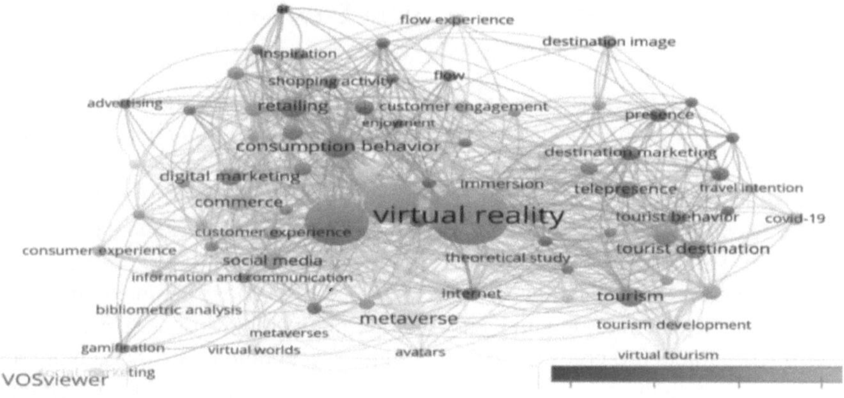

FIGURE 4.5 Bibliographic coupling of documents.

as a movie, or using avatars to represent brands. This marriage between technology advances and tourism is not only a value-add to the traveller but also helps brand extenders to present products and services in context and create a memorable immersive interaction that can help with brand building and social bonding (Dieck and Jung, 2018).

Preposition 4.1: Analyse the privacy and ethical issues towards data gathering via mobile app with AR/VR incorporation in the tourism industry.

4.4.2 Cluster 2: "Enhancing Customer Experience through Immersive Technologies"

AR/VR as customer experience is currently the most promising trend that has the potential to fundamentally change the way we perceive products and services. This fusion will create a dual-benefit paradigm: tangible benefits that get reflected in the monetary value of the goods consumed for their functionality and efficiency as well as intangible benefits that deliver pleasure and enjoyment (Guan, 2024). Soon enough, end users can walk around in virtual stores, testing a specific product in environments that reflect the client's desires. Companies can create custom training environments that help their employees become more skilled and efficient at their jobs. Such technologies are on the horizon to become more advanced leading to the convoluted interface of physical and digital spaces, which could redefine the traditional paradigm of value and satisfaction in consumer contact (Kimmel, 2010). It may result in a more engaging, closer, and effective customer experience delivering journeys in a range of sectors from retail to education and even healthcare.

Preposition 4.2: Employ AR/VR applications that will cater to fun as well as functionality to enable maximal usage by the users.

4.4.3 Cluster 3: "Immersive Omnichannel Engagement: Leveraging VR in Digital Marketing Strategies"

The cluster hence pertains to employing a concept of utilizing VR for Digital Marketing in a way that produces a harmonious and seamless client experience, bridging the gap of various paths and touchpoints. Using VR, brands can create and offer customers spaces within which they are allowed to interact with products, services, or experiences by going beyond the usual norms of advertisement (Bogicevic et al., 2019). This approach can increase customer interactions, as the customers can stay involved and get a coherent brand experience for various channels, including stationery stores, the internet, and other mobile platforms (Mosquera et al., 2017). Because consumers are now turning to be more touchpoint savvy than before and expecting more fluid and engaging experiences, VR as part of an omnichannel experience is beneficial to marketers to satisfy such needs for smooth and engaging experiences. With more interaction, those consumers tend to become loyal and thereby make more orders, thus resulting in high conversion rates (Lazaris and Sarantopoulos, 2021).

4.4.4 Cluster 4: "Avatar-Driven Marketing in the Metaverse Economy"

This theme takes a look at how branding can be done with little real-life touchpoints, how avatars can be used for marketing and product placement, and how appropriate KPIs can be developed for this new world. It also briefly discusses how these environments may give rise to novel types of endorsement, live-streamed performances, and targeted selling approaches (Dwivedi et al., 2022).

Preposition 4.3: Investigating the impact of avatar-driven privacy concerns in the metaverse economy.

4.5 FINDINGS

A number of documents were examined based on their relevance to AR and VR in the marketing context, within a time frame of 2016–2025. The bibliometric analysis performed with R software analysed the documents which showed there is a striving growth in the number of documents from 2019 to 2023, therefore it confirms that AR and VR are tremendously adopted by marketing, probably in every sector. The United States, the UK, and China were the prominent countries with larger publications in the field of AR and VR. The top trending keywords hooked up with this theme were virtual reality, marketing, and consumption behaviour. RQ1 was addressed through bibliographic coupling to identify major themes. The emerging research themes were: "Immersive Brand Experiences in Digital Tourism", Enhancing Customer Experience through Immersive Technologies", Immersive Omnichannel Engagement: Leveraging Virtual Reality in Digital Marketing Strategies, and Avatar-Driven Marketing in the Metaverse Economy.

4.6 MANAGERIAL APPLICATIONS

Data management and privacy concerns become increasingly critical as these technologies gather more detailed and personal user information. Managers must recognize that AR and VR technologies are not just passing trends, but transformative tools that will reshape marketing strategies and consumer interactions. Businesses need to invest in developing AR/VR capabilities within their organizations, either by building in-house expertise or partnering with specialized firms. Managers and researchers also need to address the potential challenges, such as high development costs and the need for consumer education, to fully leverage the potential of AR and VR in driving future marketing success.

REFERENCES

Agrawal, Rohit, Abhijit Majumdar, Kirty Majumdar, Rakesh D. Raut, and Balkrishna E. Narkhede. "Attaining sustainable development goals (SDGs) through supply chain practices and business strategies: A systematic review with bibliometric and network analyses." *Business Strategy and the Environment* 31, no. 7 (2022): 3669–3687.

Bilal, Muhammad, Yunfeng Zhang, Shukai Cai, Umair Akram, and Alrence Halibas. "Artificial intelligence is the magic wand making customer-centric a reality! An investigation into the relationship between consumer purchase intention and consumer engagement through affective attachment." *Journal of Retailing and Consumer Services* 77 (2024): 103674.

Bogicevic, Vanja, Soobin Seo, Jay A. Kandampully, Stephanie Q. Liu, and Nancy A. Rudd. "Virtual reality presence as a preamble of tourism experience: The role of mental imagery." *Tourism Management* 74 (2019): 55–64.

Bucea-Manea-Ţoniş, Rocsana, Elena Gurgu, Oliva Maria Dourado Martins, and Violeta Elena Simion. "An overview of how VR/AR applications assist specialists in developing better consumer behavior and can revolutionize our life." *Consumer Happiness: Multiple Perspectives* 1 (2021): 231–253.

Dargan, Shaveta, Shally Bansal, Munish Kumar, Ajay Mittal, and Krishan Kumar. "Augmented reality: A comprehensive review." *Archives of Computational Methods in Engineering* 30, no. 2 (2023): 1057–1080.

deRegt, Anouk, Kirk Plangger, and Stuart J. Barnes. "Virtual reality marketing and customer advocacy: Transforming experiences from story-telling to story-doing." *Journal of Business Research* 136 (2021): 513–522.

Dwivedi, Yogesh K., Laurie Hughes, Abdullah M. Baabdullah, Samuel Ribeiro-Navarrete, Mihalis Giannakis, Mutaz M. Al-Debei, Denis Dennehy et al. "Metaverse beyond the hype: Multidisciplinary perspectives on emerging challenges, opportunities, and agenda for research, practice and policy." *International Journal of Information Management* 66 (2022): 102542.

Gnoth, Juergen. "Leveraging export brands through a tourism destination brand." *Journal of Brand Management* 9 (2002): 262–280.

Guan, Zhiwei. "Utilize XR as a sustainable service design for Nordic tourism." (2024). University of Lapland, Master's Thesis, 1-61 (2024). https://urn.fi/URN:NBN:fi-fe2024061553168

Hamad, Ayah, and Bochen Jia. "How virtual reality technology has changed our lives: an overview of the current and potential applications and limitations." *International Journal of Environmental Research and Public Health* 19, no. 18 (2022): 11278.

Heller, Brittan, and Avi Bar-Zeev. "The problems with immersive advertising: in AR/VR, nobody knows you are an ad." *Journal of Online Trust and Safety* 1, no. 1 (2021).

Judijanto, Loso, and Restu Auliani. "Bibliometric analysis of biotechnology development." *West Science Nature and Technology* 2, no. 02 (2024): 108–117.

Katz, Helen. *The Media Handbook: A Complete Guide to Advertising Media Selection, Planning, Research, and Buying*. Routledge (2022).

Kimmel, Allan J. *Connecting with Consumers: Marketing for New Marketplace Realities*. Oxford University Press (2010).

Lazaris, Chris, Panagiotis Sarantopoulos, Adam Vrechopoulos, and Georgios Doukidis. "Effects of increased omnichannel integration on customer satisfaction and loyalty intentions." *International Journal of Electronic Commerce* 25, no. 4 (2021): 440–468.

Mosquera, Ana, Cristina Olarte Pascual, and Emma Juaneda Ayensa. "Understanding the customer experience in the age of omni-channel shopping." *Icono14* 15, no. 2 (2017): 4.

Riar, Marc, Jakob J. Korbel, Nannan Xi, Rüdiger Zarnekow, and Juho Hamari. "The use of augmented reality in retail: A review of literature." (2021), pp. 638–647. http://hdl.handle.net/10125/70689

Rosário, Albérico Travassos, and Anna Carolina Boechat. "Systematic literature review: Bibliometrics on ethics practised by Influencers." *Global Perspectives on Social Media Influencers and Strategic Business Communication* (2024): 292–325.

Rust, Roland T. "The future of marketing." *International Journal of Research in Marketing* 37, no. 1 (2020): 15–26.

Sarkar, Apurbo, Hongyu Wang, Airin Rahman, Waqar Hussain Memon, and Lu Qian. "A bibliometric analysis of sustainable agriculture: Based on the Web of Science (WOS) platform." *Environmental Science and Pollution Research* 29, no. 26 (2022): 38928–38949.

Shah, Denish, and B. P. S. Murthi. "Marketing in a data-driven digital world: Implications for the role and scope of marketing." *Journal of Business Research* 125 (2021): 772–779.

Storbacka, Kaj, and Ted Moser (2020). The changing role of marketing: transformed propositions, processes and partnerships. *AMS Review*, 10, 299–310.

Tom Dieck, M. Claudia, and Timothy Jung. "A theoretical model of mobile augmented reality acceptance in urban heritage tourism." *Current Issues in Tourism* 21, no. 2 (2018): 154–174.

5 Bibliometric Analysis of Financial Inclusion and Its Impact on Poverty Eradication and Reduction

Raj Kumar Singh, Soni Mitali, Vikas Agrawal, Ankita Bansal, and Vivek Anand Singh

5.1 INTRODUCTION

In a nation rich with vibrant cultures and diverse communities, access to financial services was once a privilege that many could only dream of (Barras, 1990). Rural inhabitants, agriculturalists, and small-scale entrepreneurs throughout the country had everyday challenges in meeting their basic needs, frequently depending on local lenders who imposed excessively high interest rates (Owusu-Antwi & Antwi, 2010). Within these challenges, the forces of transformation started to emerge on a momentous day – 24 July 1991. On that day, Shri Manmohan Singh, the Finance Minister of India, presented a set of ambitious economic changes to the nation, which would permanently modify the country's financial environment. His remarks reverberated like the initial precipitation during an extended period of aridity, renewing optimism for millions (Saha & Qin, 2023). The reforms established the foundation for a radical financial transformation that sought to provide financial services to all regions of India, even the most isolated and overlooked places. This development represented a significant stride in achieving financial inclusion for all individuals, rather than simply those with more privileges (Satpathy et al., 2014). In due course, the changes started to have positive results. Financial institutions expanded their presence by establishing branches in cities, towns, and villages, while automated teller machines (ATMs) emerged in previously unexplored locations (Gandhi, 2013). Over time, mobile phones, formerly associated with opulence, have become more accessible and ubiquitous, evolving into potent instruments for financial empowerment. Now, consumers can transfer funds, settle expenses, and even set aside funds for future use, all conveniently accessible through mobile devices (Au & Kauffman, 2008). However, notwithstanding the advancements, the narrative remained incomplete. The pathways to achieve financial inclusion were convoluted and encrusted with impediments. Rural regions continued to be isolated from the advantages of

DOI: 10.1201/9781003564621-5

contemporary banking. Sustained social and financial isolation, accompanied by poverty, unemployment, and inequality, cast enduring shadows over these areas (Amponsah et al., 2021).

Emerging research started to illuminate this intricate path. Analysis revealed that certain financial inclusion initiatives, such as the expansion of bank branches, had a substantial beneficial influence on the economy, while others such as the widespread use of ATMs, had less impact (Maity & Sahu, 2018). The mere provision of access to financial services was insufficient and enabling individuals to utilize these services to enhance their quality of life is the most difficult part (Machuca-Vílchez et al., 2023). As an example, a study conducted in Zimbabwe perceived that small-scale farmers faced many issues while seeking to obtain banking services (Mhlanga, 2021). The chief factors persuading the ability of these farmers to initiate a transaction account, borrow money, or get insurance were the size of their households, transaction fees, and accessibility of agricultural lengthening services (Muhongayire, 2012). Money received from sources apart from farming and the age of the household were other significant deliberations in their financial condition. The sovereign government recognized the need to prepare ventures that would encourage farmers to engage in authorized financial planning, seeing this as an important element in reducing poverty and raising social dignity (Inoue, 2024). In all places or situations, the flow of digital financial distribution, referring to the use of mobile devices to advance financial work, signalled a new era of beliefs (Gabor & Brooks, 2020). Multiple studies piloted in all locations, in addition to Europe, Africa, and Asia, have affirmed the development of digital financial incorporation to abate poverty, lessen income inequality, and promote comprehensive economic enhancement (Naceur et al., 2020). Mobile phones have become absolutely necessary tools, nurturing passage to financial services that allow individuals to save, invest, and secure themselves against financial instability (World Bank, 2018a). In spite of the progress attained in digital financial inclusion, abiding setbacks are endured. Although technological growth knocked off certain problems, others continued to exist, essentially in rural areas with less financial understanding. With the restricted knowledge of financial products and services, some people were responsive to manipulative activities (Nisa et al., 2023). The prerequisite for exhaustive financial education became apparent – instructing individuals not only on how to obtain financial services but also on how to use them to reach financial stability and autonomy (Naceur et al., 2020).

In some developing nations, the financial contribution received from relatives working overseas has become a critical means of producing revenue for homes (Saha & Qin, 2022). Empiric evidence established a strong correlation between digital financial inclusion and compensations, whereby the flow of remittances increased the dominance of digital financial services on the abetment of poverty (Machuca-Vílchez et al., 2023). The combined interaction of these two elements emerged as a force stimulus for transformation, facilitating the upgradation of families and communities. As the story of financial inclusion incremented, it became clear that this attempt incorporated more than mere accessibility of funds – it entailed permissive individuals to assume power over their financial prospects. Financial literacy, governmental regulations, technology advancements, and community espousal had crucial roles to play in this perpetual pursuit (Xie, 2023). The study endures to broaden, and other chapters were pioneered. The ramification of financial inclusion was experienced rife, ranging from the abatement of energy and poverty to the elevation of sustainable development (Li et al., 2021). The alliance among governments, legislators, and financial service providers combined data, artificial intelligence, and innovative approaches to productively engage the neediest communities (How et al., 2020). Hence, the account of financial insertion continued a story of positivity, dedication, and transformation (Afjal, 2023). Along with each new effort, policy alteration, and technical advancement, the goal of a global community where all individuals had identical access to the required financial reserves for their propensity approached realization (Omar & Inaba, 2020). The portrayal continued to unfold, with each chapter tossing the globe towards a future characterized by just economic growth and combined wealth (Omar & Inaba, 2020).

In light of the ongoing financial inclusion, depicted by significant changes and the inception of digital banking, this research recognized the need to chart its particulars. A comprehensive bibliometric analysis was conducted for the development of financial inclusion analysis by inspecting carefully selected scholarly records (Chmielewski, 1981). The investigation explored prominent writers in critical fields such as economics and finance and established the increasing influence of digital inclusion and FinTech (Atkinson & Messy, 2013). Global alliance and theme changes were proved using co-authorship and co-citation analyses, providing a procedure for future study. This academic endeavour reflects the continuous process of financial inclusiveness, aiming to establish a fairer global society.

5.2 BIBLIOMETRIC RESEARCH

5.2.1 Determining the Appropriate Search Keywords

Table 5.1 presents a comprehensive analysis of the selection criteria employed to refine the initial search results for the selected keywords and database. The process of elimination was conducted by considering the publishing era, subject area,

TABLE 5.1
Selection Criteria

Draft Criteria	Eliminated	Accepted
Database: "SCOPUS"		
Date of search: "24 June 2024"		
Initial search result (on the search term): "Financial inclusion" AND "Poverty eradication" OR "Poverty" OR "Poverty reduction" OR "Poverty alleviation"		776
Duration of Publications: 2009–2024	3	773
Subject area: "Business, management and accounting, Economics, Econometrics and Finance, decision science, Psychology, Social sciences, Astronomy, Multidisciplinary, and Arts and Humanities"	91	682
Publication type: "Articles, Conference Paper, Books, Book Chapter, and Review"	4	678
Language cloaking: Include documents issued in English only	18	660
Screening (based on content)		

publication type, and language. As a result, 660 papers were accepted for additional screening based on their substance.

5.2.2 Data Collection

For this study, the data has been extracted exclusively from the SCOPUS database as it is known for its huge collection and having high-impact-factor journals, which further includes a double peer-reviewed process (Groff et al., 2020) for the selection of the articles for final publication. This research study includes an extensive review of 660 documents in total after the completion of the refinement process, which is very much necessary and mandated to make the research authentic and reliable. The first search includes various keywords such as "Financial inclusion", "Poverty eradication", "Poverty", "Poverty reduction", or "Poverty alleviation" which left 776 results. To advance the research, the search was primarily based on the period of publications (2009–2024), Subject areas, Publication type, and Language screening. Additionally, various techniques have been incorporated in this research to complete the analysis; for visual representation of the results VOSViewer tool has been used to provide an overview of the result. This research makes a major contribution to the broader aspects of financial inclusion and poverty. Considering this fact, to reduce errors and inaccuracies, this research follows a rigorous refinement process (Donthu et al., 2021; Zupic, 2015).

5.2.3 Electing the Techniques for Analysis

A bibliometric study was performed to analyse the research scenario (Siddique et al., 2023). The bibliometric study implies scientific association by featuring a network of co-authorship institutional partnerships, displaying the key partners and joint network. A systematic methodological approach has been developed using bibliometric analysis; the analytical approach involves various methodologies (Goyal & Kumar, 2021). It helps in identifying critical works and core literature, which is vital for literature reviews and for advanced researchers approaching a field (Ellegaard & Wallin, 2015).

5.3 FINDINGS

5.3.1 Performance Analysis

Figure 5.1 depicts an examination of the documents released on financial inclusion and poverty reduction from 2009 to 2024 indicating a notable increase in research activities. Between 2009 and 2015, there was a progressive increase in the number of publications, starting with 5 in 2009 and reaching 15 in 2015. Significant growth was recorded from 2016 to 2019, with the figures increasing from 28 to 59. The trend persisted, with a significant increase to 70 publications in 2020, indicating a heightened level of interest in this field. In 2021 and 2022, there were consistently high numbers of publications, with 85 and 84, respectively. This trend reached its zenith in 2023, with a total of 142 documents. Despite a decline to 68

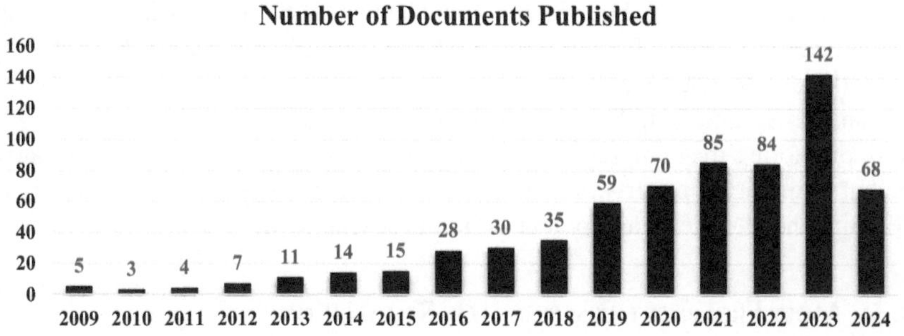

FIGURE 5.1 Performance analysis. (*Source*: https://www.scopus.com.geulsrx.lsremote.in/.)

publications thus far, the data for 2025 still suggests a strong level of scientific activity. The global rise in numbers affirms the increasing acknowledgement of financial inclusion as a vital instrument for condensing poverty and developing economic growth, accelerating the necessity for extended study and policy drive in this field.

5.3.2 Most Prominent Authors for Research on Financial Inclusion and Poverty Eradication

The investigation of the top 20 authors (Table 5.2) in the realm of financial involvement and poverty eradication underscores the notable contributions made by these researchers. At the top of the list is Koomson, I., who has nine documents, tailed by Kassim, S., and Ozili, P.K., who each have eight publications. Rastogi, S. is in second place with seven documents, while Mhlanga, D. has uncompensated six. Hassan,

TABLE 5.2
Elite 20 Authors

Author	Count of Documents	Author	Count of Documents
Koomson, I.	9	Madaleno, M.	3
Kassim, S.	8	Chitimira, H.	3
Ozili, P.K.	8	Adetiloye, K.A.	3
Rastogi, S.	7	Othman, A.H.A.	3
Mhlanga, D.	6	Singh, M.	3
Hassan, M.K.	5	Maity, S.	3
Gautam, R.S.	4	Wealth, E.	3
Mader, P.	4	Awaworyi Churchill, S.	3
Singh, J.	3	Redzuan, N.H.	3
Bernards, N.	3	Lal, T.	3

Source: https://www.scopus.com.geulsrx.lsremote.in/.

M.K. has a total of five publications, while Gautam, R.S., and Mader, P. each have four publications. Singh, J., Bernards, N., Madaleno, M., Chitimira, H., Adetiloye, K.A., Othman, A.H.A., Singh, M., Maity, S., Wealth, E., Awaworyi Churchill, S., Redzuan, N. H., and Lal, T. are with three papers each. These authors have been vital in promoting the comprehension and execution of financial inclusion initiatives aimed at eliminating poverty. Their substantial research contributions have significantly influenced policies and practices in this crucial field.

5.3.3 Most Prominent Discipline for Research on Financial Inclusion and Poverty Eradication

Figure 5.2 depicts the allocation of research efforts on financial inclusion and poverty eradication among different academic areas. Economics, Econometrics, and Finance are the most prevalent disciplines, representing 29.6% of the papers. Social Sciences closely trailed behind at 27.8%, indicating a notable interdisciplinary focus on the socioeconomic effects of financial inclusion. The fields of Business, Management, and Accounting have a significant impact, accounting for 20.3% of the research. Additional noteworthy fields of study include Computer Science (4.0%), Arts and Humanities (3.7%), Environmental Science (3.6%), Decision Science (2.7%), Energy (2.6%), Engineering (1.9%), and Psychology (1.8%). The "Other" category represents a minor fraction (1.9%) of the study, suggesting contributions from diverse supplementary disciplines. The wide range of academic participation highlights the complex and varied nature of financial inclusion and its significance in various fields of research.

5.3.4 Prominent Authors, Organizations, and Countries for Research on Financial Inclusion and Poverty Eradication

Table 5.3 illustrates prominent authors, institutions, and countries having a considerable influence on the research landscape regarding financial inclusion and poverty reduction. Isaac Koomson has the greatest number of citations, with 453 citations

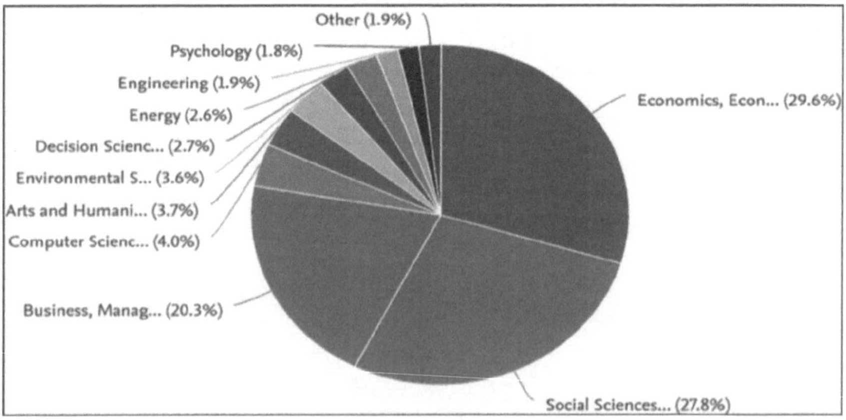

FIGURE 5.2 Prominent disciplines. (*Source*: https://www.scopus.com.geulsrx.lsremote.in/.)

TABLE 5.3
Outstanding Authors, Organizations, and Countries

TC	Author	TP	TC	Institution	TP	TC	Country	TP
453	Koomson, Isaac	9	453	"Network for Socioeconomic Research and Advancement (NESRA), Ghana"	8	1256	United Kingdom	51
370	Soederberg, Susanne	3	410	"UNE Business School, University of New England, Armidale, NSW, Australia"	4	1254	India	131
238	Mader, Philip	4	63	"University of Kentucky, United States"	2	1229	United States	75
167	Ozili, Peterson K.	8	50	"Birla Institute of Technology and Science (BITS), Pilani, India"	2	960	Australia	39
133	Madaleno, Mara	3	50	"University of South Africa, Pretoria, South Africa"	2	832	Canada	13
129	Mhlanga, David	6	36	"Institute of Development Studies, Brighton, United Kingdom"	2	832	China	46
111	Bernards, Nick	3	36	"Singapore University of Social Sciences, Singapore"	2	702	Ghana	26
93	Lal, Tarsem	3	30	"University of Johannesburg, Auckland Park, South Africa"	2	660	South Africa	48
65	Inoue, Takeshi	3	29	"Central Bank of Nigeria, Nigeria"	2	625	Nigeria	43
41	Szafarz, Ariane	3	23	"Covenant University, Nigeria"	3	516	Malaysia	42
40	Ashta, Arvind	3	18	"RMIT University, Australia"	2	483	Japan	14
34	Hassan, M. Kabir	5	18	"University of Warwick, United Kingdom"	2	446	Spain	18
33	Maity, Shrabanti	3	13	"International Islamic University Malaysia"	3	397	Bangladesh	18
30	Mpofu, Favourate Y.	4	13	"University of New England, Armidale, Australia"	2	218	Pakistan	18
29	Carrillo-Hidalgo, Isabel	3	12	"North-West University, South Africa"	2	197	Philippines	8
29	Pulido-Fernández, Juan Ignacio	3	10	"University of New Orleans, New Orleans, United States"	3	194	Germany	10
25	Awaworyi Churchill, Sefa	3	10	"University of South Africa, South Africa"	2	184	Brazil	8
13	Othman, Anwar Hasan Abdullah	3	8	"Central Bank, Nigeria"	3	168	Vietnam	8
9	Singh, Jaskirat	3	8	"Punjabi University, Patiala, India"	2	165	Turkey	10
9	Singh, Manjit	3	8	"Vocational Training Institute, Morocco"	2	142	UAE	8

Source: Authors' preparation based on VOSViewer output.
Note(S): Tc = Total Citations, Tp = Total Number of Article(S) Publications.

from nine articles. Susanne Soederberg follows with 370 citations from three publications, while Philip Mader has 238 citations from four publications. The "Network for Socioeconomic Research and Advancement" (NESRA) in Ghana and the UNE Business School in Australia have received 453 and 410 citations, respectively, highlighting their significant contribution to the advancement of this discipline. Moreover, the University of Kentucky in the United States and the Birla Institute of Technology and Science (BITS) in India are renowned for their substantial contributions. The United Kingdom and India are the most influential countries, with 1,256 and 1,254 citations, respectively. They are closely followed by the United States and Australia, which have 1,229 and 960 citations, respectively. Canada, China, Ghana, South Africa, Nigeria, and Malaysia are also significant donors, bringing vital research and viewpoints. These contributions jointly improve the comprehension and execution of financial inclusion measures, which are essential for eradicating poverty globally.

5.3.5 Most Dominant Journals for Research on Financial Inclusion and Poverty Eradication

The examination of journals uncovers the most prominent sources in the realm of financial inclusion and poverty eradication, determined by the aggregate number of citations (TC) and total number of papers published (TP) as presented in Table 5.4. The top-ranked entry is "Sustainability (Switzerland)," which has received 319 citations from 18 publications, indicating its substantial influence in the field of study. "World Development" has received 285 citations from six publications, while the "International Journal of Social Economics" has received 282 citations from 21 publications, demonstrating its significant impact on the literature. "Applied Economics" has received 257 citations and has been published nine times, while "Social Indicators Research" has received 231 citations and has been published ten times. These numbers highlight the significant contributions made by these journals. Additional prominent journals include "Forum for Social Economics" (199 citations, five publications), "Development in Practice" (198 citations, nine publications), and "Review of International Political Economy" (120 citations, five publications).

Furthermore, the journals "Geoforum" (with 76 citations and four publications), "Review of Development Economics" (with 66 citations and seven publications), and "Development Southern Africa" (with 65 citations and four publications) have had notable influences. Publications such as "Cogent Economics and Finance", "Journal of Risk and Financial Management", and "International Journal of Ethics and Systems", each having more than 50 citations, are crucial in promoting research in this field. The periodical in question mutually enhances the capacity and execution of financial inclusion methods, serving as an important archive of knowledge and indication to guide policies and practices aimed at wiping out poverty.

5.3.6 Most Influential Articles for Research on Financial Inclusion and Poverty Eradication

Table 5.5 showcases the most impactful articles on the topic of financial inclusion and poverty eradication, based on the total number of citations they have received. At the finest of the list is the report titled "Debtfare States and the Poverty Industry:

TABLE 5.4
Most Influential Journals

Journal	TC	TP
Sustainability (Switzerland)	319	18
World Development	285	6
International Journal of Social Economics	282	21
Applied Economics	257	9
Social Indicators Research	231	10
Forum for Social Economics	199	5
Development in Practice	198	9
Review of International Political Economy	120	5
Geoforum	76	4
Review of Development Economics	66	7
Development Southern Africa	65	4
Cogent Economics and Finance	55	6
Journal of Risk and Financial Management	54	6
International Journal of Ethics and Systems	53	6
Journal of International Development	52	5
Enterprise Development and Microfinance	49	7
Cogent Social Sciences	38	4
African Journal of Economic and Management Studies	27	4
Banks and Bank Systems	27	4
Sage Open	20	4

Source: Authors' preparation based on VOSViewer output.
Note: TC = total citations, TP = total number of article(s) publications.

Money, discipline and the Surplus Population" (Soederberg, 2014). This publication has admitted 267 citations, displaying its substantial effect. Omar and Inaba's (2020) paper, titled "Does Financial Inclusion Reduce Poverty and Income Inequality in Developing Countries? A Panel Data reasoning" has received 245 citations, indicating its important contribution to understanding the role of financial inclusion.

Further evident publications are by Chibba (2009) "Financial Inclusion, Poverty Reduction and the Millennium Development Goals", which has acknowledged 241 citations, and Demir et al. (2022) "Fintech, Financial Inclusion and Income Inequality: A Quantile Regression Approach", which has cumulative 231 quotations. The study done by Neaime and Gaysset (2018) on "Financial inclusion and support in the Middle East and North Africa (MENA) suburb: Experimental findings on poverty and inequality" has reunited significant effect, as seen by its 226 citations. Articles by Chao et al. (2021) as to group decision-making in financial inclusion, Koomson and Danquah (2021) on the topic of financial inclusion and energy deficit in Ghana, and Mushtaq and Bruneau (2019) on the conclusion of microfinance and ICT for poverty have received ample attention, with citations ranging from 178 to 212.

Other contributions include Mader (2018) titled "Contesting Financial Inclusion"; Koomson et al. (2020), who explored the encounter of financial inclusion on poverty

TABLE 5.5
Most Influential Articles

Author(s)	Title	TC
Soederberg (2014)	"Debtfare states and the poverty industry: Money, discipline and the surplus population"	267
Omar and Inaba (2020)	"Does financial inclusion reduce poverty and income inequality in developing countries? A panel data analysis"	245
Chibba (2009)	"Financial inclusion, poverty reduction and the millennium development goals"	241
Demir et al. (2022)	"Fintech, financial inclusion and income inequality: A quantile regression approach"	231
Neaime and Gaysset (2018)	"Financial inclusion and stability in MENA: Evidence from poverty and inequality"	226
Chao et al. (2021)	"Large-scale group decision-making with non-cooperative behaviors and heterogeneous preferences: An application in financial inclusion"	212
Koomson and Danquah (2021)	"Financial inclusion and energy poverty: Empirical evidence from Ghana"	184
Mushtaq and Bruneau (2019)	"Microfinance, financial inclusion and ICT: Implications for poverty and inequality"	178
Park and Mercado (2018)	"Financial inclusion, poverty, and income inequality"	176
Mader (2018)	"Contesting financial inclusion"	158
Koomson et al. (2020)	"Effect of financial inclusion on poverty and vulnerability to poverty: Evidence using a multidimensional measure of financial inclusion"	157
Ozili (2021)	"Financial inclusion research around the world: A review"	145
Cobb et al. (2016)	"Funding financial inclusion: Institutional logics and the contextual contingency of funding for microfinance organizations"	144
Diniz et al. (2012)	"Triggers and barriers to financial inclusion: The use of ICT-based branchless banking in an Amazon County"	141
Abor et al. (2018)	"Mobile telephony, financial inclusion and inclusive growth"	135
Mhlanga (2021)	"Artificial intelligence in the Industry 4.0, and its impact on poverty, innovation, infrastructure development, and the sustainable development goals: Lessons from emerging economies?"	119
Demirgüç-Kunt et al. (2020)	"The Global Findex Database 2017: Measuring financial inclusion and opportunities to expand access to and use of financial services"	110
Dogan et al. (2021)	"Which households are more energy vulnerable? Energy poverty and financial inclusion in Turkey"	107
Churchill and Marisetty (2020)	"Financial inclusion and poverty: A tale of forty-five thousand households"	105
Mishra and Singh Bisht (2013)	"Mobile banking in a developing economy: A customer-centric model for policy formulation"	102

Source: Authors' preparation based on VOSViewer output.
Note: TC = total citations.

and amenability; and Ozili (2021), who reviewed global research on financial inclusion. Each of these articles, in addition to the contextual belief of funding for microfinance, the application of ICT-based branchless banking, and the grant of mobile telephony to comprehensive growth, has committed valuable insights and attitude to the discussion on financial inclusion and poverty abatement. These significant works in sync enhance the comprehension of how financial inclusion approach might be efficiently utilized to mitigate poverty and feed fair economic development on a comprehensive scale. Various other articles listed in the table (Abor et al., 2018; Churchill & Marisetty, 2020; Cobb et al., 2016; Demirgüç-Kunt et al., 2020; Diniz et al., 2012; Dogan et al., 2021; Mhlanga, 2021; Mishra & Singh Bisht, 2013; Park & Mercado, 2018) have made a valuable contribution to this research era.

5.3.7 Knowledge Foundations for Research on Financial Inclusion and Poverty Eradication

Figure 5.3 presents a network visualization, representing co-authorship relationships and patterns among researchers in the research era of financial inclusion and poverty. Based on the collaborative works the nodes represent individual authors, while the size of each node indicates the influence of study within this research community. Notably five clusters are the prominent analysis of this figure, where the largest

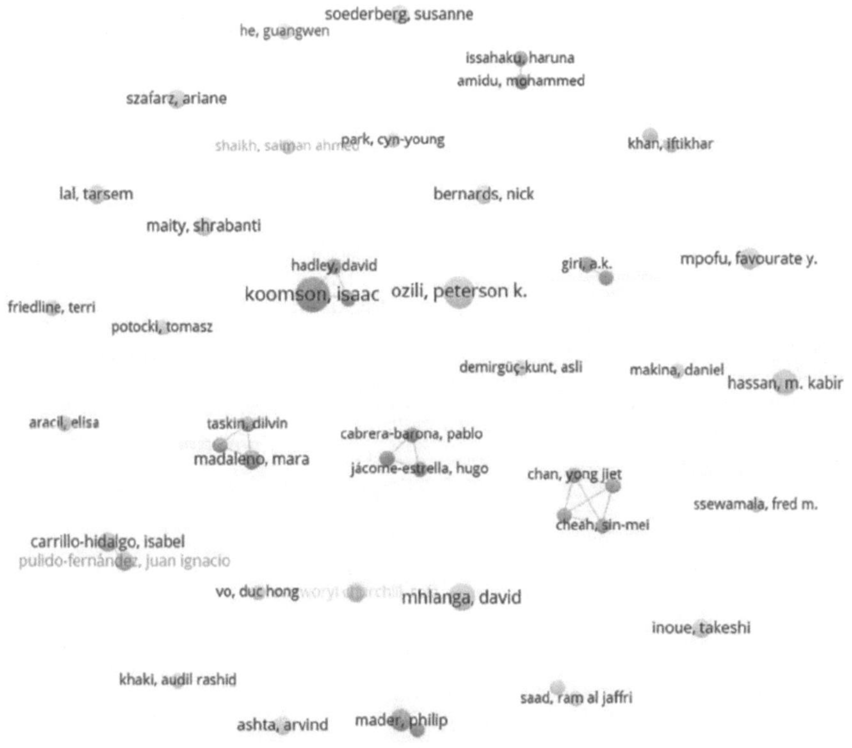

FIGURE 5.3 Co-authorship analysis. (*Source*: VOSViewer output.)

cluster includes authors "Koomson, Isaac", "Ozili, Peterson K.", and "Mhlanga, David", suggesting that these authors are key contributors to this field, with significant co-authorship impact. Clustered groups like "Chan, Yong Jiet", "Cheah, Sin-Mei", and other groups suggest they work on a similar topic together. The immediacy of nodes indicates the closeness of relationships between authors, either through direct collaborations or shared research topics. For instance, Koomson, Isaac and Ozili, Peterson K. appear closely connected, indicating a stronger or more frequent collaboration. This visualization provides insights into the intellectual landscape of financial inclusion research, highlighting the most influential scholars, the strength of collaborative networks, and the thematic structure of the field.

Figure 5.4, the co-citation network analysis, demonstrates the intellectual framework and influential publications in the domain of financial inclusion and poverty reduction. Notable groupings suggest important areas of concentration, with authors such as Asli Demirgüc-Kunt and Leora Klapper standing out as central personalities, showcasing their substantial contributions and impact. The extensive interrelationships among authors such as Beck, Levine, and Morduch emphasize a tightly knit scholarly community studying financial development and its effects. The works by Kumar N., Sarma M., and Swamy V. on financial inclusion and indices are highly interconnected, as evidenced by the significant nodes and citations. This highlights their fundamental contribution to the existing literature. The intricate nature of the network, characterized by dynamic connections between many clusters, indicates a diverse and interdisciplinary conversation that covers topics such as financial

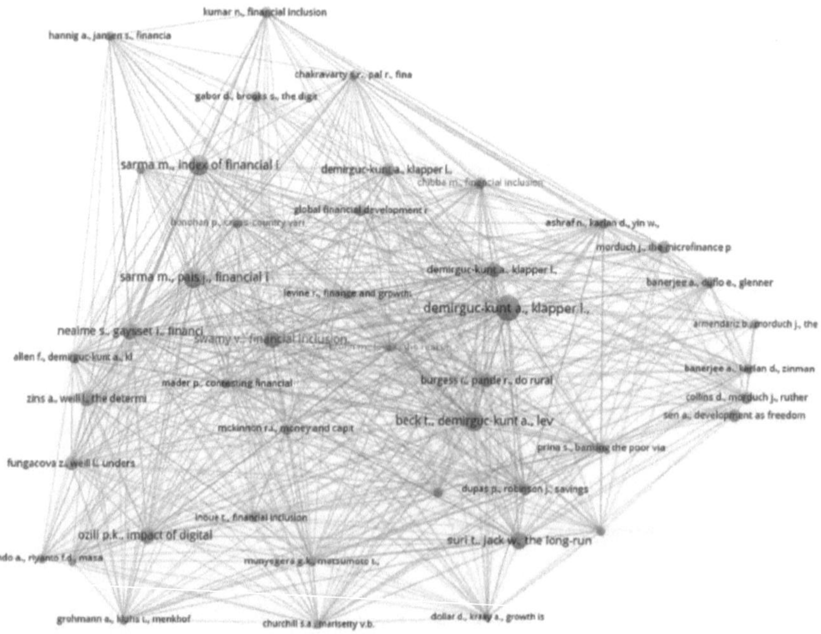

FIGURE 5.4 Co-citation analysis. (*Source*: VOSViewer output.)

technology, poverty reduction, and policy execution. This intricate network of references not only charts the essential knowledge foundation but also pinpoints influential scholars and groundbreaking works that propel the discussion on financial inclusion and economic prosperity.

5.3.8 Thematic and Influence Structure Analysis through Bibliographic Coupling

Table 5.6 analyses that the research on financial inclusion and poverty reduction is built upon a set of fundamental concepts, which cover various important topics and provide valuable perspectives. Based on five different themes, the research articles have been bifurcated to show emerging eras of research. The examination of the topic and influence structure using bibliographic coupling reveals the interconnections and significance of important research works in the field of financial inclusion and poverty reduction; research such as by Omar M.A. and Inaba K.'s examination of panel data and Chibba M.'s study on the millennium development objectives emphasizes the crucial significance of financial inclusion in diminishing poverty and inequality. Park C.-Y. and Mercado R., Jr., as well as Koomson I. et al. do research that delves further into these linkages by utilizing multidimensional measurements. Demir A. et al. investigate the effects of fintech using quantile regression, while Mushtaq R. and Bruneau C. analyse the consequences of microfinance and ICT. Technological progress is a significant topic in both studies. Diniz E. et al. examine the factors that promote and hinder financial inclusion through ICT-based branchless banking, while Mishra V. and Singh Bisht S. suggest a customer-focused framework for mobile banking in emerging economies. Koomson I. and Danquah M. in Ghana, and Dogan E. et al. in Turkey examine the correlation between financial inclusion and energy poverty, emphasizing the risks and advantages linked to financial accessibility. Stability and critical viewpoints are examined in the research. Neaime S. and Gaysset I. specifically analyse the MENA region, Ozili P. K. provides a comprehensive worldwide overview, and Adam Cobb J. et al. evaluate the impact of financial contingencies. Soederberg, Mader, Churchill, and their colleagues provide a critical analysis of the wider effects of financial inclusion, specifically examining the influence of artificial intelligence and the insights provided by the Global Findex database. This fundamental research jointly improves the comprehension and execution of financial inclusion techniques that are crucial for successful poverty eradication.

Soederberg S.'s "Debtfare states and the Poverty Industry", which has been cited 267 times, provides a thorough analysis of how financial institutions contribute to the perpetuation of poverty. The study conducted by Omar M.A. and Inaba K., which has received 245 citations, offers significant insights into the impact of financial inclusion on poverty and income disparity in developing nations using panel data analysis. Chibba M.'s research, which has been referenced 241 times, highlights the significance of financial inclusion in attaining the millennium development objectives. Additional significant publications encompass research on the effects of fintech and ICT on financial inclusion and inequality, empirical findings regarding financial inclusion and energy poverty in Ghana, and the consequences of mobile telephony for inclusive growth. These extensively referenced works create a strong interconnected

TABLE 5.6
Cluster Table

Theme	Author(s)	Title	TC
Financial inclusion and poverty/inequality reduction	Omar and Inaba (2020)	"Does financial inclusion reduce poverty and income inequality in developing countries? A panel data analysis"	245
	Chibba (2009)	"Financial inclusion, poverty reduction, and the Millennium Development goals"	241
	Park and Mercado (2018)	"Financial inclusion, poverty, and income inequality"	176
	Koomson et al. (2020)	"Effect of financial inclusion on poverty and vulnerability to poverty: Evidence using a multidimensional measure of financial inclusion"	157
Financial inclusion and technological advancements	Demir et al. (2022)	"Fintech, financial inclusion, and income inequality: A quantile regression approach"	231
	Mushtaq and Bruneau (2019)	"Microfinance, financial inclusion, and ICT: Implications for poverty and inequality"	178
	Diniz et al. (2012)	"Triggers and barriers to financial inclusion: The use of ICT-based branchless banking in an Amazon County"	141
	Mishra and Singh Bisht (2013)	"Mobile banking in a developing economy: A customer-centric model for policy formulation"	102
Financial inclusion and energy poverty	Koomson and Danquah (2021)	"Financial inclusion and energy poverty: Empirical evidence from Ghana"	184
	Dogan et al. (2021)	"Which households are more energy vulnerable? Energy poverty and financial inclusion in Turkey"	107
	Abor et al. (2018)	"Mobile telephony, financial inclusion and inclusive growth"	135
	Chao et al. (2021)	"Large-scale group decision-making with non-cooperative behaviors and heterogeneous preferences: An application in financial inclusion"	212
Financial inclusion and stability	Neaime and Gaysset (2018)	"Financial inclusion and stability in MENA: Evidence from poverty and inequality"	226
	Ozili (2021)	"Financial inclusion research around the world: A review"	145
	Cobb et al. (2016)	"Funding financial inclusion: Institutional logic and the contextual contingency of funding for microfinance organizations"	144
	Demirgüç-Kunt et al. (2020)	"The Global Findex Database 2017: Measuring financial inclusion and opportunities to expand access to and use of financial services"	110
Critical perspectives on financial inclusion	Soederberg (2014)	"Debtfare states and the poverty industry: Money, discipline and the surplus population"	267
	Mader (2018)	"Contesting financial inclusion"	158
	Churchill and Marisetty (2020)	"Financial inclusion and poverty: A tale of forty-five thousand households"	105
	Mhlanga (2021)	"Artificial intelligence in Industry 4.0, and its impact on poverty, innovation, infrastructure development, and the sustainable development goals: Lessons from emerging economies?"	119

Source: Authors' preparation based on VOSViewer output.
Note: TC = total citations.

system of knowledge, showcasing how policies for financial inclusion may successfully tackle poverty and foster fair economic development.

5.3.9 Prominent Keywords Used for Research on Financial Inclusion and Poverty Eradication

In Figure 5.5, the co-occurrence analysis of terms in research on financial inclusion and poverty eradication reveals the main topics and interconnected concepts in this field. The basic approach is "financial inclusion", which is heartily similar to expressions like "poverty", "poverty alleviation", and "income inequality", presenting the vital emphasis on the effect of financial inclusion on economic disparity. The terms "microfinance", "microcredit", and "Islamic finance" are the different financial methods used to advance inclusivity. The provision "sustainable development", "economic growth", and "human development" are also compelling, as they represent the higher socioeconomic objectives associated with initiatives adorning financial inclusion. The depiction of technological growth can be observed through remarks such as "fintech", "digital finance", "mobile money", and "technology", promoting the effect of digital solutions in the development of financial access. Terms such as "sub-Saharan Africa", "Bangladesh", "India", and "gender" are used

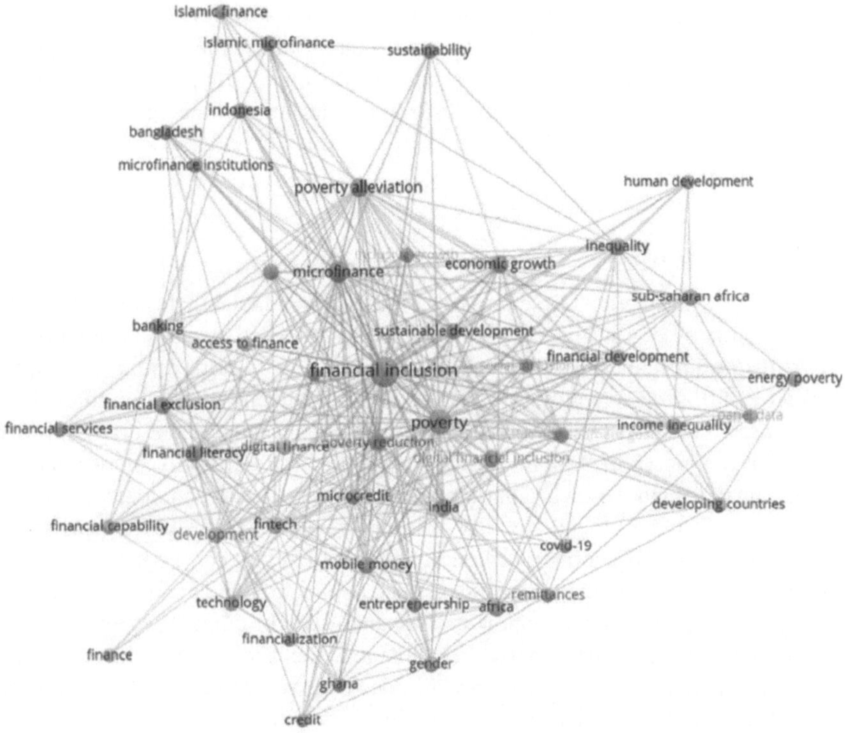

FIGURE 5.5 Co-occurrence analysis. (*Source*: VOSViewer output.)

to analyse the unique geographic and demographic focuses that are central in the study. Moreover, the clout of external prospects such as the "COVID-19" pandemic on drive to advance financial inclusion is accepted. The chain of keywords granted here demonstrates a thorough and combined research environment that seeks to employ financial inclusion to knock off poverty and feed fair economic evolution on a global scale.

5.4 FUTURE DIRECTIONS

Eventual research on financial inclusion and poverty elimination should compute many key areas to appreciate understanding and employment in this crucial sphere. First, it is necessary to consider the abiding effects of digital financial services, such as mobile banking and fintech, on both economic stability and the elimination of poverty. Research should delve into the brunt of digital literacy dynamism on improving the adequacy of these technologies among disempowered communities. Additionally, research must examine the impact of financial inclusion initiatives on definite demographic disciple, such as women and rural community, to formulate more centred policies. The COVID-19 epidemic has had a compelling impact on the endeavour to promote financial inclusion, which is an important subject of study. Clearly, it is worth wandering how digital quick fix might help reduce the economic importance of such crises. Also, it is critical to employ interdisciplinary methodologies that incorporates economics, social sciences, and technology studies to adequately tackle the many difficulties linked with financial inclusion. Thus, assessing the fusion of financial inclusion into wider sustainable development manifesto will offer an invaluable understanding of how these endeavours might contribute to sustained economic development and abatement of poverty.

5.5 CONCLUSION

The complete examination of the research outlook on financial inclusion and poverty elimination uncovers a strong and combined body of task led by eminent authors, significant institutions, and distinct geographic target. The synergic and interdisciplinary quality of this topic is presented by the co-authorship and co-citation chain. Isaac Koomson, Peterson K. Ozili, and Asli Demirgüc-Kunt have made great improvements. The bibliographic coupling study reveals the main areas of interest, such as the trappings of financial inclusion on poverty and injustice, technological progress, energy poverty, and financial stability. The research focuses on key concepts such as "financial inclusion", "poverty", "microfinance", and "fintech", which feature the main capacity and changing arrangement. The application underscores the implication of digital explanation and region-specific analysis. The cumulative findings from these studies emphasize the importance of financial inclusion in reaching sustainable development and decent economic growth. They endeavour to provide a comprehensive guide for future research and policy-making efforts to adequately tackle poverty and boost inclusive prosperity.

5.6 IMPLICATIONS

Comprehensive examination of financial inclusion and poverty elimination reveals symbolic consequence for policymakers, practitioners, and scholars. The statement affirms the need for accurate interventions that are directed to address the exclusive characteristics of distinct countries and demographics, with an appropriate focus on disempowered people in places such as sub-Saharan Africa, Bangladesh, and India. Experts are being aided to strengthen regulatory backing to strengthen digital financial services, encourage advancements in financial technology, and secure a resilient digital infrastructure. Financial institutions should weigh digital clarification, such as mobile banking and fintech, to approach unbanked people, while also ensuring acceptable digital literacy to empower efficient discharge of services. Accumulating financial inclusion drives further sustainable development goals can enormously amplify its control on reducing poverty and establishing economic stability. The study features the implication of interdisciplinary association, integrating viewpoints from economics, social sciences, technology, and development studies to locate the complicated threat of financial inclusion. Consecutive studies should continue to broaden upon significant findings, investigating innovative aspects such as the ramifications of COVID-19, to enhance compassionate and application in this crucial sphere.

REFERENCES

Abor, J. Y., Amidu, M., & Issahaku, H. (2018). Mobile telephony, financial inclusion and inclusive growth. *Journal of African Business*, *19*(3), 430–453.

Afjal, M. (2023). Bridging the financial divide: a bibliometric analysis on the role of digital financial services within FinTech in enhancing financial inclusion and economic development. *Humanities and Social Sciences Communications*, *10*(1). https://doi.org/10.1057/s41599-023-02086-y

Amponsah, M., Agbola, F. W., & Mahmood, A. (2021). The impact of informality on inclusive growth in Sub-Saharan Africa: does financial inclusion matter? *Journal of Policy Modeling,* 43(6), 1259–1286. https://doi.org/10.1016/j.jpolmod.2021.03.009

Atkinson, A., & Messy, F. A. (2013). *Promoting Financial Inclusion through Financial Education: OECD/INFE Evidence, Policies and Practice*. Organisation for Economic Co-operation and Development. https://doi.org/10.1787/5k3xz6m88smp-en

Au, Y. A., & Kauffman, R. J. (2008). The economics of mobile payments: understanding stakeholder issues for an emerging financial technology application. *Electronic Commerce Research and Applications,* 7(2), 141–164.

Barras, R. (1990). Interactive innovation in financial and business services: the vanguard of the service revolution. *Research Policy,* 19(3), 215–237.

Chao, X., Kou, G., Peng, Y., & Viedma, E. H. (2021). Large-scale group decision-making with non-cooperative behaviors and heterogeneous preferences: an application in financial inclusion. *European Journal of Operational Research*, 288(1), 271–293.

Chibba, M. (2009). Financial inclusion, poverty reduction and the millennium development goals. *The European Journal of Development Research,* 21(2), 213–230.

Chmielewski, M. A. (1981). Elliptically symmetric distributions: a review and bibliography. *International Statistical Review,* 49, 67–74.

Churchill, S. A., & Marisetty, V. B. (2020). Financial inclusion and poverty: a tale of forty-five thousand households. *Applied Economics,* 52(16), 1777–1788.

Cobb, J. A., Wry, T., & Zhao, E. Y. (2016). Funding financial inclusion: Institutional logics and the contextual contingency of funding for microfinance organizations. *Academy of Management Journal,* 59(6), 2103–2131. https://doi.org/10.5465/amj.2015.0715

Demir, A., Pesqué-Cela, V., Altunbas, Y., & Murinde, V. (2022). Fintech, financial inclusion and income inequality: a quantile regression approach. *The European Journal of Finance,* 28(1), 86–107. https://doi.org/10.1080/1351847X.2020.1772335

Demirgüç-Kunt, A., Klapper, L., Singer, D., Ansar, S., & Hess, J. (2020). The global findex database 2017: measuring financial inclusion and opportunities to expand access to and use of financial services. *The World Bank Economic Review*, 34, S2–S8.

Diniz, E., Birochi, R., & Pozzebon, M. (2012). Triggers and barriers to financial inclusion: The use of ICT-based branchless banking in an Amazon county. *Electronic Commerce Research and Applications,* 11(5), 484–494. https://doi.org/10.1016/j.elerap.2011.07.006

Dogan, E., Madaleno, M., & Taskin, D. (2021). Which households are more energy vulnerable? Energy poverty and financial inclusion in Turkey. *Energy Economics,* 99, 105306.

Donthu, N., Kumar, S., Mukherjee, D., Pandey, N., & Marc, W. (2021). How to conduct a bibliometric analysis : an overview and guidelines. *Journal of Business Research,* 133(April), 285–296. https://doi.org/10.1016/j.jbusres.2021.04.070

Ellegaard, O., & Wallin, J. A. (2015). The bibliometric analysis of scholarly production: how great is the impact? *Scientometrics*, 105, 1809–1831.

Gabor, D., & Brooks, S. (2020). The digital revolution in financial inclusion: international development in the fintech era. In D. Bryan, L. Rafferty, & R. W. H. Engle (Eds.), *Material Cultures of Financialisation* (pp. 69-82). Routledge.

Gandhi, M. M. (2013). Role of banks in financial inclusion in India: issues, challenges and strategies. *International Journal of Applied Business and Economic Research*, 11(2), 163–183.

Goyal, K., & Kumar, S. (2021). Financial literacy: a systematic review and bibliometric analysis. *International Journal of Consumer Studies,* 45(1), 80–105.

Hannig, A., & Jansen, S. (2010). *Financial Inclusion and Financial Stability: Current Policy Issues.* Alliance for Financial Inclusion. https://www.afi-global.org/wp-content/uploads/publications/2010/11/AFI_Financial_Inclusion_and_Financial_Stability.pdf

Hasan, M., Le, T., & Hoque, A. (2021). How does financial literacy impact on inclusive finance? *Financial Innovation,* 7(1). https://doi.org/10.1186/s40854-021-00259-9

Hasan, M. M., & Lu, Z. (2023). Nexus among financial inclusion and sustainability in Asia: role of banking sector. *Environmental Science and Pollution Research,* 30(22), 62481–62493.

Hendriks, S. (2019). The role of financial inclusion in driving women's economic empowerment. *Development in Practice,* 29(8), 1029–1038. https://doi.org/10.1080/09614524.2019.1660308

How, M.-L., Cheah, S.-M., Khor, A. C., & Chan, Y. J. (2020). Artificial intelligence-enhanced predictive insights for advancing financial inclusion: a human-centric ai-thinking approach. Big Data and Cognitive Computing, 4(2), 1–21. https://doi.org/10.3390/bdcc4020008

Ibrahim, F., McHugh, N., Biosca, O., Baker, R., Laxton, T., & Donaldson, C. (2021). Microcredit as a public health initiative? Exploring mechanisms and pathways to health and wellbeing. *Social Science and Medicine,* 270, 113633. https://doi.org/10.1016/j.socscimed.2020.113633

Inoue, T. (2024). Digital financial inclusion, international remittances, and poverty reduction. *Journal of Economic Structures,* 13(1), 3. https://doi.org/10.1186/s40008-024-00328-z

Jiang, Y., & Liu, Y. (2022). Does financial inclusion help alleviate household poverty and vulnerability in China? *PLoS One*, 2017, e0278564.

Kar, A. K., & Bali Swain, R. (2023). Does financial inclusion improve energy accessibility in Sub-Saharan Africa? *Applied Economics.* https://doi.org/10.1080/00036846.2023.2270227

Koomson, I., & Danquah, M. (2021). Financial inclusion and energy poverty: empirical evidence from Ghana. *Energy Economics,* 94, 105085. https://doi.org/10.1016/j.eneco.2020.105085

Koomson, I., Villano, R. A., & Hadley, D. (2020). Effect of financial inclusion on poverty and vulnerability to poverty: evidence using a multidimensional measure of financial inclusion. *Social Indicators Research,* 149(2), 613–639. https://doi.org/10.1007/s11205-019-02263-0

Lagna, A., & Ravishankar, M. N. (2022). Making the world a better place with fintech research. *Information Systems Journal,* 32(1), 61–102. https://doi.org/10.1111/isj.12333

Li, A., Gao, L., Chen, S., Zhao, J., Ujiyad, S., Huang, J., Han, X., & Bryan, B. A. (2021). Financial inclusion may limit sustainable development under economic globalization and climate change. *Environmental Research Letters, 16*(5), 054049. https://doi.org/10.1088/1748-9326/abf465

Liu, C., Chen, L., Li, Z., & Wu, D. (2023). The impact of digital financial inclusion and urbanization on agricultural mechanization: evidence from counties of China. *PLoS One,* 18, e0293910. https://doi.org/10.1371/journal.pone.0293910

Loukoianova, M. E., Yang, Y., Guo, M. S., Hunter, M. L., Jahan, M. S., Jamaludin, M. F., ... & Wu, M. Y. (2018). *Financial Inclusion in Asia-Pacific*. International Monetary Fund.

Machuca-Vílchez, J. A., Ramos-Cavero, M. J., & Cordova-Buiza, F. (2023). Knowledge management in financial education in Peruvian government programs focused on women: progress and challenges. *Knowledge and Performance Management, 7*(1), 1–14. https://doi.org/10.21511/kpm.07(1).2023.01

Mader, P. (2018). Contesting financial inclusion. *Development and Change,* 49(2), 461–483.

Maity, C. M. A., & Sahu, T. N. (2018). Bank branch expansion and financial inclusion: evidence from selected commercial banks in India. *Al-Barkaat Journal of Finance & Management,* 10(1), 48–65.

Mhlanga, D. (2021). Artificial intelligence in the industry 4.0, and its impact on poverty, innovation, infrastructure development, and the sustainable development goals: Lessons from emerging economies? *Sustainability,* 13(11), 5788. https://doi.org/10.3390/su13115788

Mhlanga, D. (2021). Factors that matter for financial inclusion: evidence from Sub-Sharan Africa—The Zimbabwe case. *Academic Journal of Interdisciplinary Studies,* 10(6), 48–59.

Mishra, V., & Singh Bisht, S. (2013). Mobile banking in a developing economy: A customer-centric model for policy formulation. *Telecommunications Policy*, 37(6–7), 503–514.

Mpofu, F. Y. (2024). Industry 4.0 in finance, digital financial services and digital financial inclusion in developing countries: opportunities, challenges, and possible policy responses. *International Journal of Economics and Financial Issues,* 14(2), 120–135.

Mushtaq, R., & Bruneau, C. (2019). Microfinance, financial inclusion and ICT: implications for poverty and inequality. *Technology in Society,* 59, 101154.

Naceur, S. B., Beck, T., Belhaj, M., & Barajas, A. (2020). Financial inclusion: What have we learned so far? What do we have to learn? *IMF Working Paper*, *2020*(157), 1. https://doi.org/10.5089/9781513553009.001

Neaime, S., & Gaysset, I. (2018). Financial inclusion and stability in MENA: evidence from poverty and inequality. *Finance Research Letters,* 24, 230–237.

Nisa, C., Viverita, V., & Chalid, D. A. (2023). Competition in microfinance institutions: does voluntary saving product have a moderating role? *Jurnal Keuangan Dan Perbankan,* *27*(2), 295–305. https://doi.org/10.26905/jkdp.v27i2.11014

Omar, M. A., & Inaba, K. (2020). Does financial inclusion reduce poverty and income inequality in developing countries? A panel data analysis. *Journal of Economic Structures,* 9(1), 37.

Owusu-Antwi, G., & Antwi, J. (2010). The analysis of the rural credit market in Ghana. *International Business & Economics Research Journal* (IBER), 9(8), 45–56.

Ozili, P. K. (2021). Financial inclusion research around the world: a review. *Forum for Social Economics,* 50(4), 457–479. https://doi.org/10.1080/07360932.2020.1715238

Park, C.-Y., & Mercado, R. (2018). Financial inclusion, poverty, and income inequality. *The Singapore Economic Review,* 63(01), 185–206. https://doi.org/10.1142/S0217590818410059

Polloni-Silva, E., da Costa, N., Moralles, H. F., & Sacomano Neto, M. (2021). Does financial inclusion diminish poverty and inequality? A panel data analysis for Latin American countries. *Social Indicators Research,* 158(3), 889–925.

Saha, S. K., & Qin, J. (2022). Financial inclusion and poverty alleviation: an empirical examination. *Economic Change and Restructuring*, *56*(1), 409–440. https://doi.org/10.1007/s10644-022-09428-x

Satpathy, I., Patnaik, B. C. M., & Das, P. K. (2014). Transformation from class banking to mass banking through inclusive finance: a paradigm shift. *Asian Social Science,* 10(19), 11–16.

Sha'ban, M., Girardone, C., & Sarkisyan, A. (2020). Cross-country variation in financial inclusion: a global perspective. *European Journal of Finance,* 26(4–5), 319–340.

Siddique, J., Shamim, A., Nawaz, M., & Abid, M. F. (2023). The hope and hype of neuromarketing: a bibliometric analysis. *Journal of Contemporary Marketing Science,* 6(1), 1–21.

World Bank (2018a). *The World Bank Annual Report 2018.* Washington, DC: World Bank eBooks. https://doi.org/10.1596/978-1-4648-1296-5

World Bank (2018b). World development report 2018: Learning to realize education's promise. *Washington, DC: World Bank. [Review of the book* World development report 2018: Learning to realize education's promise]. *Population and Development Review*, *44*(2), 404–405. https://doi.org/10.1111/padr.12164

Xie, X. (2023). Analyzing the impact of digital inclusive finance on poverty reduction: a study based on System GMM in China. *Sustainability*, *15*(18), 13331. https://doi.org/10.3390/su151813331

Zupic, I., & Čater, T. (2015). Bibliometric methods in management and organization. *Organizational Research Methods, 18*(3), 429–472.

6 Combating Deepfakes
Converging and Diverging Approaches of Media and Internet Powerhouses against High-Tech Disinformation

Larisa Mistrean, Muskan Singh, and Sachin Ghai

6.1 INTRODUCTION

New difficulties for journalistic verification arise from the use of artificial intelligence in technologically orchestrated communication processes in the networked world. It has improved several phases of news production systems at the same time. Artificial intelligence technology is evident in both the communication flows and a significant portion of the socialization dynamics. Thus, one of the most current risks to the integrity of journalism and news trustworthiness is the threat posed by the creation of deepfakes, or artificial intelligence-manipulated films. Deepfakes are a problem for journalism as well, but their existence has made people less certain about accessing news information. Likewise, experts are on guard due to the growing complexity of this type of false news (Vaccari & Chadwick, 2020).

Over the past few years, misinformation has become more relevant and has taken centre stage in the public discourse (Vargo et al., 2018). The quantity of initiatives and actions taken to combat this issue has increased significantly over time. The European Commission's Action Plan Against Disinformation might serve as an illustration of this (2018). Journalists and media professionals are mindful of how the popularity of hoaxes threatens democracies and its dependability. Consequently, companies attempt to respond by taking steps like fact-checking that promote openness and the observance of their ethical and professional standards (Lowrey, 2017). Several schools of thought have been established on this problem. They are all attempting to offset the effects of misleading political and social tendencies that have gained significance in 21st-century civilizations (McNair, 2017) in an environment where online communication is a major platform for the creation and spread of false information, as well as the associated fallout (Nelson & Taneja, 2018).

Methods that ensure the effectiveness of information verification, which has been a fundamental component of journalism since its establishment as a communication tool in the contemporary era (Kovach & Rosenstiel, 2014), are searching inside technological innovation for instruments that might assist specialists in their day-to-day work.

DOI: 10.1201/9781003564621-6

It is true that there are instances where the standards adhered to in order to produce precise informative articles are ambiguous and complex (Shapiro et al., 2013). However, in order to assess questionable material, journalism needs to upgrade its antiquated authentication procedures to computational ones instead of sticking with them (Ciampaglia et al., 2015). Evaluation systems with fact-checking methods are currently being updated. These might aid in the formulation of news articles that demonstrate the legitimacy of news in the age of technology following the application of a complex set of fundamental, societal, and technical links (Carlson, 2017, p. 13). Certain methods employed by these information verification services are possible to increase consumers' dependability even while there is still a lot of scepticism (Brandtzaeg & Følstad, 2017).

Additionally, in response to this circumstance, fresh ideas for updated professional practices and profiles have emerged (Palomo & Masip, 2020). Constructive journalism may fall under this category, with the aim of winning back the public's confidence (McIntyre & Gyldensted, 2017). The current trend in journalism takes novel approaches. It will take some time to see, though, if these novel techniques mesh with the growing Third Millennium journalism paradigm, which is distinctly committed to social responsibility, openness, and truth.

The rise of deepfake technology, which creates remarkably lifelike but entirely fake audio and visual material using advanced machine learning and artificial intelligence, poses a serious danger to digital information. The more these altered materials resemble authentic recordings, the more likely it is that lies and false information will proliferate. Major online shops and media sites have thus begun to employ a number of strategies to detect and mitigate the impact of deepfakes. These tactics, however, vary widely and encompass a variety of approaches, from cutting-edge technological solutions to collaborative initiatives with governments and civil society. This study looks at the convergence and divergence strategies employed by these organizations, as well as their effectiveness, scalability, and implications for privacy and freedom of expression. The analysis seeks to fight the spread of technologically advanced disinformation by providing readers with a comprehensive overview of the complex realm of digital legitimacy and security together with a full assessment of current and future solutions.

Deepfake technology produces phoney audio and video content that sounds remarkably genuine. It has become a potent tool for disinformation operations, endangering public trust and reasoned conversation. Internet companies and media companies have acknowledged the issue and implemented a number of strategies to deal with and counteract these sophisticated digital frauds. These efforts range from forming alliances for the sharing of information and standardization to developing unique detection techniques. This chapter examines the complimentary as well as distinctive information integrity protection strategies used by three well-known industry leaders, each of which has substantial influence over digital property. This study looks at the benefits and drawbacks of current methods as well as the larger implications for privacy and policy in the digital age. This study sheds light on the technological battle against false information while raising significant concerns about how to reconcile innovation with moral responsibility in the fight against false information disseminated online.

In the digital age, the spread of deepfake technology poses a serious threat because of its capacity to subtly modify audio and video content. This technology undermines the foundation of accurate communication and calls for a strong response from those who oversee the flow of information since it can produce persuasive but completely fraudulent material. Though their approaches and goals differ greatly, media firms and large internet platforms—the main gatekeepers of digital content—are using a number of tactics to stop the spread of these intricate forgeries.

Prominent corporations within the domain are progressing towards the development of cutting-edge instruments capable of autonomously detecting and classifying deepfake content prior to its exposure to wider audiences. Others emphasize more the formation of coalitions that bring together lawmakers, tech corporations, and academic experts to work together and pool resources in the fight against deepfakes. Regulatory and legal structures that mandate transparency and accountability from platforms hosting user-generated content are likewise receiving more and more attention.

This essay tries to analyse the variety of tactics used by media organizations and online giants to combat deepfake-driven digital misinformation. By carefully analysing these tactics, the chapter evaluates how well they work in practical settings, looks into any potential downsides, and talks about the moral ramifications of these anti-deepfake approaches. By doing this, it highlights the intricate relationship between ethical governance and advances in technology in the age of digital manipulation and offers a nuanced understanding of how various actors in the digital domain are cooperating and working independently to secure the integrity of information. This study is essential for understanding the current situation as well as for planning a multi-stakeholder, coordinated remedy to the issues that deepfake technologies may face in the future.

6.2 LITERATURE REVIEW

6.2.1 Misinformation through the Ages

False, fantasy, and fake news have emerged as a topic of discussion in communication and media, raising concerns about their impact on society (McNair, 2017). Misinformation has been around since the Roman Empire (Burkhardt, 2017), but the emergence of printing in the 15th century marked a significant shift. The ease of disseminating written knowledge also enabled the spread of falsehoods.

The introduction of new communication channels has led to a rise in purposeful misleading material, which is not necessarily detrimental. The radio transmission of Orson Welles' film The War of the Worlds on October 30, 1938, is a prime example of disinformation. According to Gorbach (2018), the narration of Welles' cast led many listeners to fear the Earth was under extraterrestrial invasion.

The radio show aimed to delight the audience by altering reality. Manipulation of the facts has historically been employed in military engagements to accuse the opponent of malevolent behaviour (Bloch, 1999, p. 182). The US Navy ship USS Maine's boiler explosion on February 15, 1898, exemplified the use of disinformation in press coverage. During the competition for the most read newspaper against

Joseph Pulitzer's The New York World, William Randolph Hearst, editor of the New York Journal, dispatched a writer to Cuba to cover a Spanish raid on a ship. Upon arrival on the island, the journalist reported that there was no purported attack. The journal disseminated inaccurate tales about the incident, fuelling hatred for Spain and support of the impending conflict. Finally, the United States declared war on Spain (Amorós, 2018, p. 34). Following the struggle, misinformation was employed against the enemy during wartime. Peters (2018) identifies techniques used in recent battles such as World Wars I and II, Vietnam, and the Gulf War.

The issue of false information, fantasy, and fake news has been a long-standing concern in the realm of communication and media, drawing attention to its potential societal impacts (McNair, 2017). While misinformation isn't a new phenomenon, its prevalence and impact have evolved over time. Historical examples, such as during the Roman Empire (Burkhardt, 2017), illustrate its enduring presence. However, the advent of printing in the 15th century ushered in a new era, significantly amplifying the spread of falsehoods alongside written knowledge. As communication mediums evolved, so did the tools for conveying intentionally erroneous information. A noteworthy example is Orson Welles' radio broadcast of The War of the Worlds on October 30, 1938, which notably provoked fear among listeners who felt Earth was under extraterrestrial attack (Gorbach, 2018). This instance demonstrates how manipulating reality for amusement reasons can result in unforeseen consequences.

Beyond entertainment, manipulation of facts has also been utilized in more serious contexts, including military engagements. In the late 19th century, the press coverage of the USS Maine's explosion in 1898 exemplified the use of disinformation to sway public opinion. William Randolph Hearst's New York Journal, in competition with Joseph Pulitzer's The New York World, fabricated stories of a Spanish attack on the ship, ultimately stoking public outrage and paving the way for the United States' declaration of war against Spain (Amorós, 2018). The utilization of misinformation in wartime has a long history, with various techniques identified in conflicts such as World Wars I and II, Vietnam, and the Gulf War (Peters, 2018). These instances highlight the enduring challenge of distinguishing truth from falsehood during conflict and the potential consequences of misinformation on both a national and global scale.

6.2.2 Fake News as a Threat to Journalism

Throughout history, the field of communication has been plagued with false information. Fake news is a new kind of disinformation that mimics journalistic content but is intentionally manipulated (Rochlin, 2017). This type of misinformation has coexisted alongside actual news. The use of high-speed, low-contrast communication methods, including social media, creates an environment conducive to spreading disinformation. Social media platforms like Facebook and Twitter have become main news sources for internet users (Bergström & Jervelycke-Belfrage, 2018). As a result, the web has become the primary avenue for fake material distribution, using the anonymity given by places (Vosoughi et al., 2018). Fake news providers have the potential to reach huge audiences as established journalistic brands (Fletcher et al., 2018). This makes verifying the falsities more complex.

Various approaches for identifying bogus news have emerged in recent years. Tandoc, Lim, and Ling developed one of these (2018), which is perhaps the most exhaustive: News satire is a popular sort of false news that may be seen in publications, websites, and radio/TV broadcasts. News parody is like news satire but does not focus on current events. Fictional components are created for specific goals, such as news stories that mimic the format of traditional media. The promoters of these pieces attempt to deceive by blending false information with truthful information, altering images and videos to create a false reality, disguising advertising as journalistic reporting, and spreading propaganda. Political groups use tales to influence individuals' opinions about them. Like earlier examples, they mimic the formal structure of news articles.

Fake news mimics the formal presentation of real news. Misinformation methods often include visual codes, headlines, photos, videos, hypertext, and journalistic-style writings (Amorós, 2018, p. 65). Fake news often attacks readers' existing opinions, particularly on contentious topics like racism, xenophobia, homophobia, and other forms of hate (Bennett & Livingston, 2018; Waisbord, 2018). This link allows for quick duplication of material via social media sites, with occurrences such as electoral processes (Lowrey, 2017) and, most recently, the COVID-19 epidemic (Salaverría et al., 2020; Shimizu, 2020). The spread of fake news has increased significantly, utilizing both simple means and powerful technologies. Misinformation spreads quickly in the digital era because of easy access to many internet platforms. Wardle and Derakhshan (2017) found that the concept of "information disorder" includes not just fake news, but also misunderstanding and misinformation. Misinformation is defined as legitimate information exploited to inflict harm, such as leaking personal information or disseminating private papers (Wardle & Derakhshan, 2017).

Furthermore, the emergence of deepfakes complicates the distinction between fact and fiction. Deepfakes are incredibly realistic movies or audios produced by artificial intelligence, frequently portraying famous personalities saying or doing things they never did. These clever alterations blur the border between fact and fiction, complicating the disinformation environment. The psychology of how bogus news spreads is also being studied. According to Pennycook and Rand (2018), people are more inclined to trust and spread information that confirms their previous ideas or prejudices. This cognitive bias leads to the spread of misinformation because individuals like to seek out and magnify news that confirms their viewpoint.

As well, algorithms have a significant impact on online conversation. Social media sites use algorithms to customize users' feeds based on their previous behaviour and interests, which can lead to filter bubbles and echo chambers. This algorithmic bias may unintentionally facilitate the propagation of incorrect information among like-minded networks while excluding disagreeing voices. Fighting the spread of false news necessitates a multifaceted strategy that includes media literacy education, technical interventions, and regulatory measures. Scholars argue for teaching critical thinking skills and digital literacy at a young age so that people can distinguish between reliable sources and disinformation. Furthermore, platforms must invest in developing content moderation algorithms and enforcing policies against the spread of misinformation.

6.2.3 Deepfake: A Novel Form of Fake News

Deepfakes, a mixture of "deep learning" and "fake news" (Westerlund, 2019), have the characteristics of "highly realistic and difficult-to-detect digital manipulations of audio or video" (Chesney & Citron, 2019). It refers to the use of computer code to modify videos (Fernandes et al., 2019, p. 1721). This involves substituting a person's voice or visage with that of another person. Artificial intelligence has increased the prevalence and quality of photo and video editing, which has been a long-standing practice. Although some of these films are humorous, the bulk are harmful (Maras & Alexandrou, 2019). This movement began in 2017 and has grown rapidly since then (Deep trace Labs, 2018, pp. 2–4). This technology utilizes Generative Adversarial Networks (GANs), which replace human features or voices in photos and videos to enhance their realistic quality (Li et al., 2018). These algorithms provide the benefit of self-improvement through deepfakes. Learning from prior experiences will enhance future inventions. This feature makes disinformation more dangerous, especially with the rise of mobile applications and computer programs that enable non-programmers to create deepfakes (Nirkin et al., 2019; Schwartz, 2018). Farid et al. (2019, pp. 4–6) classified deepfakes into four categories: (1) Face replacement, often known as face swapping, involves substituting one person's face with the targets. (2) Face re-enactment involves manipulating a person's facial characteristics, such as moving their mouth or eyes. (3) GANs can be used to create a new face. (4) Speech synthesis involves altering or creating a new discourse based on cadence and intonation.

Deepfake algorithms have potential applications in psychology, such as generating digital identities for silent people or using improved facial recognition for police enforcement (Akhtar & Dasgupta, 2019; Zhu et al., 2020). However, the usage of these technologies has been linked to fraud and crime, making it more detrimental than useful (Stupp, 2019). One of the most challenging aspects of combating deepfakes is the lack of timely development of debunking solutions for disinformation (Galston, 2020). Efforts to build technology-based methods for identifying and correcting it are ongoing in both public and commercial institutions (Deeptrace Labs, 2018, p. 2). These technologies are useful for many forms of communication, particularly journalism. Deepfakes, a combination of "deep learning" and "fake," are a big step forward in digital manipulation technology (Westerlund, 2019). These manipulations entail the fabrication of very realistic audio or visual information that is difficult to identify as fake (Chesney & Citron, 2019).

The main approach underlying deepfakes is based on advanced computer algorithms, notably GANs, which allow for the seamless replacement of faces or voices in movies and photos. This technology has democratized the creation of false material, enabling non-programmers to create convincing deepfakes using simple tools and apps (Nirkin et al., 2019; Schwartz, 2018). Researchers have classified deepfakes into numerous sorts based on their manipulation methods. These include face replacement, which swaps one person's face for another's, face re-enactment, which alters facial emotions, GAN-based development of wholly new faces, and voice synthesis, which creates or modifies speech patterns (Farid et al., 2019).

While deepfake technology has potential uses in a variety of sectors, including psychology and law enforcement (Akhtar & Dasgupta, 2019; Zhu et al., 2020), its abuse is very dangerous. Cases of fraud and criminality associated with deepfake technology have generated worries about its negative influence on society (Stupp, 2019). Combating the proliferation of deepfakes poses significant hurdles, notably in establishing effective debunking methods to counter misinformation in real time (Galston, 2020). Both public and commercial institutions are currently developing technical tools and techniques to identify and mitigate the negative consequences of deepfakes (Deeptrace Labs, 2018). In the field of journalism, the advent of deepfakes highlights the significance of media literacy and critical thinking abilities among consumers in distinguishing manipulated information from genuine sources. Furthermore, news organizations must implement comprehensive verification mechanisms to prevent the spread of false information (Deeptrace Labs, 2018).

6.2.4 Fact-Checking: Journalism's Response to the Misinformation Wave

Verified information is crucial for effective communication (Ekström et al., 2020). During disruptive events such as the coronavirus epidemic, people's intake of information increased significantly (Masip et al., 2020). Political leaders now convey their thoughts online rather than through conventional media channels (López-García & Pavía, 2019). The media has highlighted the need for verification systems to address internal and external faults (Geham, 2017). A new professional profile, the fact-checker, has evolved to dispel disinformation and prevent it from being consumed by audiences. Professionals gather and compare available data to determine the source of information or claims (Graves, 2016, p. 110). Journalists can now efficiently check photographs, videos, and websites thanks to advancements in technology (Brandtzaeg et al., 2016).

Verification is crucial in journalism, especially with the rise of fake news in recent years. According to the University of Duke Reporters' Lab's census, there are about 300 fact-checkers in over 60 nations by the middle of 2020, which is 100 more than in 2019 (Stencel & Luther, 2020). Fact-checking can be considered a multinational movement (Graves, 2018). Both legacy and independent media companies attempted to reestablish faith in the media (Bennett & Livingston, 2018). In today's information ecosystem, where disinformation abounds, fact-checking has become increasingly important (Ekström et al., 2020). During times of crisis, such as the COVID-19 pandemic, demand for verified information has increased as individuals seek reputable sources to make informed judgements (Masip et al., 2020).

The rise of online political debate emphasizes the need for verification mechanisms to prevent incorrect information from spreading (López-García & Pavía, 2019). As a result, a new class of experts known as fact-checkers has formed in the media sector. These people specialize in validating claims and assertions in order to avoid the spread of disinformation (Graves, 2016). Technological advancements have enabled journalists to do more effective fact-checking across several media types, such as images, videos, and websites (Brandtzaeg et al., 2016). This technical advancement has enabled fact-checkers to quickly analyse the veracity of material and give fast revisions as needed. The spread of fact-checking programmes throughout the world demonstrates a rising realization of their relevance in protecting journalism's integrity. According to the University of Duke Reporters' Lab, the number of fact-checkers globally surpassed 300 by mid-2020, a substantial rise from prior years (Stencel & Luther, 2020). This worldwide movement highlights the coordinated efforts of legacy and independent media companies to battle disinformation and reestablish trust with their viewers (Graves, 2018).

6.3 METHODS

This investigation will begin with a systematic literature review (Kitchenham, 2004). This method offers a strategy for addressing and studying deepfakes. This strategy provides a comprehensive understanding of the unfamiliar reality (Codina, 2017): what are researchers doing to assess this phenomenon and what efforts are being made to stop its spread?

This case study will examine how six businesses employ various techniques to detect, classify, and alert deepfakes, including protocols, technology, collaboration with universities, and financing for creative initiatives.

This will be accomplished by analysing accessible reports and statements from six organizations. Our study aims to find similarities and differences in the strategies of three major media sources and key internet platforms in combating the spread of deepfakes. This information is important to determine if six key members from these two disciplines are working together to restrict the spread of the virus and enhance existing practices.

6.4 RESULTS

6.4.1 Counteracting Deepfakes at the *Wall Street Journal*, the *Washington Post*, and Reuters

The democratization of artificial intelligence has made it possible for regular people to produce deepfakes. Given the potentially negative effects of these inventions, particularly prior to election processes, this poses a significant challenge to our society. The *Wall Street Journal* has established a section of 21 journalists with the specific goal of identifying, classifying, and disproving false material, including deepfakes, in anticipation of the US general election in 2020 (Southern, 2019). This team, which is a collaboration between the Standards and Ethics and R&D departments, works closely with technology. Journalists with backgrounds in research, photography, video, and graphics as well as news are present, and they have received training in deepfake identification (Marconi & Daldrup, 2018). Additionally, in collaboration with other scholars, The *Wall Street Journal* offers specialized training in the identification of deepfakes and disinformation. This prompted the creation of a three-step norm to locate instances of this type of misinformation: searching for older online footage versions, examining footage using photo and video editing software, and examining the source through interaction with the original source, authorship proof of identity, and the metadata check, among other activities.

The strategy used by the news organization Reuters might be a third way to look at this situation. According to Crosso (n.d.), the news services provider, it is aware of and concerned about the rise of deepfakes. As a result, it has partnered with Facebook to identify as much manipulated user-generated content as it can from among all of the images and videos that are posted on the network (Patadia, 2020).

To this end, Reuters has launched a blog in both Spanish and English with the goal of validating doctored information. All of it with the goal of disproving as much false information as possible before the US election of 2020. This demonstrates the growing cooperation between media outlets and digital platforms in attempting to curb the spread of deepfakes and fake news during crucial events like the presidential campaign.

6.4.2 Internet Giants' Strategies against Deepfakes

Nowadays, social networking sites and other online areas provide a challenge to the propagation of misinformation for companies such as Google, Facebook, and Twitter. Consequently, in the last few months, they have launched many efforts, each with the specific goal of identifying effective means of identifying and thwarting the spread of misleading data and, more recently, deepfakes.

In this context, these three businesses take distinct stances on the issue. For example, Google has made a sizeable collection of modified and unmanipulated movies available to the scientific community (Dufour & Gully, 2019). Through this endeavour, they want to contribute to the advancement of identification procedures by using the vast quantity of data stored in their archives. Furthermore, they work with the Defence Advanced Research Projects Agency to provide funding to various researchers who are creating media forensic instruments.

Conversely, Facebook is funding several research initiatives as part of its "Deepfake Detection Challenge." This programme, supported by research departments from several US colleges as well as corporations like Facebook, Microsoft, and Amazon Web Services, aims to help scientists who are developing artificial intelligence-based deepfake detection techniques. As a result, these researchers who compete to propose practical mechanisms in order to win various awards had access to a corpus of over 1,00,000 movies.

Additionally, in an effort to combat this kind of false information, Mark Zuckerberg's social media platform removes edited images or videos or designates it as fake news with the assistance of verifying facts media sources (Bickert, 2020). Given the potential impact of false news on the 2020 US presidential election campaign, this is especially crucial for everyone involved.

Twitter, at last, offers a more straightforward solution to this issue. According to Harvey (2019), their approach may be summed up in the following four rules: The altered content should be identified through a notice on Tweets that contain it, users should be warned before sharing it due to its manipulated state, links to reports or other reliable sources should be included so that users can learn more about the reasons behind the manipulation, and all potentially hazardous or dangerous content should be removed.

While official intervention in monitoring content on social networking sites and other networks is still in its early stages, these divergent tactics employed by the major online platforms are partly the result of self-regulated approaches for combating deepfakes. This raises the question of whether other governmental institutions should devote resources to addressing societal interests in this circumstance, or if the internet platforms are the only ones responsible for combating disinformation.

In 2018, the European Commission made clear that governments must fund investigations into the spread of false information and urge them to make social media corporations responsible for their actions (European Commission, 2018). To address the problem, the EU has so far launched a number of projects in the past two years, including the establishment of the Rapid Alert System, the Social Observatory for Disinformation and Social Media Analysis, a code of practice against disinformation, and other R&D projects like PROVENANCE, SocialTruth, EUNOMIA, or WeVerify (European Commission, 2020). Deepfakes are still not receiving as much attention as other academics would want, despite significant efforts being made to prevent the spread of false information within the EU. These scholars also express concern about programmes that appear to be understaffed (Bressnan, 2019). Various levels of involvement are employed by governments to encourage social media corporations to take action against false news, with the primary goal being to combat misinformation in political advertisements.

While Australia mandates that all paid electoral advertising, including those on social networking sites, be authorized and contain a permission statement, France, Germany, and other countries require platforms on the internet to set up a visible and easily accessible method for users to report inaccurate data (Levush, 2019). A few states in the United States have already taken action against deepfakes, but it's still reactive rather than preventive. For example, Texas passed a law making it illegal to release and spread deepfake videos with the intent to cause damage to an individual during the election process. California also passed a law last October making it illegal for anyone to distribute deepfakes with the intent to deceive voters or perjure a candidate (Castro, 2020).

According to Nick Clegg, the company's vice president for worldwide affairs and communications and a former deputy prime minister of the United Kingdom, Facebook's creation of the Independent Oversight Board, which "aims to make Facebook more accountable and improve the decision-making process," is evidence that significant governance issues should be addressed in the years to come (Moltzau, 2020). In the meantime, a lot of businesses are starting to re-evaluate their governance structures.

6.5 DISCUSSION

The media and internet companies have taken a number of steps to counteract disinformation, including the development of deepfakes, as previously noted. These two communication sectors' techniques have similarities and differ from one another. Collaboration between platforms and media outlets looks to be growing over time. Reuters and Facebook have partnered to detect and correct false news. Fact-checking organizations collaborate with this social network to detect and warn users about bogus material. Another coincidence is the use of technology as a weapon against deepfakes. To combat fake news, news organizations and internet platforms acknowledge the need of innovative technology and smart algorithms. Media outlets are preparing journalists and interdisciplinary teams to identify lies. The third match might be the emerging synergies between the intellectual and communicative sides. To train and use their algorithms, media outlets and platforms collaborate with experts and institutes that specialize in false news detection.

To address the issue of deepfakes, web platforms might sponsor research initiatives to create AI systems for recognizing fake news. The media lacks these opportunities due to the high cost of its activities. The media handles disinformation differently, correcting it rather than removing it. Several social media networks have implemented measures to eliminate doctored information. This is an obvious difficulty. Deleting modified movies or photographs affects future users, but not those who have already viewed them. In contrast, labelling verification media outlets' strategy of identifying fraudulent or modified materials may be useful for future users.

In the ongoing battle against misinformation, the collaboration between media outlets and internet platforms has become increasingly evident. Reuters and Facebook, for instance, have formed partnerships aimed at detecting and rectifying false news circulating on social media. This collaboration extends to the involvement of fact-checking groups, which work hand in hand with these platforms to identify and alert users about fraudulent information. Recognizing the gravity of the situation, both news organizations and internet platforms have acknowledged the indispensable role of technology in combating the proliferation of false news. Consequently, efforts are being made to leverage advanced technology and sophisticated algorithms to bolster their defence mechanisms.

Media outlets are investing in training programmes for journalists and interdisciplinary teams so that they can successfully identify and rebut lies. Furthermore, there is a rising tendency of collaboration between the intellectual and communicative realms, with media outlets and platforms teaming up with researchers and research organizations specialized in the detection of fake news. This partnership enables the development and deployment of cutting-edge algorithms and approaches for successfully combating disinformation.

To tackle the specific challenge posed by deepfakes, web platforms are increasingly supporting research initiatives aimed at developing AI systems capable of recognizing and mitigating the spread of fake news. However, the high cost associated with such endeavours often limits the involvement of traditional media outlets in similar research initiatives. Consequently, the media adopts a different approach to handling disinformation, focusing more on correction rather than removal. While several social media networks have implemented measures to eradicate doctored information, this approach presents inherent challenges. Removing modified content may impact future users, but it does little to address the issue for those who have already been exposed to it. In contrast, the strategy adopted by media outlets, which involves the identification and labelling of fraudulent or modified materials, offers a more nuanced approach that can benefit future users by providing them with crucial context and awareness.

Legal and regulatory frameworks, in addition to technology improvements and joint initiatives, are critical in the fight against disinformation. Governments all across the world are creating legislation and rules to hold platforms accountable for the dissemination of misleading information. For example, the European Union's General Data Protection Regulation (GDPR) establishes stringent rules on data processing by internet platforms, reducing their capacity to combat disinformation. Furthermore, nations such as Singapore have enacted legislation that particularly targets internet falsehoods, requiring platforms to give corrections or remove incorrect information.

Beyond technology solutions, user education and media literacy programmes are critical components in combating misinformation. Media literacy programmes attempt to provide individuals with the critical thinking skills required to assess information sources and recognize disinformation. Initiatives by both governmental and non-governmental groups aim on teaching consumers about the hazards of misleading information and equipping them with tools to distinguish between legitimate and questionable sources.

Ethical media standards: While technology is important, ethical media standards remain critical in the battle against disinformation. News organizations follow professional standards and codes of ethics, which include the concepts of truth, impartiality, and transparency. Journalistic integrity is critical to retaining public confidence and credibility, especially in this digital age. Fact-checking, source verification, and responsible reporting are all components of ethical journalism that assist in the spread of misleading information.

Public-Business Partnerships: Collaboration among governments, business sector firms, and civil society groups is widely regarded as an important tool for combating disinformation. Public-private partnerships use the talents and resources of diverse parties to create comprehensive solutions for countering misinformation. Initiatives like the Partnership on AI (PAI) bring together tech corporations, universities, and non-profits to work on AI research and development while keeping ethical aspects in mind, such as disinformation identification and mitigation.

Working Together Globally: Misinformation spreads quickly across borders, therefore governments must collaborate to combat it. The Global Alliance for Partnerships on Media and Information Literacy, coordinated by UNESCO, facilitates collaboration across nations to increase people's capacity to absorb media and information. Furthermore, internet platforms and governments share information to detect and prevent emerging problems, such as the spread of misleading information or online scams.

6.6 CONCLUSION

Deepfakes are now a common occurrence in our communication infrastructure. Internet service providers and media outlets attempt to offset it with opposing viewpoints. Still, the methods for spreading false information appear to be developing more quickly than those for refuting it. In this regard, the majority of the research that is now accessible focuses on two areas: first, detection technologies; and second, the consequences of this type of false news for democracy and national security. The fact that only large technological corporations have been able to introduce high-tech, costly solutions to combat deepfakes thus far encourages the vast majority of research on the topic to be devoted to answering the primarily technological questions posed by these corporations. However, journalism-focused media are unable to forward their worries onto the research agenda since they cannot afford to make significant financial investments in deepfake detection. Because of this, doing research on the effects of deepfakes on journalism and under journalistic principles, as well as more study on how the media prepares its personnel to recognize sophisticated disinformation, seems essential.

This deceptive technique's novelty contributes to its overlooked status, but the ongoing expansion of research on the subject indicates that it will become a significant area of study for disinformation and media forensics specialists in the years to come. The study can, however, demonstrate to some extent how the techniques of the media and digital platforms differ and have striking parallels. This could be because of how differently their business models operate, but occasionally it still appears like an investment issue. Digital platforms have collaborated with partners in technology, academia, and entrepreneurship, investing significant financial resources in this area—something that many media outlets are unable to do.

These partnerships, meanwhile, can extend outside the business world. Around the world, there is increasing demand from states and governments to impose varying degrees of regulation on these social media corporations. Nations have begun implementing procedures that enable a closer examination of these firms since there is a growing body of information about how their internal social interaction instruments might impact fundamental democratic components (elections, political advertising, etc.). Facebook's recent hiring of Nick Clegg, a former UK deputy prime minister, is illustrative of this, since it appears the company is prepared to change its governance structures in order to implement more common-sense rules. The trend appears to characterize a conflict between these corporations' attempts to uphold their own autonomy and the states' and governments' goals of regulating.

However, we must point out the limits of our sample and our study. Our sample exhibits a Western-oriented cultural bias, as the viewer may have noted, for both the media and online medium examination. Because of this, extrapolating our results to analogous organizations in other regions could prove challenging. This study, however, takes a descriptive method with the aim of analysing the strategies used by large communication firms to stop the spread of deepfakes, the most recent and technologically advanced type of deception. Furthermore, only major companies—mostly headquartered in the United States—can now afford the level of technology and human resources necessary for the debunking process due to the novelty and high degree of complexity of this new method of manufacturing false news. It will be fascinating to watch how deepfakes are produced and identified in the future, as well as the courses that academics in this field choose to take. Although this is a relatively new phenomenon, its rapid expansion will need the establishment of procedures to control its dissemination in print and digital media.

Prospective avenues for investigation on this subject matter might encompass an in-depth evaluation of how journalists interpret deepfakes in their day-to-day activities, as well as the difficulty of verification for journalists, including concerns about technology proficiency and best practices guidelines.

REFERENCES

Akhtar, Z., & Dasgupta, D. (2019). A comparative evaluation of local feature descriptors for deepfakes detection. In *2019 IEEE International Symposium on Technologies for Homeland Security, HST 2019*. Woburn, MA: IEEE. https://doi.org/10.1109/hst47167.2019.9033005

Amorós, M. (2018). *Fake News: La verdad de las Noticias Falsas* [Fake news: The truth of false news]. Barcelona: Plataforma Editorial.

Bennett, W. L., & Livingston, S. (2018). The disinformation order: Disruptive communication and the decline of democratic institutions. *European Journal of Communication*, *33*(2), 122–139. https://doi.org/ 10.1177/0267323118760317

Bergström, A., &Jervelycke-Belfrage, M. (2018). News in social media. *Digital Journalism*, *6*(5), 583–598. https://doi.org/10.1080/21670811.2018.1423625

Bickert, M. (2020). Enforcing against manipulated media. *Facebook*. Retrieved from https://about.fb.com/news/2020/01/enforcing-against-manipulated-media

Bloch, M. (1999). *Historia E Historiadores* [History and Historians]. Madrid: Akal.

Brandtzaeg, P. B., &Følstad, A. (2017). Trust and distrust in online fact-checking services. *Communications of the ACM*, *60*(9), 65–71. https://doi.org/10.1145/3122803

Brandtzaeg, P. B., Lüders, M., Spangenberg, J., Rath-Wiggins, L., &Følstad, A. (2016). Emerging journalistic verification practices concerning social media. *Journalism Practice*, *10*(3), 323–342. https://doi.org/ 10.1080/17512786.2015.1020331

Bressnan, S. (2019). Can the EU prevent deepfakes from threatening peace? Carnegie Europe. Retrieved from https://carnegieeurope.eu/ strategiceurope/79877

Burkhardt, J. M. (2017). Chapter 1: History of fake news. *Library Technology Reports*, *53*(8), 5–9.

Carlson, M. (2017). *Journalistic Authority: Legitimating News in the Digital Era*. New York, NY: Columbia University Press.

Castro, D. (2020). Deepfakes are on the rise: How should government respond? *Government Technology*. Retrieved from https://www.govtech.com/policy/DeepfakesAre-on-the-Rise-How-Should-Government-Respond.html

Chesney, R., & Citron, D. (2019). Deepfakes and the new disinformation war. *Foreign Affairs*. Retrieved from https://www.foreignaffairs.com/articles/world/2018-12-11/deepfakes-and-new-disinformation-war

Ciampaglia, G. L., Shiralkar, P., Rocha, L. M., Bollen, J., Menczer, F., &Flammini, A. (2015). Computational fact checking from knowledge networks. *PLoS One*, *10*(6). https://doi.org/10.1371/journal.pone.0128193

Codina, L. (2017). Revisionessistematizadas y cómollevarlas a cabo con garantías: Systematic reviews y SALSA Framework [Systematized reviews and how to conduct it with guarantees: Systematic reviews and SALSA Framework]. *Lluiscodina*. Retrieved from https://www.lluiscodina.com/revision-sistematica-salsa-framework

Crosse, G. (n.d.). Truth and authentication: The frightening world of video fakes. *Reuters*. Retrieved from https://agency.reuters.com/en/insights/articles/articles-archive/truth-and-authentication-the-frightening-world-of-video-fakes.html

Deeptrace Labs. (2018). *The State of Deepfakes: Reality Under Attack*. Amsterdam: Deeptrace Labs. Retrieved from https://s3.eu-west-2.amazonaws.com/rep2018/2018-the-state-of-deepfakes.pdf

Dufour, N., & Gully, A. (2019). Contributing data to deepfake detection research. *Google Blog*. Retrieved from https://ai.googleblog.com/2019/09/contributing-data-to-deepfake-detection.html

Ekström, M., Lewis, S. C., &Westlund, O. (2020). Epistemologies of digital journalism and the study of misinformation. *New Media & Society*, *22*(2), 205–212. https://doi.org/10.1177/1461444819856914

European Commission. (2018). *Action Plan against Disinformation*. Brussels: European Commission. Retrieved from https://eeas.europa.eu/sites/eeas/files/action_plan_against_disinformation.pdf

European Commission. (2020). *Tackling Online Disinformation*. Brussels: European Commission. Retrieved from https://ec.europa.eu/digital-single-market/en/tackling-online-disinformation

Farid, H., Davies, A., Webb, L., Wolf, C., Hwang, T., Zucconi, A., &Lyu, S. (2019). *Deepfakes and audiovisual disinformation*. London: Centre for Data Ethics and Innovation. Retrieved from www. gov.uk/government/publications/cdei-publishes-its-first-series-of-three-snapshot-papers-ethical-issues-in-ai/snapshot-paper-deepfakes-and-audiovisual-disinformation

Fernandes, S., Raj, S., Ortiz, E., Vintila, I., Salter, M., Urosevic, G., & Jha, S. (2019). Predicting heart rate variations of deepfake videos using neural ODE. In *Proceedings: 2019 International Conference on Computer Vision Workshop, ICCVW 2019* (pp. 1721–1729). Seoul: IEEE. https://doi.org/10.1109/iccvw.2019.00213

Fletcher, R., Cornia, A., Graves, L., & Nielsen, R. K. (2018). *Measuring the Reach of Fake News and Online Disinformation in Europe*. Oxford: Reuters Institute for the Study of Journalism. Retrieved from https://reutersinstitute.politics.ox.ac.uk/our-research/measuring-reach-fake-news-and-online-disinformation-europe

Galston, W. A. (2020). *Is seeing still Believing? The Deep-Fake Challenge to Truth in Politics*. Washington, DC: Brookings. Retrieved from www.brookings.edu/research/is-seeing-still-believing-the-deepfake-challenge-to-truth-in-politics

Gorbach, J. (2018). Not your grandpa's hoax: A comparative history of fake news. *American Journalism*, *35*(2), 236–249. https://doi.org/10.1080/08821127.2018.1457915

Graves, L. (2016). *Deciding What's True*. New York, NY: Columbia University Press.

Graves, L. (2018). Boundaries not drawn. *Journalism Studies*, *19*(5), 613–631. https://doi.org/10.1080/1461670X.2016.1196602

Harvey, D. (2019). Help us shape our approach to synthetic and manipulated media. *Twitter.* Retrieved from https://blog.twitter.com/en_us/ topics/company/2019/synthetic_manipulated_media_policy_feedback.html

Kitchenham, B. (2004). *Procedures for Performing Systematic Reviews*. Keele: Keele University. Retrieved from www.inf.ufsc.br/~aldo.vw/ kitchenham.pdf

Kovach, B., & Rosenstiel, T. (2014). *The Elements of Journalism, Revised and Updated: What Newspeople Should know and the Public Should Expect* (3rd ed.). New York, NY: Three Rivers Press.

Levush, R. (2019). Government responses to disinformation on social media platforms. Library of Congress. Retrieved from www.loc.gov/law/help/social-media-disinformation/compsum.php

Li, Y., Chang, M.-C., &Lyu, S. (2018). In ICTU oculi: Exposing AI created fake videos by detecting eye blinking. In *2018 10TH IEEE International Workshop on Information Forensics and Security (WIFS)* (pp. 1–5). Hong Kong: IEEE. https://doi.org/10.1109/wifs.2018.8630787

Lowrey, W. (2017). The emergence and development of news fact-checking sites. *Journalism Studies*, *18*(3), 376–394. https://doi.org/10.1080/1461670X. 2015.1052537

Maras, M.-H., & Alexandrou, A. (2019). Determining authenticity of video evidence in the age of artificial intelligence and in the wake of deepfake videos. *The International Journal of Evidence & Proof*, *23*(3), 255–262. https://doi.org/10.1177/1365712718807226

Marconi, F., & Daldrup, T. (2018). How The Wall Street Journal is preparing its journalists to detect deepfakes. *NiemanLab*. Retrieved from https://www.niemanlab.org/2018/11/how-the-wall-street-journal-is-preparing-its-journalists-to-detect-deepfakes

Masip, P., Aran-Ramspott, S., Ruiz-Caballero, C., Suau, J., Almenar, E., &Puertas-Graell, D. (2020). News consumption and media coverage during the confinement by Covid-19: Information overload, ideological bias and sensationalism. *El Profesional de la Información*, *29*(3). https://doi.org/10.3145/epi.2020.may.12

McIntyre, K., &Gyldensted, C. (2017). Constructive journalism: An introduction and practical guide for applying positive psychology techniques to news production. *The Journal of Media Innovations*, *4*(2), 20–34. https://doi.org/10.5617/jomi.v4i2.2403

McNair, B. (2017). *Fake News: Falsehood, Fabrication and Fantasy in Journalism*. London: Routledge.

Moltzau, A. (2020). What strategy does Europe have to tackle deepfakes? *Medium*. Retrieved from https://medium.com/dataseries/what-strategy-does-europe-have-to-tackle-deepfakes-fb159040f0c

Nelson, J. L., & Taneja, H. (2018). The small, disloyal fake news audience: The role of audience availability in fake news consumption. *New Media &Society*, *20*(10), 3720–3737. https://doi.org/10.1177/1461444818758715

Nirkin, Y., Keller, Y., &Hassner, T. (2019). FSGAN: Subject agnostic face swapping and reenactment. *Nirkin*. Retrieved from https://nirkin.com/fsgan

Palomo, B., &Masip, P. (2020). Journalistic reinvention for an automated and polarized scenario. In C. Toural Bran, Á. Vizoso, S. Pérez-Seijo, M. Rodríguez-Castro, & M.-C. Negreira-Rey (Eds.), *Information visualization in the era of innovative journalism* (pp. 161–170). New York, NY: Routledge.

Patadia, D. (2020). Reuters launches fact-checking initiative to identify misinformation, in partnership with Facebook. *Reuters*. Retrieved from www.reuters.com/article/rpb-fbfact checking/reuters-launches-fact-checking-initiative-to-identify-misinformation-in-partnership-with- facebook-idUSKBN2061TG

Pennycook, G., & Rand, D. G. (2020). Who falls for fake news? The roles of bullshit receptivity, overclaiming, familiarity, and analytic thinking. *Journal of Personality*, *88*(2), 185–200. https://onlinelibrary.wiley.com/doi/abs/10.1111/jopy.12476

Peters, M. A. (2018). The information wars, fake news and the end of globalisation. *Educational Philosophy and Theory*, *50*(13), 1161–1164. https://doi.org/10.1080/00131857.2017.1417200

Rochlin, N. (2017). Fake news: Belief in post-truth. *Library Hi Tech*, *35*(3), 386–392. https://doi.org/10.1108/LHT-03-2017-0062

Schwartz, O. (2018). You thought fake news was bad? Deep fakes are where truth goes to die. *The Guardian*. Retrieved from www.theguardian.com/technology/2018/nov/12/deep-fakes-fake-news-truth

Shimizu, K. (2020). 2019-nCoV, fake news, and racism. *The Lancet*, *395*(10225), 685–686. https://doi.org/10.1016/S0140-6736(20)30357-3

Southern, L. (2019). 'A perfect storm': The Wall Street Journal has 21 people detecting 'deep- fakes.' *Digiday*. Retrieved from https://digiday.com/media/the-wall-street-journal-has-21-people- detecting-deepfakes

Stencel, M., & Luther, J. (2020). Annual census finds nearly 300 fact-checking projects around the world. *Duke Reporters' Lab*. Retrieved from https://reporterslab.org/annual-census-finds-nearly-300-fact-checking-projects-around-the-world

Stupp, C. (2019, August 30). Fraudsters used AI to mimic CEO's voice in unusual cybercrime case. *The Wall Street Journal*. Retrieved from www. wsj.com/articles/fraudsters-use-ai-to-mimic-ceos-voice-in-unusual-cybercrime-case-11567157402

Tandoc, E. C., Lim, Z. W., & Ling, R. (2018). Defining "fake news." *Digital Journalism*, *6*(2), 137–153. https://doi.org/10.1080/21670811.2017.1360143

Vaccari, C., & Chadwick, A. (2020). Deepfakes and dis- information: Exploring the impact of synthetic political video on deception, uncertainty, and trust in news. *Social Media + Society*, *6*(1). https://doi.org/ 10.1177/2056305120903408

Vargo, C. J., Guo, L., &Amazeen, M. A. (2018). The agenda-setting power of fake news: A big data analysis of the online media landscape from 2014 to 2016. *New Media & Society*, *20*(5), 2028–2049. https://doi.org/10.1177/1461444817712086

Vosoughi, S., Roy, D., & Aral, S. (2018). The spread of true and false news online. *Science*, *359*(6380), 1146–1151. https://doi.org/10.1126/science.aap9559

Waisbord, S. (2018). Truth is what happens to news. *Journalism Studies*, *19*(13), 1866–1878. https://doi.org/ 10.1080/1461670X.2018.1492881

Wardle, C., & Derakhshan, H. (2017). *Information Disorder: Toward an Interdisciplinary Framework for Research and Policymaking*. Vol. 27. Strasbourg: Council of Europe. pp. 1–109. https://tverezo.info/wp-content/uploads/2017/11/PREMS-162317-GBR-2018-Report-desinformation-A4-BAT.pdf

Westerlund, M. (2019). The emergence of deepfake technology: A review. *Technology Innovation Management Review*, *9*, 40–53. https://doi.org/10.22215/timreview/1282

Zhu, B., Fang, H., Sui, Y., & Li, L. (2020). Deepfakes for medical video de-identification: Privacy protection and diagnostic information preservation. In *AIES 2020: Proceedings of the AAAI/ACM conference on AI, ethics, and society* (pp. 414–420). New York, NY: ACM. https://doi.org/10.1145/3375627.3375849

7 Bibliometric Insights
Emerging Trends in AI Adoption and Integration in Educational Institutions

Diksha Ahuja, Ramneek Ahluwalia, and Rasna Sharma

7.1 INTRODUCTION

In the past few decades, the education sector has experienced significant changes due to globalization, technological advancements, and social transformations, which have been irreversible (Iñiguez & Lorange, 2022). The rise of digital transformation in education has ushered in a new era marked by improved accessibility and increased autonomy (Ülkü, 2023). The higher education sector has undergone significant changes in recent decades due to advancements in information and communication technology (ICT). This change will help institutions meet the changing needs of students, provide excellent education, and ready them for the digital age's challenges.

As online learning continues to evolve, artificial intelligence (AI) techniques can be integrated into various devices, such as chatbots, dashboards, automated assessment systems, and intelligent tutoring systems, to achieve different objectives (Amini et al., 2024). In our post-digital era, universities are encountering a new phase of the Fourth Industrial Revolution, where General Artificial Intelligence (GAI) introduces higher levels of machine innovation and independence. Human monitoring is crucial to ensure socially responsible education and leadership in this context (Ellis, 2024). With the advent of the Fourth Industrial Revolution, or Industry 4.0, it has fundamentally changed our world, giving rise to a modern era of economic and societal change. It is marked by a network of cutting-edge technologies like robotics, the internet of things, automation, and the integration of AI into various scientific, technological, and business operations. AI, in particular, is considered the cornerstone of Industry 4.0, often associated with machine learning, a subset of AI predicted to be crucial for future business operations (Bukartaite & Hooper, 2023).

New teaching and learning solutions that integrate intelligent tutoring systems, technological architecture, and AI capabilities are being developed and tested worldwide. These innovative approaches use AI technologies to improve educational experiences, customize instruction, and assist learners and educators (Grover et al., 2024).

The topic of integrity has gained more attention due to the extensive use of educational resources for teaching and assessment in universities. Scholars have extensively

 DOI: 10.1201/9781003564621-7

studied the convergence of technology and education. The inclusion of advanced technology, particularly AI, in educational environments has added complexity to this discussion (Ülkü, 2023).

Machine learning is a crucial element of AI. The focus is on instructing computers to learn from data and enhance their performance through experience, rather than being specifically programmed to do tasks. By utilizing machine learning in education, educators can acquire valuable insights into how students learn, enabling them to make informed decisions based on data and customize their teaching methods to suit each student's needs (Grover et al., 2024).

AI is demonstrating notable promise in the field of text-generating chatbots, which have gained popularity not only among university students but also among various users in recent months (Ellis, 2024). Chatbots can simplify the process of accessing information for students by responding to their queries and requests, whether through text or audio input. However, they do not assist in understanding the information. Typically, students rely on text for obtaining information, but merely acquiring the relevant information is not sufficient (Ellis, 2024).

Some optimistic individuals believe that AI will greatly enhance the lives of historically marginalized groups. Conversely, pessimists fear that AI will negatively impact society through mass unemployment, heightened bias, and increased surveillance, particularly affecting these marginalized groups (Novozhilova et al., 2024).

In the realm of tests and tasks, ChatGPT-4 stands as an advanced language model created by OpenAI, showcasing its ability to produce a narrative with a human-like structure. This phenomenon brings forth various possibilities and ethical challenges (Ülkü, 2023). Relying too heavily on AI could impede the cultivation of critical thinking abilities and raise concerns about AI models creating deceptive academic content. Educators face difficulties in discerning genuine student work from material produced by AI, creating a complex and challenging situation.

7.2 RESEARCH METHODOLOGY

The present research utilizes the systematic procedural technique (Tranfield et al., 2003) to guarantee deliberate, replicable, and comprehensive coverage of the literature. Structured reviews facilitate the assessment of the present state of knowledge and help identify potential research deficiencies in a particular field of study (Cai & Lo, 2020). The data for this research has been extracted from the Scopus database. Scopus has emerged as the predominant, esteemed, and authenticated database for scholarly writings. Several bibliometric studies have shown the reliability of the Scopus database for research analysis. The present research utilizes VOSviewer 1.6.17 for doing bibliometric analysis. The bibliometric technique is characterized by its simplicity and ability to be replicated, as shown by van Eck and Waltman (2010). The VOSviewer software facilitates the process of doing bibliometric mapping and network analysis. The analysis has been carried out at many levels: The topics covered in the text include the co-occurrence of terms, overlay visualization, source growth, and the relation between countries. A compilation of research on AI at educational institutions spanning from 1985 to 2024 was collected from the Scopus database. Until now, there has been little bibliometric study undertaken to examine the impact of AI in educational institutions. The objective of this study is to identify the theoretical foundations, prominent researchers, and academic journals in the area of AI and educational institutions. This study aims to analyze the framework of AI utilization and provide potential directions for future research in the domains of AI and educational institutions.

7.3 DATA AND METHOD

7.3.1 Data Selection and Search Criteria

The final search criteria are summarized below:

This research employs bibliometric analysis as its chosen methodology. The bibliometric analysis methodology addresses research inquiries by examining advancements and literature in the field. The employed analysis method encompasses both evaluative and descriptive aspects of bibliometric analysis. The study's sample consisted of 224 publications retrieved from the Scopus database, aligning with the specified keywords. Utilizing the expansive interdisciplinary coverage of the Scopus database, researchers conducted a search for data sources pertaining to AI in management educational institutes. Refining the collected data involves a series of steps, starting with identification, followed by screening, assessing eligibility, and, ultimately, the inclusion step. Figure 7.1 represents the process of data collection.

Figure 7.1 illustrates the initial stage of the data collection process, where the identification process is the first step. The researcher inputs keywords into the search field of the Scopus database. The keywords entered are ("Artificial Intelligence") AND ("Management") AND ("Education"). The identification process yielded publication data for 339 articles. Following the identification process, the subsequent step involves screening based on specific criteria. In this case, the researcher screens publications, ensuring they are in English and take the form of journal articles. Out of the screening results, 249 publications were identified as meeting the specified criteria, while 90 publications were excluded and will not proceed to the next phase.

Following the screening, the researcher conducts a feasibility process to determine the eligibility of publications for inclusion in the subsequent stage. This manual assessment is carried out to ensure the suitability of the identified publications. The

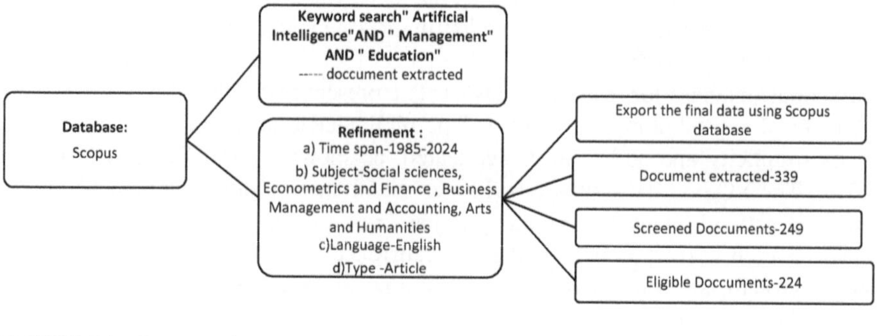

FIGURE 7.1 Process of data collection.

researcher examined the titles and abstracts of 249 publications, evaluating those that incorporated the variable of AI in management students. Upon the completion of the third phase, 224 publications were identified as meeting the criteria for inclusion in the subsequent stage. Descriptive analysis of the trend in publications concerning the application of AI to management students is conducted through bibliometric analysis, extracting data from the Scopus database. A graph depicting the annual publication trends from 2000 to 2024, along with a linear trendline, will be generated using Microsoft Excel software.

Figure 7.2 illustrates the clusters that reflect the research focus. There are several different colors indicating research clusters connected to Machine Intelligence in management educational institutions. There are five clusters, with the largest being Cluster 1, followed by Cluster 2, Cluster 3, Cluster 4, and Cluster 5. This indicates that the research focus is segregated into five clusters. There are six items in the first cluster (in red). If seen from the size of the circle in this first cluster, the keywords including AI, technology, knowledge management and online learning represents the largest diameter. However, the second cluster includes four items. The keywords including education and higher education are the center of the second research together with machine learning as these represent the largest circle size in the cluster. There are five items found in the third cluster having the same occurrence, whereas the fourth cluster consist of three items, where all the three keywords also have the same number of occurrences. Lastly, the fifth cluster constitutes five items, all five keywords have the same occurrence.

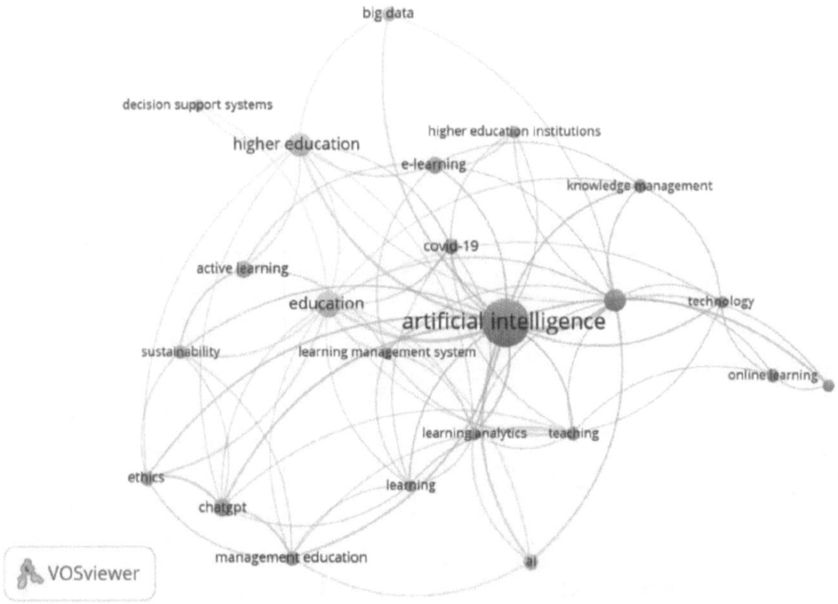

FIGURE 7.2 Keyword co-occurrence.

According to Figure 7.3, circles represent keywords that signify a new theme. Figure 7.3 highlights ChatGPT, online learning, learning analytics, ethics, and management education as the new thematic keywords in this field (Table 7.1).

Figure 7.4 illustrates the source dynamics of the top five journals, utilizing LOESS (locally estimated scatterplot smoothing) to demonstrate the publication counts over time. According to the figure, Sustainability (Switzerland) exhibited a rapid increase in publications in 2022. However, in 2023, the Sustainability (Switzerland) journal experienced a significant decline compared to the higher number of publications in the Education Services and Education and Information Technologies journals.

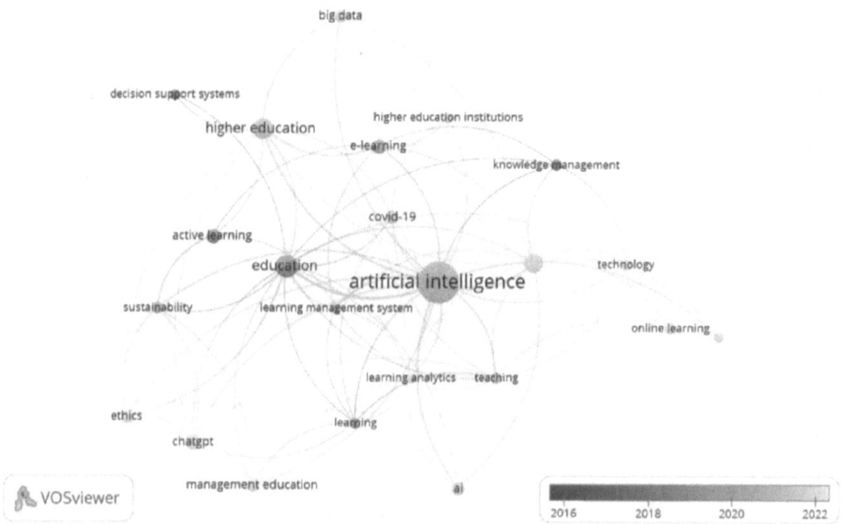

FIGURE 7.3 Overlay visualization.

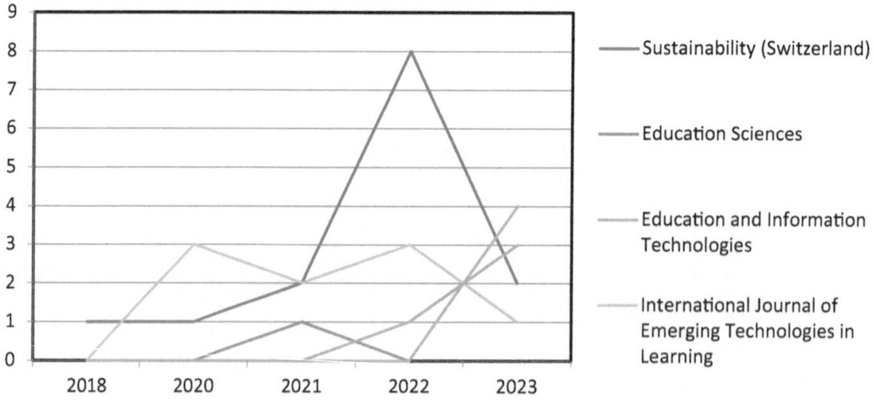

FIGURE 7.4 Source growth.

TABLE 7.1
Source Growth of Five Journals

Year	Source	No of Publications
2018	*Sustainability (Switzerland)*	1
2020	*Sustainability (Switzerland)*	1
2021	*Sustainability (Switzerland)*	2
2022	*Sustainability (Switzerland)*	8
2023	*Sustainability (Switzerland)*	2
2021	*Education Sciences*	1
2023	*Education Sciences*	4
2023	*Education and Information Technologies*	4
2020	*International Journal of Emerging Technologies in Learning*	3
2021	*International Journal of Emerging Technologies in Learning*	2
2022	*International Journal of Emerging Technologies in Learning*	3
2023	*International Journal of Emerging Technologies in Learning*	1
2022	*Computers and Education: Artificial Intelligence*	1
2023	*Computers and Education: Artificial Intelligence*	3

The journal Computers and Education: AI experienced a notable increase in publications, rising from 1 publication in 2022 to 3 in 2023.

Figure 7.5 clearly signifies that the US circle has the largest circle diameter in contrast to other countries. This depicts that the country has the highest level of harmonious and cooperative relations accompanied by China. The display of VOS viewer exhibits that UK has allied with 11 other countries, followed by China with a total of 7 links.

7.4 LIMITATIONS

That being said, even though it is thorough research, it does have some limitations. While the search was conducted using the Scopus it is possible that other databases may contain supplementary information that was not incorporated in the study like Web of Science. Consequently, even bibliometric evaluations are limited by particular characteristics that are primarily attributable to their methodology. It is crucial to select appropriate search criteria in order to locate a database containing the maximum number of papers pertinent to the subject under consideration. Additionally, it is important to mention that the source growth is only considered from 2018 onward, disregarding the growth of journals in prior years. However, in an effort to address these constraints, this study conducted an exhaustive search and verified the presence

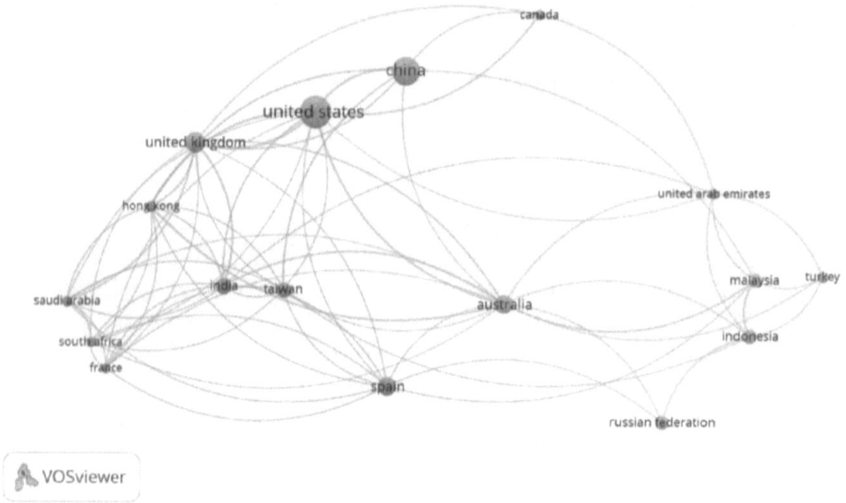

FIGURE 7.5 Relation between countries.

of an exact correspondence between the names retrieved from the database and the authors referenced in the articles being examined.

7.5 RECOMMENDATIONS FOR FUTURE RESEARCH

The goal of the research was to examine the existing research on the implementation of AI in educational institutions. The objective of this study was to only examine papers and documents that establish a connection between AI and educational institutions. However, the majority of the research found are mostly focused on AI and its relationship with elements such as online learning, management education, machine learning, and so on. No article has been found on the role of AI tools and trends. Primarily, publications pertaining to AI mostly focus on the topic of learning rather than trends. It affirms that AI has become prevalent in marketing, but there is less study on the use of AI in education to understand how AI tools or trends might assist students or instructors in decision-making. Therefore, it is essential to go into this area of study.

7.6 CONCLUSION

This chapter presents a bibliometric examination of AI in the literature of educational institutions. As a comprehensive analysis, our purpose is to gather existing information on the connection between AI and educational institutions. This chapter does a systematic evaluation of the existing literature in the topic, using both bibliometric and content analytic methods. We used bibliometric analysis to ascertain yearly publishing patterns, the geographic origins of all contributing sources, the interrelationship among countries, source expansion, and study of keyword occurrence.

The present study showcases the chronicle of significant publications and enhances the field by clarifying the connections between impactful investigations. The research makes significant contributions since it is the first to use a systematic approach, namely bibliometric analysis, within the field of AI and educational institutions. Furthermore, bibliometric analysis facilitates the identification of both well-established and emergent tendencies within the area of study. In addition, the study provides suggestions for further investigations. Utilizing these resources may provide advantageous perspectives.

REFERENCES

Amini, A., Firouzkouhi, N., Nazari, M., Ghareeb, N., Cheng, C., & Davvaz, B. (2024). Artificial intelligence global online learning (GOL) theory by generalized n-ary fuzzy relation. *Artificial Intelligence Review,* Vol. 57(3), *68*. https://doi.org/10.1007/s10462-023-10691-1

Bukartaite, R., & Hooper, D. (2023). Automation, artificial intelligence and future skills needs: an Irish perspective. *European Journal of Training and Development*, *47*(10), 163–185. https://doi.org/10.1108/EJTD-03-2023-0045

Cai, Y. J., & Lo, C. K. Y. (2020). Omni-channel management in the new retailing era: a systematic review and future research agenda. *International Journal of Production Economics*, *229*(April), 107729. https://doi.org/10.1016/j.ijpe.2020.107729

Ellis, R. A. (2024). The education leadership challenges for universities in a postdigital age. *Postdigital Science and Education*, *0123456789*. https://doi.org/10.1007/s42438-024-00461-9

Grover, V., Nandal, M., Balusamy, B., Sahu, D., & Dogra, M. (2024). Proposed hybrid model in online education. *EAI Endorsed Transactions Proposed hybrid Model in Online Education*. *10*, 1–7. https://doi.org/10.4108/eetiot.4770

Novozhilova, E., Mays, K., & Katz, J. E. (2024). Looking towards an automated future: U.S. attitudes towards future artificial intelligence instantiations and their effect. *Humanities and Social Sciences Communications*, *11*(1), 1–11. https://doi.org/10.1057/s41599-024-02625-1

Tranfield, D., Denyer, D., & Smart, P. (2003). Towards a methodology for developing evidence-informed management knowledge by means of systematic review. *British Journal of Management*, *14*, 207–222. https://doi.org/10.1007/978-3-030-92836-0_28

Ülkü, A. (2023). Artificial intelligence-based large language models and integrity of exams and assignments in higher education: the case of tourism courses. *Tourism and Management Studies*, *19*(4), 21–34. https://doi.org/10.18089/tms.2023.190402

van Eck, N. J., & Waltman, L. (2010). Software survey: VOSviewer, a computer program for bibliometric mapping. *Scientometrics*, *84*(2), 523–538. https://doi.org/10.1007/s11192-009-0146-3

8 Employee Training, Educating Staff on Recognizing and Reporting Deepfakes, Building a Vigilant Corporate Culture

Sami Ullah, Mohit Kukreti, and Muhammad Rehan Shaukat

8.1 INTRODUCTION

In early 2020, an authentic-looking video started to make the rounds on social media of one of the prominent CEOs with a shocking message: the company was going to file for bankruptcy. Staff, shareholders, and even media personnel scrambled to react, only to realize later that it was a deepfake video crafted to mimic the executive visually and audibly convincingly. Although the conspiracy theory was soon proved a hoax, the damage had already been done; stock prices fell, and trust began to bleed away. This incident and others like it mark just the beginning of a growing threat of deepfakes to corporations.

Businesses are growing increasingly interlinked and digitalized, but risks in deepfakes are no longer theoretical (Kietzmann et al., 2020). A UK-based energy company lost more than $2,00,000 in 2019 after one of its employees was impersonated by a deepfake by someone assuming the guise of the company's chief executive, who thus duped him into transferring the amount. The examples highlight that organizations must train employees to identify and promptly report deepfakes. Building a vigilant corporate culture would go a long way to mitigate this growing threat (Hight, 2022; Mustak et al., 2023). The leaders need to create an atmosphere where the workforce is trained and empowered to act when suspicious content presents itself in view (Abbas & Taeihagh, 2024). The risks of inaction or ignorance are too high to ignore as the technology continues rapidly to outpace corporate defenses (Deepfake Conundrum).

This chapter will guide managers and leaders through the essential steps of offering a robust response to deepfakes in organizational contexts. It starts with explaining deepfakes and their potential growing threat to organizations. It then discusses

DOI: 10.1201/9781003564621-8

why employee vigilance is required, alerting staff to their role in helping protect their organizations. The chapter then moves on to practical steps to develop such a holistic training program, detailing why it is necessary and how it fits into the wider security frameworks. The chapter concludes with strategies for cultivating a vigilant corporate culture with a discourse on leadership's influence on that culture and overcoming resistance to training.

8.2 DEEPFAKE AWARENESS IN CORPORATE CULTURE

Deepfake combines "deep learning" and "fake," referring to synthetic media in which the person's likeness, voice, or action is convincingly changed to represent false narratives (Seow et al., 2022; Westerlund, 2019). The concept started as a niche within artificial intelligence (AI) and first came to public notice with political propaganda, but now it has emerged as one of the most recent emerging threats to businesses worldwide (Mariani & Dwivedi, 2024; Whittaker et al., 2023). It poses threats to security, distrust, and reputation problems by precisely mimicking hyper-realistic digital forgeries related to the person's speech or actions. The seamless nature of deepfakes makes it dangerous with the increasing digitalization of businesses. Remote work and virtual meetings have become norms, increasing the risk of deepfakes spreading through corporate communication channels. Deepfakes can be used to mislead employees, customers, and shareholders (Van der Sloot & Wagensveld, 2022; Wach et al., 2023), the effects of which might be disastrous for a firm. Therefore, companies should take urgent measures to determine deepfake awareness, which is now a critical component of corporate culture, given the rising accessibility and sophistication of this technology.

8.3 IMPORTANCE OF EMPLOYEE VIGILANCE

Employees are the first line of defense in the fight against deepfakes (Schreiber & Schreiber, 2024). Since deepfakes can penetrate all aspects of an organization—from email communications to video conferences—the organizations must encourage a culture of caution in the workplace (Ahmed, 2021). Employees should be able to identify anomalies, report them to someone, and realize that the threat of deepfakes is real and evolving (Levine, 2020). For instance, in 2020, a global bank almost succumbed to a cyberattack. The bank's finance department received an email from its CFO requesting approval for a wire transfer. However, one employee noticed minute inconsistencies in the tone and wording of the email and decided to delay the transaction processing. The employee tried another way to confirm the authenticity of the request. The email was found to be part of a phishing attack through deepfake audio manipulation. If the employee had not raised an alarm, the bank would have suffered a severe loss of money. This highlights the need for awareness and vigilance on the part of the employee. Solutions that come from technology include AI-driven detection systems, but these are still in development and are far from foolproof. Employees should be taught to detect visual and audio abnormalities and follow internal procedures, such as verifying unusual requests through other channels (Byeon et al., 2024). The same can be applied to customers and external parties; transparency and clarity in verification procedures could help prevent the exploitation of deepfakes (Gu et al., 2024).

Vigilance is beyond just recognizing abnormalities. The employees need to understand the larger scheme that deepfakes fit into. Such deepfake attacks target those with decision-making authority, the executives, the managers, and the financial officers. Training such individuals to identify the face of manipulation is critical since they are usually on the receiving end of the fraudulent actors' activities (Gowrisankar & Thing, 2024; Jayashre & Amsaprabhaa, 2024). However, rank-and-file employees should not be ignored; they, too, play a vital role in maintaining the integrity of corporate operations. A culture of collective vigilance ensures that every member—from the most junior to even the highest-ranking officer—contributes to the organization's security (Hight, 2022; Jacobsen & Simpson, 2024).

8.4 CONNECTING TO CORPORATE VULNERABILITIES

Deepfakes represent a new threat to corporate vulnerabilities because they exploit trust, which is the lifeline of any organization (Ayata, 2024; Cover, 2022). Trust exists in corporate structures between employees and executives, the entity and its shareholders, and the business and customers. Deepfakes can destroy such trust immediately and in the worst form: They take advantage of human nature since we tend to believe our eyes and ears; they have become a powerful tool for fraud and misinformation (Kolagati et al., 2022; Liz-Lopez et al., 2024). Authenticity of communication is the very essence in the fast-moving world of business, where many decisions are made in a flash. A financial disaster would result if employees did not know this could all be a deepfake mimicry of an incredibly well-trusted executive (de Rancourt-Raymond & Smaili, 2023; Whittaker et al., 2023).

Deepfakes can also influence public opinion. In 2021, a fake video circulated online allegedly showing a CEO making derogatory comments about customers and employees. The video surfaced easily and was quickly discredited, but it was too late, as the damage had already been done. The company faced a brief crisis wherein there was severe backlash on social media, media inquiries began pouring in, and stock prices dipped. While the company was eventually cleared of the allegations, the incident highlighted how susceptible companies can be to deepfakes, especially in today's social media era, where misinformation spreads quickly.

Deepfakes also coincide with other corporate vulnerabilities relating to intellectual property (IP) and trade secrets (Beuve et al., 2023). Consider a deepfake video of an executive where the individual is disclosing confidential information concerning a new product launch or merger. Even if the video is fake, the fact that such a video exists can erode investor confidence, harm relationships with key stakeholders, or diminish customers' trust in the company. Such vulnerabilities of deepfakes can make them a particularly insidious threat. Therefore, companies should include deepfake awareness in risk management and security efforts (Cochran & Napshin, 2021; Schreiber & Schreiber, 2024). In this case, it would consolidate the IT, cybersecurity, HR, and legal departments to ensure vulnerability exposure. In this regard, training programs should be created to inform employees of the rampant deepfakes and the space they hold in the larger picture of corporate vulnerabilities. Employees should realize that deepfakes are not just a concern of a technology issue; they are essentially business issues impacting every activity within a company's operations—from financial transactions to reputation management and everything in between (Abbas & Taeihagh, 2024; Ayata, 2024).

8.5 DEVELOPING A COMPREHENSIVE TRAINING PROGRAM

As deepfake technology becomes increasingly widespread, businesses need to prepare themselves ahead of time against the growing threat through defensive measures (Cochran & Napshin, 2021). A defense would then mean a wide-ranging training course to equip the workforce with the insight and practice of seeing a possible deepfake and reporting the same. Failure to train staff on this issue exposes organizations to risks that go far beyond financial loss and thus put their corporate security, reputation, and trust entirely at risk, with debilitating consequences in the long term (Masood et al., 2023). Hence, deepfakes must be included in broad cybersecurity frameworks, ensuring the defenses built on such awareness are resilient and adaptive.

8.6 WHY TRAINING IS CRUCIAL

A recent report by the cybersecurity firm Deeptrace shows that the rate of deepfake videos circulating online is increasing quickly, doubling every six months since 2018. Considering this growing presence, businesses must be prepared for deepfakes in all their manifestations, not limited to manipulated emails but also fabricated video calls (Abbas & Taeihagh, 2024). The organization remains vulnerable if employees are not adequately trained to detect these threats and respond accordingly. Suppose in an internal meeting in a multinational pharmaceutical company, a deepfake video of a senior executive circulates the news that production had to be halted due to so-called regulatory violations. Panicked employees quickly spread the rumor to partners and other industry stakeholders through the grapevine. By the time the deepfake was identified and circulated, stock in the company had taken a hit. This could have all been minimized or prevented if employees had been trained to be sensitive to deepfake manipulation. Comprehensive training is critical for early detection.

Training, in this sense, serves two purposes. First, it mitigates threats already at hand and minimizes the immediate threats. Second, it fosters the essential long-term culture of vigilance. Deepfakes will only get more complex over time, as will an organization's detection approach (Ahmed, 2021; Ayata, 2024). Employee training on the latest forgery techniques and AI-driven technology can ensure companies act quickly against these new threats. The training, thus, should also cover the technicalities of identifying deepfakes and the protocols for maintaining verification procedures and reporting any suspicious activity to the proper channels (Beuve et al., 2023; Cochran & Napshin, 2021).

8.7 RISKS OF UNTRAINED STAFF

The untrained staff poses multidimensional risks. When the employees are not trained, they succumb to any deepfake scheme dished out to them, thereby bringing a host of disastrous consequences for the organization (Bray et al., 2023). Beyond financial loss, deepfakes can compromise corporate security, damage reputation, and gradually erode trust within the company and among other external stakeholders (Gowrisankar & Thing, 2024; Kietzmann et al., 2020). One classic example happened in 2021 when some middle management at one of the world's largest automotive firms became victims of a deepfake email scam. Cybercrime sent an AI-generated deepfake email from the company's CEO, authorizing a cybercriminal to transfer proprietary research data on a grand scale to an external server. When the breach was finally discovered, valuable IP had long since disappeared. An attack like this highlights the risk involved with an untrained employee, particularly those in sensitive positions, who have no recognition of the slightest indicators of a deepfake attack.

The reputational damage that such an incident can cause is immense. In an even more recent example in 2022, a deepfake of the founder of a tech company posted an admission of illegal business practices that never occurred. The company was able to prove it was faked, and its case was overturned, but for weeks in legal battles and media rebuttals, the damage remained unreversed. Without the ability to detect deepfake attacks, employees can be too slow to prevent or minimize damage from such videos; thereby, unbridled misinformation spreads. Trust is the foundation of all successful business relationships, whether it is between employees, clients, or investors. If staff lacks information about deepfake threats, then there is an easy breach in trust.

Similarly, a financial giant was engulfed in a public relations crisis when a leaked internal email containing the deepfake audio of its board member maltreating other employees by pretending to be talking about unethical financial practices damaged the reputation of a board member. The situation got out of hand before the organization could regain control because the failure of the organization to suspect the deepfake in the first instant of the leaked email posed a question over the authenticity of the communicated idea in its organization. It defines how untrained staff can inadvertently expose an organization to substantial risks to the reputation and stakeholder confidence.

8.8 LINK TO BROADER SECURITY FRAMEWORKS

Integrating deepfake training into security frameworks enhances corporate resilience and strengthens defenses across multiple fronts (Westerlund, 2019; Whittaker et al., 2023). A comprehensive deepfake training program should not be considered in isolation but as part of a broader effort to reinforce cybersecurity and compliance with regulations (Masood et al., 2023). Many organizations are increasingly structured with comprehensive approaches to cybersecurity, with protection aspects for data, phishing, and access. Training for deepfakes is an organic extension of these programs, as the threats that deepfakes pose overlap significantly with many of the same vulnerabilities.

For example, the financial services sector is already highly regulated and strongly focuses on ensuring data security. In 2021, one major investment firm implemented training for deepfakes as part of its overall cybersecurity strategy. They were trained to detect deepfakes and how these forgeries could be used with traditional cyberattacks, such as phishing schemes or social engineering tactics. The company integrated deepfake awareness into ongoing cybersecurity drills so the employees understood the connections between deepfakes and other cyber threats. Consequently, the firm was much more likely to identify and neutralize complex cyberattacks synchronized by deepfakes with other forms of malicious exploitation.

Besides cybersecurity capabilities, deepfake training can help firms comply with regulatory obligations. The financial, healthcare, and energy sectors have broad regulatory requirements regarding data integrity, privacy, and communications. Failure to protect against deepfake manipulation will result in probable regulatory sanctions, particularly if a deepfake leads to compromised customer data or internal communications. Sensitizing employees through awareness training on deepfakes enhances an organization's compliance with such regulatory standards, thus reducing legal risk (Ahmed, 2021; Beuve et al., 2023). For instance, a hospital in the United States has engaged in deepfake awareness training to enhance HIPAA compliance. They realized that deepfakes can be manipulated to tamper with a patient's medical information or even details about the patient, which puts privacy and data integrity at risk. Because the hospital included deepfake detection training in its compliance program, the risk of any occurrence was mitigated.

The introduction of deepfake training into a broader security framework encourages departmental collaboration (Mariani & Dwivedi, 2024; Rabhi et al., 2024). Threats related to deepfakes run across entire lines of business, from HR and legal teams through IT to communications. Training programs should encourage cross-functional collaboration so that deepfake risks are handled holistically. In 2022, a large multinational corporation created a cross-departmental task force to manage its deepfake response strategies. This task force comprised IT, legal, HR, and corporate communications representatives, each bringing a unique perspective on how deepfakes could impact their respective areas. The company then collaborated and rolled out a comprehensive approach to deepfake threats following its broader cybersecurity and compliance frameworks.

8.9 EDUCATING STAFF ON REPORTING MECHANISMS

Half the battle in organizations trying to protect themselves from the emergent threat would be training employees to detect and report suspected deepfakes (Masood et al., 2023). Businesses will also need to establish clear reporting mechanisms that allow access to such mechanisms and provide staff with sufficient time to assume responsibility and take swift action upon encountering suspicious content (de Rancourt-Raymond & Smaili, 2023; Schreiber & Schreiber, 2024). A robust approach to reporting will help empower the employees to act and enable the organization to respond to deepfakes in a way that is fair, reasonable, and consistent with the legal and regulatory mandated compliance requirements. The following section discusses the importance of robust reporting channels, details the alignment requirement of the channels with the broader compliance framework, and so on.

8.10 ESTABLISHING CLEAR REPORTING CHANNELS

Reporting channels are the basic ingredient in any deepfake awareness program. Clear lines and well-communicated reporting channels mean that employees know exactly what to do if they feel or suspect that a deepfake is being used within the organization, which may be a video call scam, email scam, or social media post (Romero Moreno, 2024). By providing an easy reporting pathway, businesses can streamline their response processes and identify and control any threats from deepfakes as quickly as possible. The reporting system is open to suspected deepfakes, which was thus introduced to a global insurance company in 2021 (Kaate et al., 2023). The tool allowed employees to submit potential deepfake materials such as emails, video content, or voice recordings directly to a cybersecurity team for further analysis. The reporting tool was integrated into the company's IT help desk software, making it easy and intuitive for employees. The company also conducted training sessions to ensure that staff were trained on accessing the tool and what types of content could cause red flags. With this centralized reporting, the company could respond quickly and effectively to such instances while limiting the chances of false alarms or delaying the identification of actual threats.

However, having a reporting mechanism in place is not enough without it being effective. All staff members should be informed about the procedures for reporting suspected deepfakes, irrespective of different departments and job roles (Schreiber & Schreiber, 2024). Most organizations leave the mentioning of cybersecurity threats to the expertise of the IT departments or security specialists, while others in the organization stay uninformed (Thomson et al., 2024). To avoid such a pitfall, reporting channels for deepfakes must be included in the business's security awareness program. That communication must integrate with other regular information measures. These reporting mechanisms need to be fused with an observant and effective cybersecurity team to back up these mechanisms. Employees should feel that their reporting will be taken seriously and acted upon promptly. Such a dedicated team would be specifically tasked with investigating incidents of deepfakes or carrying this responsibility as part of the existing security team, which would build the desired trust in the reporting system. With this, businesses can address deepfake threats promptly and effectively, potentially limiting their harm.

8.11 INCORPORATING LEGAL AND COMPLIANCE CONSIDERATIONS

Deepfake reporting mechanisms must also be carefully designed to align with legal and regulatory frameworks. Deepfakes often pertain to sensitive threats, especially impersonation cases, financial fraud, or altering proprietary information (Kietzmann et al., 2020). The reporting processes should be compliant with the laws and industry standards. This might entail getting oneself on the wrong side of the law, aside from possible reputational consequences and loss of regulatory relations. For example, Europe's general data protection regulation (GDPR) and California's Consumer Privacy Act (CCPA) are the latest stringent data protection acts. The reporting mechanisms of the businesses must comply with those regulations if the suspected deepfake contains personal data, including customer information, employee records, or private communications (Schreiber & Schreiber, 2024). This establishes rules for collecting, storing, and sharing data on suspected deepfake incidents. It is, therefore, a move toward enlightening workers on the legal dimensions of dealing with deepfake material, especially within sectors such as healthcare and finance that require more confidentiality on data (de Rancourt-Raymond & Smaili, 2023).

In highly monitored sectors, such as finance or contracting with the government, there usually is a need to report incidents that could affect financial integrity, security, or public trust. For instance, incidents related to deepfakes, such as impersonating senior executives or falsifying official documents, may lead to detrimental and far-reaching regulatory actions if left inadequately reported (Fui-Hoon Nah et al., 2023). A global bank sets an excellent example of how to bring regulatory considerations into deepfake reporting. When one of the regional offices fell victim to a deepfake scam, the bank started a reporting system that allowed employees to report suspected deepfakes internally and flagged instances that may warrant reporting to financial regulators. By including regulatory compliance under the reporting mechanism, the bank ensured that any deepfake-related incidents would be dealt with in adherence to the bank's internal protocols and regulatory expectations of the industry. In this manner, the bank covered itself for fines and reputation loss while instilling a culture of accountability and transparency among the employees.

Lastly, companies must also consider risks to employee legal privacy when developing reporting mechanisms (Romero Moreno, 2024). Deepfakes could also constitute an assault on an employee's image or voice, for instance, leading to defamation claims, invasion of privacy, or claims of identity theft (Bray et al., 2023; Kaate et al., 2023). Organizational procedures must be defined to respect employees' rights during such investigations and responses. This could include attorneys assisting those employees or advising them on pursuing their personal privacy issues. Thus, through streamlining reporting channels along compliance frameworks, businesses are protected from legal risks while fostering a culture of transparency and accountability that supports strengthening their security.

8.12 BUILDING A VIGILANT CORPORATE CULTURE

Building a solid training program and clear reporting mechanisms are defense weapons for an organization that appears to deal with deepfakes. But this is not enough. Resilience in an organization requires vigilance and responsibility to such an extent that learning is a constant process, and security is proactively placed for survival and existence (Hancock & Bailenson, 2021). Such a culture goes beyond tick-the-box training sessions and recurrences of isolated, one-off reminders; instead, it requires sustained engagement, shared responsibility, and strong leadership to model and reinforce the necessary behaviors to protect the organization.

8.13 FOSTERING A PROACTIVE SECURITY MINDSET

A vigilant corporate culture will begin at the grassroots level, as it will ensure all organization members are proactive about security. It should be a continuous way of learning, awareness, and alertness rather than a one-time initiative. Deepfake technology keeps evolving, and so should the organization's defenses (Westerlund, 2019). Employees need to be kept abreast of the current AI-driven forgery techniques and their impact on their respective roles in the company (Schreiber & Schreiber, 2024). One of the best ways to foster such an attitude is through non-stop, scenario-based learning. A global telecommunications company created a "cybersecurity challenge" program with their employees regularly facing simulated phishing, deepfake attacks, and other cyber threats. Scenarios were molded into the specific department so employees could learn to recognize real-world threats. This made learning engaging and relevant, and it developed an inquisitive mindset about security that went beyond traditional onboarding. The idea began to sink into employees that cybersecurity was not some box-checking activity but was, instead, an integral part of their work.

Cybersecurity risks like deepfakes should be openly discussed and advocated to support such a mentality (Bray et al., 2023; Cochran & Napshin, 2021). Employees should be able to raise their concerns or ask questions without fear of judgment or reprisal (Schreiber & Schreiber, 2024). Two-way communication can also be fostered through regular internal discussion forums or cybersecurity "office hours," where employees can seek advice from security specialists. When the employee base is actively involved and has a forum to air their issues, employees will remain vigilant and motivated enough to safeguard the organization (Masood et al., 2023). Furthermore, organizations can encourage the staff to be vigilant by rewarding those who ensure excellent vigilance. This strategy will motivate the staff to be vigilant and will also be a reminder that everybody's part is to keep the company's security intact.

8.14 LEADERSHIP'S ROLE IN SHAPING CULTURE

Everybody has a stake in a vigilant corporate culture; however, leadership is uniquely influential in shaping and maintaining this environment. Leaders must lead by example and set expectations for the teams; they must express their interest in security and vigilance at every turn (Cover, 2022). Leaders taking cybersecurity seriously and deepfake awareness visibly set an example for the whole organization. For example, in 2021, the chief executive officer of a large tech company initiated a company-wide cybersecurity week, which included personal participation by the CEO in employee training sessions on deepfake detection. Here, the CEO shared his experience of facing threats to his security and stated that everyone needs to play their part to protect the organization. This has helped cybersecurity convey that deepfake is not just an IT issue but a responsibility for everyone. When the CEO is actively involved, the employees will treat such training seriously and take a security-first way of thinking.

Leaders can also foster a culture of vigilance by anchoring deepfake awareness and cybersecurity to the organization's broader strategic goals (Gowrisankar & Thing, 2024; Van der Sloot & Wagensveld, 2022). For example, during quarterly meetings or company-wide updates, executives could discuss key cybersecurity metrics, uncover new deepfake threats, and various company efforts to strengthen defenses. Such positioning of cybersecurity as a strategic initiative signals that vigilance is a long-term commitment, not a passing fancy. Moreover, leadership can empower employees to report security concerns. On the other hand, if a leadership team is response-indifferent or inconsistent about security-related concerns, employees might also be hesitant to report suspected threats. This would naturally increase the vulnerability of the organization.

Leadership teams should give adequate heed to cross-departmental collaboration over issues of deepfake awareness and cybersecurity initiatives. Deepfake threats often tend to be associated with deep penetrations into several areas of the business: IT, legal, HR, and communications (Mustak et al., 2023; Seow et al., 2022). Collaboration of these departments will ensure that an organizational approach to deepfake threats is holistic and relevant to the rest of its business objectives. The cross-functional security teams should regularly review deepfake threats, share information, and coordinate the organization's response. This leadership-driven collaboration will dethrone information silos and have the entire organization working toward protecting against emerging threats.

8.15 CHALLENGES IN TRAINING AND CULTURAL TRANSFORMATION

Building a vigilant corporate culture against deepfake threats is not a small feat. Even with a holistic training program promoting continuous learning, the difficulty will originate from strong resistance from employees on how not to let the value of training programs be lost as technology continues advancing (Nguyen et al., 2022). Moreover, ensuring security without breaching employee privacy is another ethical challenge that must be handled delicately (Romero Moreno, 2024). With all these in mind, in the following few pages, we discuss key challenges and what can be done to beat them to ensure long-term success for deepfake awareness initiatives.

8.16 OVERCOMING RESISTANCE

The biggest obstacle to adopting deepfake awareness programs is resistance from the employees. This can be in the form of skepticism over the seriousness of threats of deepfakes, unwillingness to seek additional training, or simply becoming complacent (Ayata, 2024; Beuve et al., 2023). Employees may feel that deepfake training has no relation to their immediate work and that deepfakes are such a far-off concern, as they "will never happen here" or are only of concern for celebrities or politicians (Schreiber & Schreiber, 2024). A real example of the challenge is a multinational retail company that tried to implement deepfake training in its broader cybersecurity awareness program. The program was communicated as an important initiative, but most staff showed little interest in it, especially among management. Most managers dismissed the threat, stating that deepfakes would not threaten the company's operations. This resulted in low participation rates in the training program and allowed the new threats to expose the organization to vulnerability.

To mitigate such resistance, businesses must show the implications of deepfake threats regarding risks specific to a particular industry or job position. Deepfakes should not be projected as a hypothetical cause but applied through case studies or scenarios to show how deepfakes have affected companies in similar sectors (Cochran & Napshin, 2021; Hight, 2022; Jacobsen & Simpson, 2024). For example, banks can describe scenarios of legitimate fraudulent transactions with deepfakes, whereas manufacturing companies can focus on cases related to IP attack incidents. The type of threat depiction makes the threat concrete and relevant in overcoming the initial resistance by illustrating the consequences of taking no action.

Another approach to overcome this resistance is to make the training interactive and fun. Engaging them more in learning the process increases the chances of material being accepted. For example, trainees might enjoy learning through gamification techniques such as interactive simulations or role-playing scenarios. Leadership engagement is important because if the training is proven to be supported or endorsed by top leaders, the employees will have them take it more seriously (Mustak et al., 2023; Nguyen et al., 2022). Thirdly, organizations must recognize and reward employees who have shown vigilance, creating a culture of security being valued and accomplished.

8.17 ENSURING CONTINUOUS IMPROVEMENT

Deepfakes are rapidly evolving, and each innovation creates forgeries that are almost realistic and hard to detect (Kietzmann et al., 2020; Seow et al., 2022). Deepfake awareness training needs to be updated and enhanced regularly to stay current with the required knowledge and skills. Training without following technological advancements might be outdated, keeping employees unaware of the recent threats and best practices (Schreiber & Schreiber, 2024). For example, a media company first developed the deepfake training program but failed to update it over time since the technology changed. The workers became disengaged, and the training could not keep up with new forms of sophisticated deepfakes over time. This left the company vulnerable to new attack vectors missed by the outdated training.

Companies must take a dynamic approach to training to further drive continuous improvement. One effective way to manage this is by collaborating with cybersecurity firms or experts specializing in deepfake detection for constant updates on the latest trends and threats (Cochran & Napshin, 2021; Jacobsen & Simpson, 2024). It is also crucial for organizations to implement AI and machine learning systems that track and predict future developments of deepfakes, thus ensuring that training programs along with technology are moved forward. Regular inspections through quizzes or simulation exercises can even test the awareness and preparedness of employees, thus providing valuable points for enhancing the training content.

Another significant strategy to remain relevant is creating a feedback loop in the organization (Masood et al., 2023). Workers should be aware of potential threats and vulnerabilities in deepfakes; they can openly share experiences and insights (Hight, 2022). Employee feedback can be valuable for improving the training program, ensuring that it represents real-world challenges that employees may face. Organizations can also create a cross-functional team to continually review and update the content to ensure it stays aligned with the latest security standards and addresses emerging risks.

8.18 BALANCING VIGILANCE AND PRIVACY

As organizations strive to develop a vigilant corporate culture, employee privacy is one of the primary ethical considerations behind their security measures. Most surveillance tools tracking deepfakes must employ data analytics and reporting systems that could breach employees' privacy (Hancock & Bailenson, 2021;

Kietzmann et al., 2020). A balance between vigilance and control over privacy ensures trust within the organization while remaining secure. For instance, a technology company was criticized for using an employee-monitoring system, which keeps tabs on an organization's potential deepfake attacks. It did correctly signal the suspicious content, although it didn't have the entire impact of gaining the employees' trust against the management. This wasn't effective, as the loss of trust weakened the company's security.

These risks force organizations to monitor and report with transparency and ethical considerations (Romero Moreno, 2024). For example, one way to balance vigilance with privacy is by clearly communicating to employees the purpose and scope of monitoring tools. Employees should be told what is being monitored, why it is necessary, and how their data will be protected. Further, organizations can set strict access and usage guidelines for employees' data that are only used in the context of deepfake detection and not for anything else. The second one is that the implemented monitoring systems should comply with the local and international data privacy laws. For example, GDPR in Europe or CCPA in the USA govern how personal data is collected, stored, and used, and failure to adhere to such rules can attract a very hefty penalty. As such, businesses should collaborate closely with their legal teams to comply as they tackle deepfake monitoring practices, respecting the privacy of the rights bestowed on employees.

8.19 CONCLUSION

The deepfake technology continues to advance and become ever more available, increasing the risk and threat to corporate environments. The ability of deepfake to mimic trust figures and force changes in corporate communications seriously threatens financial integrity, reputation, and business security. Therefore, training employees is the first and most important defense against such risks. The employees themselves, who can identify the threats of deepfakes and indicate them promptly to the proper channels, would prove to be effective for the organization. Beyond training, though, organizations need to promote a culture of alertness that goes beyond different levels in the organizational hierarchy. A corporate culture of alertness, through constant learning, open communication, and leadership excellence, ensures timely identification and elimination. Engaged, informed, and motivated employees to remain vigilant are strong defenses against deepfake exploitation. Meanwhile, as a vital component of culture, leadership enables consolidation of the importance of security in an organization through their actions and priorities. AI offers resilience against deepfakes that companies can implement by integrating issue awareness into broader strategies and making it an integral part of business operations.

As we move into the future, it will be readily apparent that deepfakes will be even more sophisticated and advanced than ever and far less detectable. And with the final product of deepfakes consisting of manipulated audio, video, and pictures coming together with alarming accuracy, future versions may take it to an all-new level. The making of deepfakes might be more commonly used in real-time video communications or instead be invented through AI to be used along with other modes of forgery. That is because organizations will not easily trace such scams. Secondly, an advanced AI model would provide easier access to people who are not necessarily technically sound in developing the means of making deepfakes, thereby spurring attacks on the organization.

These trends require changes in companies' training programs and cybersecurity strategies. Employee training continually needs to be updated based on the shifts in deepfake technology. Scenario-based learning, real-time simulations, and continuous feedback loops will prepare staff for these threats' dynamic nature. Businesses will have to invest in technical tools through which AI can detect deepfakes; those investments will complement human vigilance nurtured by training. Culturally, organizations must remain proactive, where security is not a goal but an ongoing adaptive process. This would include a lifelong learning attitude where employees are kept updated on new cyber threats and encouraged to keep abreast of the latest emerging technologies. Again, leadership will be important in maintaining this thinking, even though the nature of the corporate threat landscape is constantly evolving.

REFERENCES

Abbas, F., & Taeihagh, A. (2024). Unmasking deepfakes: A systematic review of deepfake detection and generation techniques using artificial intelligence. *Expert Systems with Applications*, *1*, 124260.

Ahmed, S. (2021). Fooled by the fakes: Cognitive differences in perceived claim accuracy and sharing intention of non-political deepfakes. *Personality and Individual Differences*, *182*, 111074.

Ayata, O. (2024). Artificial Realities: Mitigations against Deepfakes. *Artificial Intelligence*, *1*, 22–33.

Beuve, N., Hamidouche, W., & Déforges, O. (2023). Hierarchical learning and dummy triplet loss for efficient deepfake detection. *ACM Transactions on Multimedia Computing, Communications and Applications*, *20*(3), 1–18.

Bray, S. D., Johnson, S. D., & Kleinberg, B. (2023). Testing human ability to detect 'deepfake' images of human faces. *Journal of Cybersecurity*, *9*(1), tyad011.

Byeon, H., Shabaz, M., Shrivastava, K., Joshi, A., Keshta, I., Oak, R., Singh, P. P., & Soni, M. (2024). Deep learning model to detect deceptive generative adversarial network generated images using multimedia forensic. *Computers and Electrical Engineering*, *113*, 109024.

Cochran, J. D., & Napshin, S. A. (2021). Deepfakes: awareness, concerns, and platform accountability. *Cyberpsychology, Behavior, and Social Networking*, *24*(3), 164–172.

Cover, R. (2022). Deepfake culture: the emergence of audio-video deception as an object of social anxiety and regulation. *Continuum*, *36*(4), 609–621.

deRancourt-Raymond, A., & Smaili, N. (2023). The unethical use of deepfakes. *Journal of Financial Crime*, *30*(4), 1066–1077.

Fui-Hoon Nah, F., Zheng, R., Cai, J., Siau, K., & Chen, L. (2023). *Generative AI and ChatGPT: Applications, Challenges, and AI-human Collaboration* (Vol. 25, pp. 277–304). Taylor & Francis.

Gowrisankar, B., & Thing, V. L. (2024). An adversarial attack approach for explainable AI evaluation on deepfake detection models. *Computers & Security*, *139*, 103684.

Gu, C., Jia, S., Lai, J., Chen, R., & Chang, X. (2024). Exploring consumer acceptance of AI-generated advertisements: from the perspectives of perceived eeriness and perceived intelligence. *Journal of Theoretical and Applied Electronic Commerce Research*, *19*(3), 2218–2238.

Hancock, J. T., & Bailenson, J. N. (2021). *The Social Impact of Deepfakes* (Vol. 24, pp. 149–152). Mary Ann Liebert, Inc.

Hight, C. (2022). Deepfakes and documentary practice in an age of misinformation. *Continuum*, *36*(3), 393–410.

Jacobsen, B. N., & Simpson, J. (2024). The tensions of deepfakes. *Information, Communication & Society*, *27*(6), 1095–1109.

Jayashre, K., & Amsaprabhaa, M. (2024). Safeguarding media integrity: A hybrid optimized deep feature fusion based deepfake detection in videos. *Computers & Security*, *142*, 103860.

Kaate, I., Salminen, J., Santos, J., Jung, S.-G., Olkkonen, R., & Jansen, B. (2023). The realness of fakes: Primary evidence of the effect of deepfake personas on user perceptions in a design task. *International Journal of Human-Computer Studies*, *178*, 103096.

Kietzmann, J., Lee, L. W., McCarthy, I. P., & Kietzmann, T. C. (2020). Deepfakes: Trick or treat? *Business Horizons*, *63*(2), 135–146.

Kolagati, S., Priyadharshini, T., & Rajam, V. M. A. (2022). Exposing deepfakes using a deep multilayer perceptron–convolutional neural network model. *International Journal of Information Management Data Insights*, *2*(1), 100054.

Levine, A. J. (2020). *Dollars, Deception, and Deepfakes: An Analysis of Deepfakes and Synthetic Media Fraud.* Utica College.

Liz-Lopez, H., Keita, M., Taleb-Ahmed, A., Hadid, A., Huertas-Tato, J., & Camacho, D. (2024). Generation and detection of manipulated multimodal audiovisual content: Advances, trends and open challenges. *Information Fusion*, *103*, 102103.

Mariani, M., & Dwivedi, Y. K. (2024). Generative artificial intelligence in innovation management: A preview of future research developments. *Journal of Business Research*, *175*, 114542.

Masood, M., Nawaz, M., Malik, K. M., Javed, A., Irtaza, A., & Malik, H. (2023). Deepfakes generation and detection: State-of-the-art, open challenges, countermeasures, and way forward. *Applied intelligence*, *53*(4), 3974–4026.

Mustak, M., Salminen, J., Mäntymäki, M., Rahman, A., & Dwivedi, Y. K. (2023). Deepfakes: Deceptions, mitigations, and opportunities. *Journal of Business Research*, *154*, 113368.

Nguyen, T. T., Nguyen, Q. V. H., Nguyen, D. T., Nguyen, D. T., Huynh-The, T., Nahavandi, S., Nguyen, T. T., Pham, Q.-V., & Nguyen, C. M. (2022). Deep learning for deepfakes creation and detection: A survey. *Computer Vision and Image Understanding*, *223*, 103525.

Rabhi, M., Bakiras, S., & Di Pietro, R. (2024). Audio-deepfake detection: Adversarial attacks and countermeasures. *Expert Systems with Applications*, *250*, 123941.

Romero Moreno, F. (2024). Generative AI and deepfakes: A human rights approach to tackling harmful content. In *International Review of Law, Computers & Technology* (pp. 1–30). Taylor and Francis , available at https://doi.org/10.1080/13600869.2024.2324540

Schreiber, A., & Schreiber, I. (2024). Bridging knowledge gap: the contribution of employees' awareness of AI cyber risks comprehensive program to reducing emerging AI digital threats. *Information & Computer Security*, *32*, 613–635.

Seow, J. W., Lim, M. K., Phan, R. C., & Liu, J. K. (2022). A comprehensive overview of Deepfake: Generation, detection, datasets, and opportunities. *Neurocomputing*, *513*, 351–371.

Thomson, T., Thomas, R. J., & Matich, P. (2024). Generative visual AI in news organizations: Challenges, opportunities, perceptions, and policies. *Digital Journalism*, 1–22.

Van der Sloot, B., & Wagensveld, Y. (2022). Deepfakes: regulatory challenges for the synthetic society. *Computer Law & Security Review*, *46*, 105716.

Wach, K., Duong, C. D., Ejdys, J., Kazlauskaitė, R., Korzynski, P., Mazurek, G., Paliszkiewicz, J., & Ziemba, E. (2023). The dark side of generative artificial intelligence: A critical analysis of controversies and risks of ChatGPT. *Entrepreneurial Business and Economics Review*, *11*(2), 7–30.

Westerlund, M. (2019). The emergence of deepfake technology: A review. *Technology Innovation Management Review*, *9*(11), 1–20.

Whittaker, L., Mulcahy, R., Letheren, K., Kietzmann, J., & Russell-Bennett, R. (2023). Mapping the deepfake landscape for innovation: A multidisciplinary systematic review and future research agenda. *Technovation*, *125*, 102784.

9 Deepfake Prescriptions
When AI Plays Doctor through Data-Driven Techniques

Elif Hasret Kumcu, Mukul Bhatnagar, and Rupa Khanna

9.1 INTRODUCTION

The advent of artificial intelligence (AI) in healthcare is reshaping traditional practices, particularly in areas like diagnosis, treatment, and prescription management. AI-generated prescriptions, powered by deepfake technology, represent a significant innovation aimed at improving healthcare efficiency (Akhtar et al., 2024). However, this integration also raises complex questions about the balance between technological advancement and ethical responsibility. The concept of deepfake technology, which uses AI to generate hyper-realistic synthetic media, is expanding beyond its initial use in entertainment and social media into more critical sectors, including healthcare. The potential benefits of using AI-generated prescriptions include streamlining the prescription process, reducing human error, and improving efficiency in medical decision-making (Nikolaev et al., 2024).

However, AI-generated prescriptions also present unique challenges, especially in terms of trust, privacy, and the ethical implications of relying on algorithms to make healthcare decisions. The central issue revolves around whether patients and healthcare professionals can trust AI-generated content as much as human-generated prescriptions (Kashif et al., 2024). This study is grounded in the Technology Acceptance Model (TAM), which suggests that trust, perceived usefulness, and perceived ease of use are pivotal in the adoption of new technologies. Additionally, it considers ethical concerns about AI in healthcare, particularly in relation to patient safety, the potential for misuse, and the preservation of medical autonomy. The research also addresses the role of regulatory frameworks in shaping the ethical use of AI in healthcare and ensuring that AI technologies are used responsibly (Kerner & Risse, 2021).

As AI in healthcare continues to evolve, understanding the interplay between AI adoption, patient trust, ethical concerns, and regulatory measures is crucial to unlocking the full potential of AI-generated prescriptions. This study aims to explore these dimensions, providing a comprehensive analysis of how deepfake technology

DOI: 10.1201/9781003564621-9

is perceived in the healthcare sector and how it can be ethically and effectively integrated into medical practice.

9.2 THEORETICAL FRAMEWORK

The theoretical framework for this study on "Deepfake Prescriptions: When AI Plays Doctor" draws upon the integration of AI in healthcare, focusing on the intersections of technology, ethics, and regulatory considerations. This research is grounded in the TAM, which posits that users' trust, perceived ease of use, and perceived usefulness are crucial factors influencing the adoption of new technologies, such as AI, in healthcare (Balaskas et al., 2022). Additionally, the framework incorporates elements of ethical decision-making theories, particularly in the context of AI-generated content and the potential risks of deepfake technology, which may compromise patient trust and medical accuracy. The research also examines the regulatory landscape, emphasizing how policies can mitigate potential harms while fostering innovation. By exploring these factors, the study seeks to understand the dynamic relationships between AI adoption, trust, ethical concerns, and regulatory effectiveness in the context of AI-generated prescriptions, highlighting both opportunities and challenges in the evolving healthcare ecosystem (Zhao et al., 2022).

9.3 LITERATURE REVIEW

The literature surrounding AI in healthcare highlights the significant potential of deepfake technology to transform medical practices, particularly in the realm of prescription generation. AI technologies, such as natural language processing and machine learning, are increasingly being utilized to automate various aspects of healthcare, including diagnosing diseases, recommending treatments, and prescribing medications. The application of deepfake technology in these processes offers several advantages, including increased efficiency and accuracy in medical decision-making (Malik et al., 2024). However, despite the promising capabilities of AI in healthcare, its integration faces substantial resistance due to concerns about trust and ethics. Patients and healthcare professionals alike express reservations about the reliability of AI in critical healthcare decisions, given its lack of human oversight and the potential risks of algorithmic biases or errors (Galyashina & Nikishin, 2022).

Furthermore, the ethical concerns related to deepfake AI in healthcare primarily revolve around data security, privacy, and the potential for misuse. The idea of AI-generated prescriptions raises questions about accountability—if an AI algorithm makes an incorrect prescription, who is responsible? This has significant implications for medical malpractice and the regulation of AI technologies (Amezaga & Hajek, 2022). Studies have also examined the role of public perception and the need for transparency in AI systems to build trust among patients and healthcare providers. Finally, regulatory frameworks have yet to catch up with the pace of technological advancement, which further complicates the widespread adoption of AI in healthcare. As such, while the potential for AI-generated prescriptions is evident, it is essential that regulatory and ethical considerations be addressed to ensure safe, fair, and effective implementation (Akbar et al., 2023).

9.4 RESEARCH METHODOLOGY

This study employs a quantitative research approach using a survey methodology to assess the perceptions of healthcare professionals and patients regarding AI-generated prescriptions. The survey includes Likert scale items designed to measure factors such as perceived trust in AI, awareness of deepfake technology, ethical concerns, patient willingness to use AI, and professional perceptions of AI's role in medical decision-making. The sample size consists of 246 respondents, including doctors, pharmacists, and patients. Data analysis was conducted using ANOVA to examine the relationships between the awareness of deepfake technology and other variables. Additionally, confidence intervals were calculated for Kendall's tau-b and Spearman's rho to assess the strength and significance of correlations among the variables.

Table 9.1 outlines the research instrument and presents various variables associated with perceptions of AI-generated prescriptions in healthcare. The Likert scale descriptions for each variable range from 1 (Strongly disagree/Not at all aware) to 7 (Strongly agree/Extremely aware), assessing factors such as perceived trust in AI, awareness of deepfake technology, ethical concerns, patient willingness to use AI in healthcare, and professional perception of AI. This table categorizes the key factors influencing attitudes toward AI integration into healthcare, emphasizing the critical role of trust, knowledge, and ethical considerations in shaping acceptance.

TABLE 9.1
Research Instrument

Variable Name	Variable Question	Likert Scale Description
Perceived trust in AI-generated prescriptions	I trust AI-generated prescriptions as much as human doctors' prescriptions.	1 = Strongly disagree, 7 = Strongly agree
Awareness and knowledge of deepfake technology in healthcare	I am aware of how deepfake technology is used in healthcare.	1 = Not at all aware, 7 = Extremely aware
Ethical and legal concerns	I am concerned about the ethical and legal risks of AI-generated prescriptions.	1 = Not concerned at all, 7 = Extremely concerned
Patient willingness to use AI in healthcare	I would be comfortable receiving an AI-generated prescription.	1 = Strongly disagree, 7 = Strongly agree
Professional perception (for doctors and pharmacists)	AI-generated prescriptions can enhance medical decision-making.	1 = Strongly disagree, 7 = Strongly agree
Risk perception of deepfake AI in healthcare	I believe deepfake AI in healthcare could lead to increased prescription fraud.	1 = Very unlikely, 7 = Very likely
Regulatory and policy factors	Current regulations are sufficient to prevent misuse of deepfake AI in prescriptions.	1 = Strongly disagree, 7 = Strongly agree

9.5 DATA ANALYSIS

Tables 9.2 and 9.3 present the ANOVA results for two main interactions: *Perceived Trust in AI Generated Prescriptions* and *Awareness and Knowledge of Deepfake Technology in Healthcare*, and *Patient Willingness to Use AI in Healthcare* with the same awareness factor. The significant F-values (52.797 and 70.518, respectively) and p-values (<0.001) indicate that both interactions are statistically significant, highlighting strong relationships between trust, knowledge, and willingness to adopt

TABLE 9.2
ANOVA Table

			Sum of Squares	df	Mean Square	F	Sig.
Perceived_trust_in_AI_generated_prescriptions * awareness_and_knowledge_of_deepfake_technology_in_healthcare	Between Groups	(Combined)	273.787	6	45.631	52.797	0.000
		Linearity	259.203	1	259.203	299.907	0.000
		Deviation from Linearity	14.584	5	2.917	3.375	0.006
	Within Groups		207.427	240	0.864		
	Total		481.215	246			
Patient_willingness_to_use_AI_in_healthcare * awareness_and_knowledge_of_deepfake_technology_in_healthcare	Between Groups	(Combined)	286.469	6	47.745	70.518	0.000
		Linearity	272.339	1	272.339	402.238	0.000
		Deviation from Linearity	14.130	5	2.826	4.174	0.001
	Within Groups		162.494	240	0.677		
	Total		448.964	246			

TABLE 9.3
Measures of Association

	R	R Squared	Eta	Eta Squared
Perceived_trust_in_AI_generated_prescriptions * awareness_and_knowledge_of_deepfake_technology_in_healthcare	0.734	0.539	0.754	0.569
Patient_willingness_to_use_AI_in_healthcare * awareness_and_knowledge_of_deepfake_technology_in_healthcare	0.779	0.607	0.799	0.638

AI prescriptions. The measures of association, with R-squared values of 0.539 and 0.607, further affirm that these variables collectively explain a substantial portion of the variance in trust and willingness to use AI, suggesting the profound influence of awareness on perceptions of AI in healthcare.

Table 9.4 reports the Kendall's tau-b correlation coefficients, revealing robust positive relationships between variables such as *Perceived Trust in AI Generated Prescriptions* and *Awareness of Deepfake Technology* ($\tau = 0.590$, $p < 0.001$), and between *Ethical and Legal Concerns* and *Patient Willingness to Use AI in Healthcare* ($\tau = 0.733$, $p < 0.001$). These correlations demonstrate that higher awareness and trust in AI are significantly associated with greater comfort in using AI-generated prescriptions, while ethical concerns and risk perceptions also strongly influence willingness to adopt such technologies, signaling that ethical and regulatory factors are critical in fostering public trust in AI-powered healthcare solutions.

Table 9.5 presents Spearman's rho values, showing significant positive correlations across multiple variables, particularly between Perceived Trust in AI-Generated Prescriptions and Patient Willingness to Use AI in Healthcare ($\rho = 0.753$, $p < 0.001$), as well as between Ethical and Legal Concerns and Patient Willingness to Use AI in Healthcare ($\rho = 0.805$, $p < 0.001$). These strong associations underscore the influence of both trust and ethical considerations on the public's willingness to embrace AI technologies in healthcare, indicating that fostering trust and addressing ethical concerns are pivotal to increasing acceptance of AI-generated prescriptions and mitigating potential resistance from patients and healthcare professionals alike.

9.6 DISCUSSION

The study reveals that both healthcare professionals and patients are hesitant to fully embrace AI-generated prescriptions, with trust being a major influencing factor. While AI's potential to enhance medical decision-making is acknowledged, concerns regarding the ethical implications and risks of misuse, such as prescription fraud, remain significant barriers. This study underscores the importance of addressing ethical concerns and improving transparency to foster trust in AI-generated prescriptions. Moreover, the findings suggest that greater awareness of deepfake technology can positively influence attitudes toward AI, emphasizing the need for educational initiatives to enhance understanding and alleviate concerns.

9.7 MANAGERIAL IMPLICATIONS

For healthcare managers, the results of this study provide valuable insights into how AI technologies, specifically deepfake-generated prescriptions, are perceived by both professionals and patients. The findings emphasize the need for managers to foster a culture of trust and transparency when integrating AI into medical practices. Healthcare organizations should invest in educating patients and healthcare providers about the benefits and limitations of AI technologies, ensuring that both groups feel confident in their reliability and safety. Furthermore, it is crucial to address ethical concerns by implementing clear guidelines and regulatory standards to safeguard patient welfare. Managers should advocate for continuous training and collaboration

TABLE 9.4
Confidence Intervals of Kendall's tau-b

	Kendall's tau-b	Significance (2-tailed)	95% Confidence Intervals (2-tailed)[a]	
			Lower	Upper
Perceived_trust_in_AI_generated_prescriptions - awareness_and_knowledge_of_deepfake_technology_in_healthcare	0.590	0.000	0.533	0.641
Perceived_trust_in_AI_generated_prescriptions - ethical_and_legal_concerns	0.623	0.000	0.570	0.671
Perceived_trust_in_AI_generated_prescriptions - patient_willingness_to_use_AI_in_healthcare	0.678	0.000	0.630	0.720
Perceived_trust_in_AI_generated_prescriptions - professional_perception_for_doctors_and_pharmacists	0.323	0.000	0.247	0.396
Perceived_trust_in_AI_generated_prescriptions - risk_perception_of_deepfake_AI_in_healthcare	0.150	0.004	0.068	0.230
Perceived_trust_in_AI_generated_prescriptions - regulatory_and_policy_factors	0.114	0.030	0.031	0.195
Awareness_and_knowledge_of_deepfake_technology_in_healthcare - ethical_and_legal_concerns	0.741	0.000	0.701	0.776
Awareness_and_knowledge_of_deepfake_technology_in_healthcare - patient_willingness_to_use_AI_in_healthcare	0.629	0.000	0.576	0.677
Awareness_and_knowledge_of_deepfake_technology_in_healthcare - professional_perception_for_doctors_and_pharmacists	0.405	0.000	0.334	0.472

(Continued)

TABLE 9.4 (*Continued*)
Confidence Intervals of Kendall's tau-b

	Kendall's tau-b	Significance (2-tailed)	95% Confidence Intervals (2-tailed)[a]	
			Lower	Upper
Awareness_and_knowledge_of_deepfake_technology_in_healthcare - risk_perception_of_deepfake_AI_in_healthcare	0.239	0.000	0.159	0.316
Awareness_and_knowledge_of_deepfake_technology_in_healthcare - regulatory_and_policy_factors	0.163	0.002	0.081	0.243
Ethical_and_legal_concerns - patient_willingness_to_use_AI_in_healthcare	0.733	0.000	0.692	0.769
Ethical_and_legal_concerns - professional_perception_for_doctors_and_pharmacists	0.402	0.000	0.330	0.470
Ethical_and_legal_concerns - risk_perception_of_deepfake_AI_in_healthcare	0.222	0.000	0.141	0.299
Ethical_and_legal_concerns - regulatory_and_policy_factors	0.166	0.002	0.084	0.245
Patient_willingness_to_use_AI_in_healthcare - professional_perception_for_doctors_and_pharmacists	0.374	0.000	0.300	0.443

(*Continued*)

TABLE 9.4 (*Continued*)
Confidence Intervals of Kendall's tau-b

	Kendall's tau-b	Significance (2-tailed)	95% Confidence Intervals (2-tailed)[a]	
			Lower	Upper
Patient_willingness_to_use_AI_in_healthcare - risk_perception_of_deepfake_AI_in_healthcare	0.217	0.000	0.137	0.295
Patient_willingness_to_use_AI_in_healthcare - regulatory_and_policy_factors	0.192	0.000	0.111	0.271
Professional_perception_for_doctors_and_pharmacists - risk_perception_of_deepfake_AI_in_healthcare	0.360	0.000	0.286	0.430
Professional_perception_for_doctors_and_pharmacists - regulatory_and_policy_factors	0.289	0.000	0.211	0.364
Risk_perception_of_deepfake_AI_in_healthcare - regulatory_and_policy_factors	0.378	0.000	0.305	0.447

[a] Estimation is based on Fisher's r-to-z transformation.

TABLE 9.5
Confidence Intervals of Spearman's rho

	Spearman's rho	Significance (2-tailed)	95% Confidence Intervals (2-tailed)[a,b]	
			Lower	Upper
Perceived_trust_in_AI_generated_prescriptions - awareness_and_knowledge_of_deepfake_technology_in_healthcare	0.676	0.000	0.599	0.740
Perceived_trust_in_AI_generated_prescriptions - ethical_and_legal_concerns	0.707	0.000	0.636	0.766
Perceived_trust_in_AI_generated_prescriptions - patient_willingness_to_use_AI_in_healthcare	0.753	0.000	0.691	0.803
Perceived_trust_in_AI_generated_prescriptions - professional_perception_for_doctors_and_pharmacists	0.373	0.000	0.257	0.479
Perceived_trust_in_AI_generated_prescriptions - risk_perception_of_deepfake_AI_in_healthcare	0.184	0.004	0.057	0.305
Perceived_trust_in_AI_generated_prescriptions - regulatory_and_policy_factors	0.142	0.025	0.014	0.266
Awareness_and_knowledge_of_deepfake_technology_in_healthcare - ethical_and_legal_concerns	0.808	0.000	0.758	0.848
Awareness_and_knowledge_of_deepfake_technology_in_healthcare - patient_willingness_to_use_AI_in_healthcare	0.713	0.000	0.644	0.771
Awareness_and_knowledge_of_deepfake_technology_in_healthcare - professional_perception_for_doctors_and_pharmacists	0.461	0.000	0.354	0.557
Awareness_and_knowledge_of_deepfake_technology_in_healthcare - risk_perception_of_deepfake_AI_in_healthcare	0.288	0.000	0.166	0.402
Awareness_and_knowledge_of_deepfake_technology_in_healthcare - regulatory_and_policy_factors	0.193	0.002	0.066	0.313

(*Continued*)

TABLE 9.5 (*Continued*)
Confidence Intervals of Spearman's rho

	Spearman's rho	Significance (2-tailed)	95% Confidence Intervals (2-tailed)[a,b]	
			Lower	Upper
Ethical_and_legal_concerns - patient_willingness_to_use_AI_in_healthcare	0.805	0.000	0.754	0.846
Ethical_and_legal_concerns - professional_perception_for_doctors_and_pharmacists	0.464	0.000	0.357	0.559
Ethical_and_legal_concerns - risk_perception_of_deepfake_AI_in_healthcare	0.266	0.000	0.142	0.381
Ethical_and_legal_concerns - regulatory_and_policy_factors	0.194	0.002	0.067	0.315
Patient_willingness_to_use_AI_in_healthcare - professional_perception_for_doctors_and_pharmacists	0.433	0.000	0.323	0.532
Patient_willingness_to_use_AI_in_healthcare - risk_perception_of_deepfake_AI_in_healthcare	0.265	0.000	0.141	0.380
Patient_willingness_to_use_AI_in_healthcare - regulatory_and_policy_factors	0.229	0.000	0.103	0.347
Professional_perception_for_doctors_and_pharmacists - risk_perception_of_deepfake_AI_in_healthcare	0.430	0.000	0.319	0.529
Professional_perception_for_doctors_and_pharmacists - regulatory_and_policy_factors	0.343	0.000	0.224	0.451
Risk_perception_of_deepfake_AI_in_healthcare - regulatory_and_policy_factors	0.442	0.000	0.332	0.540

[a] Estimation is based on Fisher's r-to-z transformation.
[b] Estimation of standard error is based on the formula proposed by Fieller, Hartley, and Pearson.

between AI developers and healthcare professionals to ensure that AI systems are designed with ethical considerations in mind. Finally, healthcare leaders must work with policymakers to develop robust regulatory frameworks that address the unique challenges posed by AI in healthcare, ensuring that AI technologies are deployed in a manner that is both ethically responsible and beneficial to patients.

9.8 CONCLUSION

The study concludes that AI-generated prescriptions, particularly those powered by deepfake technology, hold significant potential for improving healthcare efficiency and decision-making. However, the adoption of such technologies is hindered by trust-related concerns, ethical dilemmas, and regulatory challenges. The findings indicate that patients and healthcare professionals are more likely to embrace AI-generated prescriptions when they have a higher awareness of deepfake technology and when they perceive the technology to be ethically sound and transparent. Trust in AI is the most significant determinant in the willingness to adopt AI-generated prescriptions, with patients and healthcare professionals alike expressing concerns over the potential for misuse, such as prescription fraud or the loss of human oversight in medical decision-making.

Ethical concerns regarding data privacy, algorithmic transparency, and the accountability of AI systems are also critical issues that must be addressed to ensure the safe use of AI in healthcare. These concerns highlight the need for stringent regulatory frameworks that can provide clear guidelines for the responsible use of AI technologies in medical practices. The study also underscores the importance of educating both healthcare professionals and patients about the potential benefits and risks of AI-generated prescriptions to build trust and promote acceptance.

In conclusion, while AI-generated prescriptions offer a promising avenue for enhancing medical decision-making, their widespread adoption will depend on overcoming significant challenges related to trust, ethics, and regulation. It is essential for healthcare organizations, regulators, and AI developers to collaborate in addressing these challenges to ensure that AI technologies are deployed in a manner that benefits patients and the healthcare system as a whole.

REFERENCES

Akbar, M., Suaib, M., & Hussain, M. S. (2023). The rise of deepfake technology: issues, challenges, and countermeasures. In M. S. Husain, M. Faisal, H. Sadia, T. Ahmad, & S. Shukla (Eds.), *Advances in Cyberology and the Advent of the Next-Gen Information Revolution*. IGI Global. https://doi.org/10.4018/978-1-6684-8133-2.ch010

Akhtar, Z., Pendyala, T. L., & Athmakuri, V. S. (2024). Video and audio deepfake datasets and open issues in deepfake technology: being ahead of the curve. *Forensic Sciences*, *4*(3), 289–377. https://doi.org/10.3390/forensicsci4030021

Amezaga, N., & Hajek, J. (2022). Availability of voice deepfake technology and its impact for good and evil. *SIGITE 2022 – Proceedings of the 23rd Annual Conference on Information Technology Education*, 23–28. https://doi.org/10.1145/3537674.3554742

Balaskas, S., Panagiotarou, A., & Rigou, M. (2022). The influence of trustworthiness and technology acceptance factors on the usage of e-government Services during COVID-19: a case study of post COVID-19 Greece. *Administrative Sciences*, *12*(4), 129. https://doi.org/10.3390/admsci12040129

Galyashina, E. I., & Nikishin, V. D. (2022). The protection of megascience projects from deepfake technologies threats: information law aspects. *Journal of Physics: Conference Series*, 2210, (1), 012007. https://doi.org/10.1088/1742-6596/2210/1/012007

Kashif, M., Garg, H., Weqar, F., & David, A. (2024). Regulatory strategies and innovative solutions for deepfake technology. In G. Lakhera, S. Taneja, E. Ozen, M. Kukreti, & P. Kumar (Eds.), *Navigating the World of Deepfake Technology*. IGI Global. https://doi.org/10.4018/979-8-3693-5298-4.ch013

Kerner, C., & Risse, M. (2021). Beyond porn and discreditation: epistemic promises and perils of deepfake technology in digital lifeworlds. *Moral Philosophy and Politics*, *8*(1), 81–108. https://doi.org/10.1515/mopp-2020-0024

Malik, S., Surbhi, A., & Roy, D. (2024). Blurring boundaries between truth and illusion: analysis of human rights and regulatory concerns arising from abuse of deepfake technology. In R. Desai & F. Gupta (Eds.), *AIP Conference Proceedings* (Vol. 3220, Issue 1). American Institute of Physics. https://doi.org/10.1063/5.0234995

Nikolaev, A. S., Kokorina, O. A., Kirillov, N. N., & Antipov, A. A. (2024). Analysis of deepfake technology development trends based on patent landscape. *Lecture Notes in Networks and Systems*, *1119 LNNS*, 490–503. https://doi.org/10.1007/978-3-031-70300-3_36

Zhao, X., Li, T., & Hu, H. (2022). Analysis and enlightenment of international deepfake technology policy texts: taking the United States and the European union as examples; [国外深度伪造技术政策文本分析与启示—以美国尧 欧盟为例]. *Journal of Library and Information Science in Agriculture*, *34*(9), 60–71. https://doi.org/10.13998/j.cnki.issn1002-1248.22-0501

10 The Economic Implications of SDG 3 and Healthcare 3.0

A Multivariate Regression Analysis of AI Adoption, Accessibility, and Regulatory Factors

Cyril Crasto, Amol Ranadive, Dhananjay Kulkarni, and Amar Yedake

10.1 INTRODUCTION

The economic impact of healthcare systems is a multifaceted and dynamic domain that intertwines various technological, regulatory, and accessibility factors (A. Kaur et al., 2023; Rani et al., 2024; Tina et al., 2024). In an era marked by rapid advancements in artificial intelligence (AI), coupled with an increasing demand for efficient and equitable healthcare delivery, understanding the determinants that drive healthcare's economic outcomes is both crucial and complex (Abbas et al., 2022; Khan et al., 2023). This study seeks to elucidate the underlying factors contributing to the economic impact of healthcare by examining key variables such as AI adoption, healthcare accessibility and affordability, regulatory frameworks, and public trust in AI-driven healthcare systems (S. Sharma, Gupta, Khanna Malhotra, et al., 2024). By employing a rigorous multivariate regression analysis, this research aims to provide a comprehensive understanding of how these factors collectively shape the economic landscape of healthcare (S. Sharma, Gupta, Khanna Malhotra, et al., 2024).

AI adoption has emerged as a transformative force within the healthcare sector, offering the promise of enhanced operational efficiency, improved patient outcomes, and cost-effective healthcare delivery (Choi et al., 2022; Tandon et al., 2023; Westerlund, 2019). However, the integration of AI in healthcare systems is not without its challenges, including concerns regarding data privacy, algorithmic bias,

DOI: 10.1201/9781003564621-10

and the displacement of human labor (Bhardwaj et al., 2024). Despite these challenges, the positive potential of AI to drive economic efficiencies warrants a closer investigation into its role as a determinant of healthcare's economic impact (Saikumar et al., 2024; P. Singh et al., 2024; Thirumalaivasan et al., 2024; Tomar et al., 2024). This study evaluates how varying levels of AI adoption influence the economic outcomes within healthcare systems, hypothesizing a positive relationship between the widespread implementation of AI technologies and improved economic metrics (Malhotra et al., 2024).

Furthermore, healthcare accessibility and affordability remain critical determinants of healthcare's economic viability (P. Sharma, Malhotra, & Kumar, 2024). As healthcare systems across the globe grapple with the pressure of providing affordable care to a growing and aging population, the economic consequences of accessibility issues become ever more pronounced (Bhatnagar & Sehajpal, 2024). The interplay between the cost of healthcare services and the accessibility of such services for marginalized populations can create significant disparities in economic outcomes (Kaushik, 2022). By analyzing how perceived accessibility and affordability shape the economic impact of healthcare, this study seeks to provide insights into the broader social and economic implications of healthcare policy decisions (Srivastava et al., 2024).

Regulatory frameworks also play a pivotal role in shaping the economic dynamics of healthcare. Policies that govern the healthcare industry not only influence the operational aspects of healthcare organizations but also dictate the scope and scale of technological advancements, including AI (Upreti et al., 2024). Stringent regulations may restrict innovation, while lenient regulations may foster rapid technological progress but at the cost of ethical and safety considerations (Clarke et al., 2023). This study explores the relationship between regulatory frameworks and the economic impact of healthcare, with a particular focus on how regulations pertaining to AI adoption affect healthcare costs and efficiencies (P. Sharma, Malhotra, Altybekova, et al., 2024).

Lastly, the role of public trust in AI-driven healthcare is an essential yet often underexplored factor. Trust in technology is foundational to its adoption and success, and healthcare is no exception (Kukreti et al., 2025; Sehajpal & Kaur, 2025). Public confidence in AI systems, particularly in sensitive domains such as healthcare, may significantly influence the speed and extent of AI integration (R. Kumar et al., 2023). This research delves into the correlation between trust in AI adoption and the perceived economic impact of healthcare, seeking to determine whether greater trust fosters more significant economic benefits or whether distrust acts as a hindrance to the economic optimization of healthcare services (V. Kumar et al., 2021).

Through this study, we aim to provide a nuanced understanding of the multifactorial influences on healthcare's economic landscape, employing advanced statistical techniques to discern the relative contributions of AI adoption, healthcare accessibility, regulatory frameworks, and trust in driving the economic outcomes of healthcare systems. By doing so, we hope to offer actionable insights for policymakers, healthcare providers, and technology developers seeking to optimize healthcare's economic potential in an increasingly digital world.

10.2 REVIEW OF LITERATURE

The economic impact of healthcare is influenced by an intricate web of factors, with technological adoption, accessibility, regulatory frameworks, and public trust standing out as key determinants. As healthcare systems worldwide are undergoing rapid transformation through the integration of advanced technologies such as AI, a growing body of literature has sought to explore the ramifications of these changes on economic outcomes. AI's potential to optimize operational efficiency, enhance diagnostic accuracy, and reduce costs has garnered considerable attention; however, the broader economic implications remain an area of significant inquiry. Studies indicate that while AI can drive efficiency, its economic benefits are contingent on the extent of its integration and the regulatory landscape that governs its application.

In parallel, healthcare accessibility and affordability are paramount to understanding economic outcomes in healthcare systems. Research has consistently shown that disparities in access to care and the rising costs of healthcare services have profound economic consequences, not only for individual patients but also for the broader healthcare economy. A nuanced understanding of how these factors interplay with AI adoption is essential in evaluating their combined effect on economic impact (Bansal et al., 2023).

Regulatory policies represent another critical layer in this discussion, with varying degrees of regulation affecting the pace and success of AI implementation. While stringent regulations may mitigate potential risks associated with AI, they can also inhibit innovation, thereby affecting the broader economic performance of healthcare systems. Conversely, lenient regulations may promote technological advancement but at the potential cost of safety and ethical concerns (Dangwal et al., 2022).

Lastly, public trust in AI-driven healthcare technologies has emerged as a crucial determinant in the successful adoption and economic realization of these innovations (Liu et al., 2024). The relationship between public confidence and the economic impact of AI in healthcare remains an underexplored yet vital avenue for research (Reepu et al., 2023). As the landscape continues to evolve, the interconnectedness of these factors necessitates a comprehensive and multidimensional approach to understanding their collective economic implications (Kukreti et al., 2025; Sehajpal & Kaur, 2025).

10.3 RESEARCH METHODOLOGY

This study employs a quantitative research methodology, utilizing multiple regression analysis to investigate the determinants influencing the economic impact of healthcare. The model incorporates key independent variables: AI Adoption, Healthcare Accessibility and Affordability, Trust in AI-driven Healthcare, and Regulatory Frameworks, with Economic Impact Healthcare as the dependent variable. Data were gathered from a sample of 273 respondents, and diagnostic tests, including ANOVA, coefficient analysis, and residual diagnostics, were conducted to ensure model validity, robustness, and the fulfillment of underlying assumptions.

10.4 DATA ANALYSIS

Table 10.1 presents the descriptive statistics for five variables related to healthcare and AI adoption across a sample of 273 observations, revealing that the mean scores for "Economic_Impact_healthcare" (4.8095), "Adoption_AI" (5.0659), "Healthcare_Accessibility_Affordability" (4.7033), "Trust_Adoption_AI_DRIVEN" (4.5531), and "Regulatory_Ethical" (4.6007) are all moderately above the midpoint of the scale (1–7), with relatively low skewness values indicating a near-normal distribution for most variables, though "Trust_Adoption_AI_DRIVEN" and "Regulatory_Ethical" exhibit slight negative skewness, suggesting a slight tendency toward more positive responses, while the kurtosis values suggest distributions that are relatively close to normal with only slight deviations, particularly for the "Adoption_AI" variable, which shows a positive peak.

The model summary reveals in Table 10.2 a robust linear relationship, with the predictor variables—Regulatory_Ethical, Adoption_AI, Trust_Adoption_AI_DRIVEN, and Healthcare_Accessibility_Affordability—explaining approximately 79.3% of the variance in the Economic_Impact_healthcare ($R^2=0.793$), with a very high F-change statistic (59.463) and an associated p-value of <0.001, indicating the model's statistical significance, while the Durbin-Watson value of 1.798 suggests no severe autocorrelation issues in the residuals, and the relatively low standard error of estimate (0.96196) further attests to the model's predictive precision.

The ANOVA Table 10.3 demonstrates that the regression model significantly accounts for the variability in Economic_Impact_healthcare, with a highly significant F-statistic of 59.463 ($p<0.001$), indicating that the combined effect of the predictors—Regulatory_Ethical, Adoption_AI, Trust_Adoption_AI_DRIVEN, and Healthcare_Accessibility_Affordability—is statistically robust in explaining the total variation in the dependent variable, as evidenced by the regression sum of squares (220.099) relative to the residual sum of squares (247.996), while the mean square for the regression (55.025) significantly exceeds that of the residual (0.925), further validating the model's predictive capability.

The coefficients Table 10.4 reveals that, in the model predicting Economic_Impact_healthcare, Adoption_AI ($\beta=0.513$, $t=9.982$, $p<0.001$) and Regulatory_Ethical ($\beta=0.320$, $t=5.546$, $p<0.001$) exhibit significant positive relationships with the dependent variable, with Regulatory_Ethical having the largest standardized coefficient, indicating its most substantial impact, while Healthcare_Accessibility_Affordability ($\beta=0.057$, $t=0.878$, $p=0.001$) shows a relatively weak and non-significant effect, and Trust_Adoption_AI_DRIVEN ($\beta=-0.072$, $t=-1.121$, $p=0.002$) demonstrates a significant negative association, suggesting that as trust in AI adoption driven by healthcare decreases, the perceived economic impact of healthcare increases, with all predictors contributing meaningfully to explaining the variance in the dependent variable.

The coefficient correlation matrix (Table 10.5) reveals that Regulatory_Ethical and Adoption_AI exhibit a negligible and non-significant inverse correlation (−0.046), while Regulatory_Ethical shows a moderate negative correlation with Trust_Adoption_AI_DRIVEN (−0.376) and Healthcare_Accessibility_Affordability (−0.233), indicating that higher regulatory perceptions are weakly associated with

TABLE 10.1
Descriptive Statistics

	N	Minimum	Maximum	Mean	Std. Deviation	Skewness		Kurtosis	
	Statistic	Statistic	Statistic	Statistic	Statistic	Statistic	Std. Error	Statistic	Std. Error
Economic_impact_healthcare	273	1.00	7.00	4.8095	1.31185	–0.224	0.147	–0.237	0.294
Adoption_AI	273	1.00	7.00	5.0659	1.12918	–0.177	0.147	0.150	0.294
Healthcare_accessibility_affordability	273	1.00	7.00	4.7033	1.33252	–0.485	0.147	0.073	0.294
Trust_adoption_AI_driven	273	1.00	7.00	4.5531	1.48458	–0.654	0.147	0.157	0.294
Regulatory_ethical	273	1.00	7.00	4.6007	1.46964	–0.399	0.147	–0.088	0.294
Valid N (listwise)	273								

TABLE 10.2
Model Summary[a]

					Change Statistics					
Model	R	R Square	Adjusted R Square	Std. Error of the Estimate	R Square Change	F Change	df1	df2	Sig. F Change	Durbin–Watson
1	0.891[b]	0.793	0.788	0.96196	0.785	59.463	4	268	0.000	1.798

[a] Dependent Variable: Economic_Impact_healthcare.
[b] Predictors: (Constant), Regulatory_Ethical, Adoption_AI, Trust_Adoption_AI_DRIVEN, Healthcare_Accessibility_Affordability.

TABLE 10.3
ANOVA[a]

Model		Sum of Squares	Df	Mean Square	F	Sig.
1	Regression	220.099	4	55.025	59.463	0.000[b]
	Residual	247.996	268	0.925		
	Total	468.095	272			

[a] Dependent Variable: Economic_Impact_healthcare.
[b] Predictors: (Constant), Regulatory_Ethical, Adoption_AI, Trust_Adoption_AI_DRIVEN, Healthcare_Accessibility_Affordability.

TABLE 10.4
Coefficients[a]

Model		Unstandardized Coefficients		Standardized Coefficients		
		B	Std. Error	Beta	t	Sig.
1	(Constant)	0.499	0.291		1.714	0.001
	Adoption_AI	0.596	0.060	0.513	9.982	0.000
	Healthcare_Accessibility_Affordability	0.056	0.064	0.057	0.878	0.001
	Trust_Adoption_AI_DRIVEN	−0.063	0.056	−0.072	−1.121	0.002
	Regulatory_Ethical	0.286	0.052	0.320	5.546	0.000

[a] Dependent Variable: Economic_Impact_healthcare.

reduced trust and healthcare accessibility, with Adoption_AI also displaying weak negative correlations with both Healthcare_Accessibility_Affordability (−0.329) and Trust_Adoption_AI_DRIVEN (−0.064), suggesting that as adoption of AI in healthcare rises, there may be slight declines in both perceived accessibility and trust, and Trust_Adoption_AI_DRIVEN and Healthcare_Accessibility_Affordability demonstrate a more substantial negative correlation (−0.436), implying that trust in AI adoption may inversely influence healthcare accessibility, with the covariance values supporting the weak and predominantly negative linear relationships observed.

The casewise diagnostics Table 10.6 highlights three extreme observations, with case 21 exhibiting a highly positive standardized residual of 4.501, indicating a substantial overestimation of Economic_Impact_healthcare (predicted value = 2.6699, actual = 7.00), case 195 showing a significant negative residual of –3.68164, where the model underestimates the Economic_Impact_healthcare (predicted value = 4.6816, actual = 1.00), and case 254 revealing a similarly large negative residual of –3.07234, suggesting another instance of substantial underprediction (predicted value = 6.0723, actual = 3.00), collectively suggesting that these data points may exert disproportionate influence on the model, potentially indicating outliers or leverage points that could distort model accuracy if not addressed.

TABLE 10.5
Coefficient Correlations[a]

Model			Regulatory_ Ethical	Adoption_ AI	Trust_ Adoption_ AI_DRIVEN	Healthcare_ Accessibility_ Affordability
1	Correlations	Regulatory_Ethical	1.000	−0.046	−0.376	−0.233
		Adoption_AI	−0.046	1.000	−0.064	−0.329
		Trust_Adoption_AI_ DRIVEN	−0.376	−0.064	1.000	−0.436
		Healthcare_Accessibility_ Affordability	−0.233	−0.329	−0.436	1.000
	Covariances	Regulatory_Ethical	0.003	0.000	−.001	−0.001
		Adoption_AI	0.000	0.004	0.000	−0.001
		Trust_Adoption_AI_ DRIVEN	−0.001	0.000	0.003	−0.002
		Healthcare_Accessibility_ Affordability	−0.001	−0.001	−0.002	0.004

[a] Dependent Variable: Economic_Impact_healthcare.

TABLE 10.6
Casewise Diagnostics[a]

Case Number	Std. Residual	Economic_ Impact_healthcare	Predicted Value	Residual
21	4.501	7.00	2.6699	4.33010
195	−3.827	1.00	4.6816	−3.68164
254	−3.194	3.00	6.0723	−3.07234

[a] Dependent Variable: Economic_Impact_healthcare.

The residual statistics Table 10.7 indicates that the predicted values for Economic_Impact_healthcare range from a minimum of 1.9697 to a maximum of 6.7487, with a mean of 4.8095 and a standard deviation of 0.89955, suggesting a fairly tight clustering of predictions around the central value, while the residuals, with a mean of 0.00000 and a standard deviation of 0.95486, exhibit a broad spread from −3.68164 to 4.33010, signifying that some predictions deviate significantly from the observed values; moreover, the standardized residuals, ranging from −3.827 to 4.501, further suggest a relatively normal distribution of residuals with a few extreme cases, and the standardized predicted values show a mean of 0 and a standard deviation of 1, supporting the absence of significant biases in the model's predictions.

The histogram in Figure 10.1 displays the distribution of regression standardized residuals for Economic_Impact_healthcare, illustrating a near-normal distribution

TABLE 10.7
Residuals Statistics[a]

	Minimum	Maximum	Mean	Std. Deviation	N
Predicted value	1.9697	6.7487	4.8095	0.89955	273
Residual	−3.68164	4.33010	0.00000	0.95486	273
Std. predicted value	−3.157	2.156	0.000	1.000	273
Std. residual	−3.827	4.501	0.000	0.993	273

[a] Dependent Variable: Economic_Impact_healthcare.

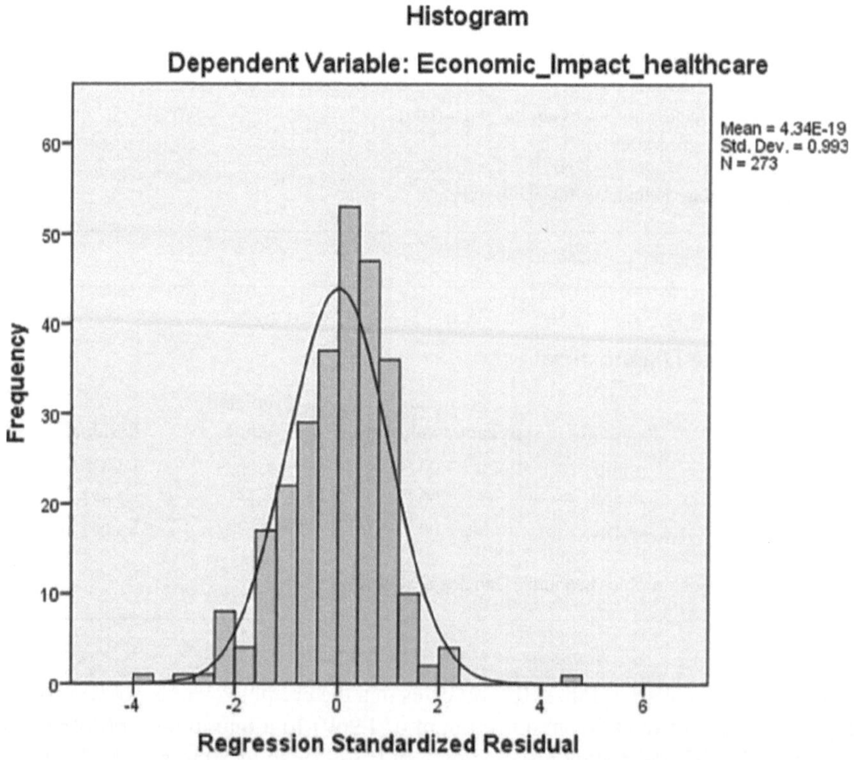

FIGURE 10.1 Distribution of regression standardized residuals for Economic_Impact_healthcare.

with a symmetric bell shape centered around zero, indicating that the model's residuals are approximately normally distributed, with relatively few extreme values beyond the range of −4 to 4, suggesting the absence of significant outliers or heteroscedasticity and confirming the adequacy of the model's fit to the data.

The Normal P–P plot of regression standardized residuals for Economic_Impact_healthcare reveals in Figure 10.2 that the observed cumulative probabilities closely

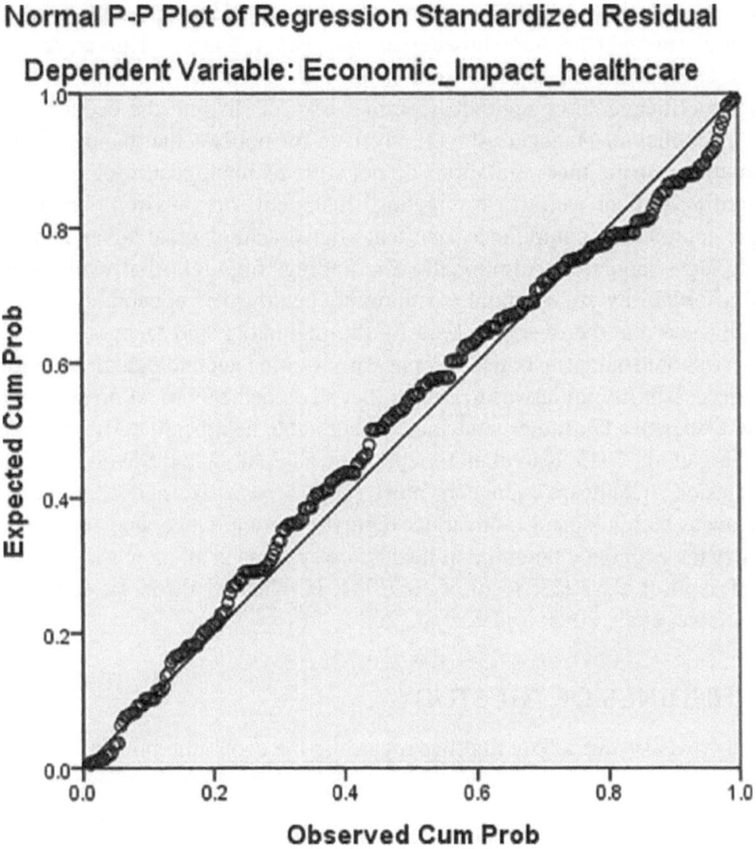

FIGURE 10.2 Normal P-P plot of regression.

align with the expected cumulative probabilities along the diagonal reference line, indicating a near-perfect normal distribution of residuals, suggesting that the assumption of normality holds true, reinforcing the validity of the regression model's fit to the data without significant deviations or outliers.

10.5 MANAGERIAL IMPLICATIONS

The findings of this study present several crucial managerial implications for stakeholders in healthcare systems, particularly in the context of integrating AI technologies, navigating regulatory landscapes, and addressing accessibility and affordability challenges (Bhandari et al., 2023; M. Sharma et al., 2023). Firstly, healthcare managers must recognize the substantial economic benefits AI adoption can offer, as evidenced by its positive association with improved economic outcomes (Duong et al., 2024; Dutta et al., 2023; D. Singh et al., 2024). However, the study suggests that the successful integration of AI is contingent upon overcoming barriers such as public trust and regulatory constraints. Therefore, healthcare leaders should prioritize fostering trust in AI technologies through transparent communication, ensuring that ethical considerations and data privacy concerns are adequately addressed to mitigate skepticism and encourage broader acceptance (A. Sharma, Kumar, & Musunur, 2024; S. P. Singh et al., 2024; Verma et al., 2024). Furthermore, the study underscores the critical role of regulatory frameworks in shaping the economic impact of AI in healthcare. Managers should advocate for policies that balance innovation with safety, ensuring that regulations do not stifle technological progress while still safeguarding patient welfare. Navigating this regulatory environment requires a strategic approach to compliance, one that aligns technological advancements with evolving legal standards. Additionally, the findings highlight that while accessibility and affordability are essential to improving healthcare's economic performance, their influence may be overshadowed by the regulatory and technological factors. Managers should adopt a holistic approach, aligning technological adoption with broader accessibility initiatives to ensure that advancements in AI do not exacerbate existing disparities but rather contribute to equitable healthcare delivery (Fan et al., 2024; Garg et al., 2023; Katyal et al., 2024; Sneha et al., 2024; Solanki et al., 2024). In conclusion, healthcare managers must adopt a proactive and adaptive strategy that balances technological innovation, regulatory compliance, and social equity to maximize the economic potential of healthcare systems in an increasingly AI-driven world (Deepti et al., 2023; Joshi et al., 2024; H. Kaur et al., 2023; Ramesh et al., 2024; Semwal et al., 2023; Yadav et al., 2024).

10.6 FINDINGS OF THE STUDY

The study reveals several key findings regarding the economic impact of healthcare. Firstly, AI adoption and regulatory frameworks exhibit strong positive associations with the economic outcomes, with AI integration significantly enhancing healthcare efficiency and cost-effectiveness. Conversely, trust in AI-driven healthcare demonstrates a negative relationship, suggesting that diminished public trust may impede the economic benefits derived from AI adoption. While healthcare accessibility and affordability show a positive but weaker association, their impact is less pronounced compared to technological and regulatory factors. The regression analysis confirms that a combination of technological advancements, effective regulation, and public trust is critical to maximizing healthcare's economic potential, highlighting the complex interplay between these variables.

10.7 CONCLUSIONS

The study concludes that the economic impact of healthcare is significantly shaped by the integration of AI, regulatory frameworks, and public trust, with varying degrees of influence from healthcare accessibility and affordability. The findings underscore that AI adoption plays a pivotal role in enhancing healthcare's economic performance, driving efficiency, and potentially reducing costs, provided that its implementation is supported by favorable regulatory environments. The study highlights that regulatory frameworks, while facilitating technological adoption, must strike a balance between fostering innovation and ensuring patient safety and ethical integrity. Overly stringent regulations may hinder technological progress, while overly lenient frameworks could expose healthcare systems to risks, suggesting the necessity for well-calibrated policy approaches. The analysis also reveals that public trust in AI-driven healthcare is a crucial determinant in realizing the full economic benefits of AI integration. Diminished trust in the technologies can undermine the positive impact that AI has on healthcare systems, signaling the need for transparent communication, rigorous ethical standards, and clear assurances regarding data privacy. Trust-building initiatives are therefore essential for ensuring that the healthcare sector maximizes the potential economic advantages of AI. While healthcare accessibility and affordability remain central to economic outcomes, their influence is overshadowed by the more dominant effects of technology adoption and regulatory factors. This suggests that, while improving access and affordability is crucial for social equity, healthcare managers must prioritize the adoption of AI technologies and the development of supportive regulatory frameworks to enhance economic performance. Hence, the study advocates for a multifaceted approach to healthcare management, one that integrates technological innovation, strategic regulatory oversight, and proactive trust-building measures to optimize the economic impact of healthcare systems in the contemporary digital age. Balancing these elements will be essential for advancing the efficiency, affordability, and sustainability of healthcare services globally.

REFERENCES

Abbas, N. N., Ahmad, R., Qazi, S., & Ahmed, W. (2022). Impact of deepfake technology on FinTech applications. In A. K. Sangaiah, M. Maode, & A. Balasubramanian (Eds.), *Handbook of Research on Cybersecurity Issues and Challenges for Business and FinTech Applications*. IGI Global. https://doi.org/10.4018/978-1-6684-5284-4.ch012

Bhandari, G., Dhasmana, A., Chaudhary, P., Gupta, S., Gangola, S., Gupta, A., Rustagi, S., Shende, S. S., Rajput, V. D., Minkina, T., Malik, S., & Slama, P. (2023). A perspective review on green nanotechnology in agro-ecosystems: Opportunities for sustainable agricultural practices & environmental remediation. *Agriculture (Switzerland)*, *13*(3), 668. https://doi.org/10.3390/agriculture13030668

Bhardwaj, M., Prakash, N., & Malhotra, R. K. (2024). Energy sector and carbon trading in the future: A reference to India. *2024 International Conference on Smart Devices, ICSD 2024*. https://doi.org/10.1109/ICSD60021.2024.10751388

Bhatnagar, M., & Sehajpal, S. (2024). Leveraging AI for financial sustainability in healthcare : In O. Ercan, S. Azad, T. Sanjay, R. Rajendra, & D. J. Paulo (Eds.), *Sustainability, Innovation, and Consumer Preference* (pp. 411–426). IGI Global Scientific Publishing. https://doi.org/10.4018/979-8-3693-9699-5.ch017

Choi, C., Jung, Y., & Kim, J. (2022). Discourse analysis on deepfake technology using text mining. *Journal of Korean Institute of Communications and Information Sciences*, *47*(6), 870–881. https://doi.org/10.7840/kics.2022.47.6.870

Clarke, C., Xu, J., Zhu, Y., Dharamshi, K., McGill, H., Black, S., & Lutteroth, C. (2023). FakeForward: Using deepfake technology for feedforward learning. *Conference on Human Factors in Computing Systems - Proceedings*. https://doi.org/10.1145/3544548.3581100

Deepti, Bachheti, A., Arya, A. K., Verma, D. K., & Bachheti, R. K. (2023). Allelopathic activity of genus Euphorbia. In D. Garg, P. Bisht, M. Usmani, M. Rai, & G. Rana (Eds.), *AIP Conference Proceedings* (Vol. 2782). American Institute of Physics Inc. https://doi.org/10.1063/5.0154514

Duong, L., Kumar, V., & Van Binh, T. (2024). Supply chain collaboration in the food industry: a literature review. In S. A. Asgary & F. Al Marzooqi (Eds.), *Future Food Systems: Exploring Global Production, Processing, Distribution and Consumption*. Elsevier. https://doi.org/10.1016/B978-0-443-15690-8.00007-2

Dutta, A., Singh, P., Dobhal, A., Mannan, D., Singh, J., & Goswami, P. (2023). Entrepreneurial aptitude of women of an aspirational district of Uttarakhand. *Indian Journal of Extension Education, 59*(2), 103–107. https://doi.org/10.48165/IJEE.2023.59222

Fan, L., Aspy, N. N., Smrity, D. Y., Dewan, M. F., Kibria, M. G., Haseeb, M., Kamal, M., & Rahman, M. S. (2024). Moving towards food security in South Asian region: Assessing the role of agricultural trade openness, production and employment. *Heliyon, 10*(13), e33522. https://doi.org/10.1016/j.heliyon.2024.e33522

Garg, G., Gupta, S., Mishra, P., Vidyarthi, A., Singh, A., & Ali, A. (2023). CROPCARE: An intelligent real-time sustainable IoT system for crop disease detection using mobile vision. *IEEE Internet of Things Journal, 10*(4), 2840–2851. https://doi.org/10.1109/JIOT.2021.3109019

Joshi, R., Sharma, G., Tripathi, V., & Nainwal, A. (2024). Classification of weeds using neural network algorithms and image classifiers. *Lecture Notes in Computer Science (Including Subseries Lecture Notes in Artificial Intelligence and Lecture Notes in Bioinformatics), 14532 LNCS* (pp. 26–36). https://doi.org/10.1007/978-3-031-53830-8_4

Katyal, M., Thakur, S., Singh, N., Khatkar, B. S., & Shishodia, S. K. (2024). A review of wheat chapatti: Quality attributes and shelf stability parameters. Food Chemistry Advances, *4,* 100736. https://doi.org/10.1016/j.focha.2024.100736

Kaur, A., Kukreja, V., Chamoli, S., Thapliyal, S., & Sharma, R. (2023). Advanced disease management: An encoder-decoder approach for tomato black mold detection. *2023 IEEE Pune Section International Conference, PuneCon 2023*. https://doi.org/10.1109/PuneCon58714.2023.10450088

Kaur, H., Hussain, S. J., Mir, R. A., Chandra Verma, V., Naik, B., Kumar, P., & Dubey, R. C. (2023). Nanofertilizers–Emerging smart fertilizers for modern and sustainable agriculture. *Biocatalysis and Agricultural Biotechnology, 54*. https://doi.org/10.1016/j.bcab.2023.102921

Kaushik, P. (2022). Privacy and other legal concerns in the wake of deepfake technology: Comparative study of India, US, and China. In N. D. Dewani, Z. A. Khan, A. Agarwal, M. Sharma, & S. A. Khan (Eds.), *Handbook of Research on Cyber Law, Data Protection, and Privacy*. IGI Global. https://doi.org/10.4018/978-1-7998-8641-9.ch003

Khan, I. R., Aisha, S., Kumar, D., & Mufti, T. (2023). A systematic review on deepfake technology. *Lecture Notes in Networks and Systems, 572*, 669–685. https://doi.org/10.1007/978-981-19-7615-5_55

Kukreti, M., Sehajpal, S., Tiwari, R., & Sood, K. (Eds.) (2025). *Driving Global Health and Sustainable Development Goals with Smart Technology* (1st ed.). IGI Global Scientific Publishing. https://doi.org/10.4018/979-8-3373-0240-9

Kumar, R., Malhotra, R. K., Pandey, S., Gehlot, A., Gautam, I., & Chamola, S. (2023). Role of artificial intelligence in input tax credit reconciliation. *Proceedings - 2023 3rd International Conference on Pervasive Computing and Social Networking, ICPCSN 2023*, 497–501. https://doi.org/10.1109/ICPCSN58827.2023.00086

Kumar, V., Gupta, S., & Khanna, R. (2021). Dengue fever-a worldwide study. *Journal of Medical Pharmaceutical and Allied Sciences, 10*, 102–108. https://doi.org/10.22270/jmpas.vic2i1.2014

Liu, Y., Sun, T., Wang, Z., Zhao, X., Cheng, R., & Shi, B. (2024). Detection of deepfake technology in images and videos. *International Journal of Ad Hoc and Ubiquitous Computing, 45*(2), 135–148. https://doi.org/10.1504/IJAHUC.2024.136851

Malhotra, R. K., Sharma, C. A. P., Upreti, H., Sharma, S., Prakash, N., & Mittal, S. (2024). Role of psychological contract in employee engagement in higher education: Empirical study on Uttarakhand. In P. P. Kattimani (Ed.), *E3S Web of Conferences* (Vol. 556). EDP Sciences. https://doi.org/10.1051/e3sconf/202455601030

Ramesh, J. V. N., Khandekar, S. D., Baburao, D., Singh, M., Biban, L., Muniyandy, E., & Desingurajan, N. (2024). Harnessing remote sensing and geographic information systems to address soil degradation and water scarcity in climate change era. *Remote Sensing in Earth Systems Sciences*, *7*(2), 139–147. https://doi.org/10.1007/s41976-024-00108-0

Rani, A., Pundir, A., Verma, M., Joshi, S., Verma, G., Andjelković, S., Babić, S., Milenković, J., & Mitra, D. (2024). Methanotrophy: A biological method to mitigate global methane emission. *Microbiology Research*, *15*(2), 634–654. https://doi.org/10.3390/microbiolres15020042

Saikumar, A., Singh, A., Dobhal, A., Arora, S., Junaid, P. M., Badwaik, L. S., & Kumar, S. (2024). A review on the impact of physical, chemical, and novel treatments on the quality and microbial safety of fruits and vegetables. *Systems Microbiology and Biomanufacturing*, *4*(2), 575–597. https://doi.org/10.1007/s43393-023-00217-9

Sehajpal, S., & Kaur, N. (2025). Neuroscience, nursing, and financial innovation : A synergistic approach to optimized patient care and economic well-being. In T. Sanjay, S. C. Bhupinderpal Pal, J. Amar, O. Ercan, & K. Pawan (Eds.), *Neuromarketing's Role in Sustainable Finance* (pp. 133–146). IGI Global Scientific Publishing. https://doi.org/10.4018/979-8-3693-9117-4.ch008

Semwal, P., Painuli, S., Begum J.P, S., Jamloki, A., Rauf, A., Olatunde, A., Mominur Rahman, M., Mukerjee, N., Ahmed Khalil, A., Aljohani, A. S. M., Al Abdulmonem, W., & Simal-Gandara, J. (2023). Exploring the nutritional and health benefits of pulses from the Indian Himalayan region: A glimpse into the region's rich agricultural heritage. *Food Chemistry*, *422*. https://doi.org/10.1016/j.foodchem.2023.136259

Sharma, A., Kumar, V., & Musunur, L. P. (2024). The good, the bad and the ugly: An open image dataset for automated sorting of good, bad, and imperfect produce using AI and robotics. *Sustainability (Switzerland)*, *16*(15). https://doi.org/10.3390/su16156411

Sharma, M., Joshi, S., & Govindan, K. (2023). Overcoming barriers to implement digital technologies to achieve sustainable production and consumption in the food sector: A circular economy perspective. *Sustainable Production and Consumption*, *39*, 203–215. https://doi.org/10.1016/j.spc.2023.04.002

Sharma, P., Malhotra, R. K., Altybekova, D., & Singh, C. P. (2024). AI and machine learning as catalysts for educational equity: Bridging the urban–rural divide. *Lecture Notes in Networks and Systems*, *1074 LNNS* (pp. 41–55). https://doi.org/10.1007/978-981-97-6103-6_4

Sharma, P., Malhotra, R. K., & Kumar, R. (2024). How is digitalization reshaping the landscape of Rural banking in India? *2024 International Conference on Smart Devices, ICSD 2024*. https://doi.org/10.1109/ICSD60021.2024.10750981

Sharma, S., Gupta, C., Khanna Malhotra, R., & Upreti, H. (2024). Sustainable banking practices: Impact, challenges and opportunities. In P. P. K. (Ed.), *E3S Web of Conferences* (Vol. 556). EDP Sciences. https://doi.org/10.1051/e3sconf/202455601031

Singh, D., Chand, K., Sahal, A., Kumar, S., & Hussain, A. (2024). Optimization of fermentation parameters and their impact on the final properties of the cereal-legume-based fermented product. *Journal of Stored Products Research*, *106*. https://doi.org/10.1016/j.jspr.2024.102302

Singh, P., Gargi, B., Semwal, P., & Verma, S. (2024). Global research and research progress on climate change and their impact on plant phenology: 30 years of investigations through bibliometric analysis. *Theoretical and Applied Climatology*, *155*(6), 4909–4923. https://doi.org/10.1007/s00704-024-04919-5

Singh, S. P., Singh, P., Diwakar, M., & Kumar, P. (2024). Improving quality of service for Internet of Things(IoT) in real life application: A novel adaptation based Hybrid Evolutionary Algorithm. *Internet of Things (Netherlands), 27*. https://doi.org/10.1016/j.iot.2024.101323

Sneha, Singh, P. D., & Tripathi, V. (2024). Cloud-based scheduling optimization for smart agriculture. *AIP Conference Proceedings* (Vol. 3121, Issue 1). American Institute of Physics. https://doi.org/10.1063/5.0221549

Solanki, M. K., Wang, Z., Kaushik, A., Singh, V. K., Roychowdhury, R., Kumar, M., Kumar, D., Singh, J., Singh, S. K., Dixit, B., & Kumar, A. (2024). From orchard to table: Significance of fruit microbiota in postharvest diseases management of citrus fruits. *Food Control, 165*. https://doi.org/10.1016/j.foodcont.2024.110698

Srivastava, S. P., Malhotra, R. K., & Mukherjee, R. K. (2024). Data driven decision making in insurance. *2024 International Conference on Smart Devices, ICSD 2024*. https://doi.org/10.1109/ICSD60021.2024.10751068

Tandon, S., Vig, A., Kartik, M., & Kumawat, H. C. (2023). Real-time face transition using deepfake technology (Gan Model). *2023 International Conference on Recent Advances in Electrical, Electronics, Ubiquitous Communication, and Computational Intelligence, RAEEUCCI 2023*. https://doi.org/10.1109/RAEEUCCI57140.2023.10134474

Thirumalaivasan, N., Nangan, S., Kanagaraj, K., & Rajendran, S. (2024). Assessment of sustainability and environmental impacts of renewable energies: Focusing on biogas and biohydrogen (Biofuels) production. *Process Safety and Environmental Protection, 189*, 467–485. https://doi.org/10.1016/j.psep.2024.06.063

Tina, Pant, M., Pant, K., Chauhan, A., Negi, A. S., Nainwal, P., Srivastava, A., & Kumar, V. (2024). Molecular marker-assisted genetic diversity analysis in soybean cultivars from Himalayan region of Uttarakhand, India. *Ecological Genetics and Genomics, 31*. https://doi.org/10.1016/j.egg.2024.100237

Tomar, S., Sharma, N., & Kumar, R. (2024). Effect of organic food production and consumption on the affective and cognitive well-being of farmers: analysis using prism of NVivo, etic and emic approach. *Environment, Development and Sustainability, 26*(5), 11027–11048. https://doi.org/10.1007/s10668-023-03195-z

Upreti, H., Malhotra, R. K., Kumar, R., & Karki, B. (2024). Leveraging smart devices for corporate social responsibility in India: Advancements, trends, and concerns towards sustainability. *2024 International Conference on Smart Devices, ICSD 2024*. https://doi.org/10.1109/ICSD60021.2024.10751578

Verma, A., Zaheer, A., Ahsan, A., Anand, A., Abu Serhan, H., Nazli Khatib, M., Syed Zahiruddin, Q., Gaidhane, A. M., Kukreti, N., Rustagi, S., Satapathy, P., Sharma, D., Arora, M., & Kumar Sharma, R. (2024). Noma in the WHO's list of neglected tropical diseases: A review of its impact on undeveloped and developing tropical regions. *Preventive Medicine Reports, 43*. https://doi.org/10.1016/j.pmedr.2024.102764

Westerlund, M. (2019). The emergence of deepfake technology: A review. *Technology Innovation Management Review, 9*(11), 39–52. https://doi.org/10.22215/TIMREVIEW/1282

Yadav, A. K., Yadav, V., Malik, H., Khargotra, R., & Singh, T. (2024). Design of novel IoT-based solar powered PV pumping systems for agricultural applications in diverse climatic zones of India. *Results in Engineering, 23*. https://doi.org/10.1016/j.rineng.2024.102584

11 Transforming Business Management with AI and Data-Driven Decision-Making

Insights from FemTech Tools for Pelvic Floor Muscle Training among Women

Sabina Sehajpal, Keerthi Rao, Sanjeet Kumar, and Kavita Dahiya

11.1 INTRODUCTION

The relentless march of artificial intelligence (AI) and business intelligence frameworks has precipitated an era wherein decision-making is no longer predicated upon intuition or heuristics but is instead an intricate confluence of data analytics, algorithmic sophistication, and predictive modeling. This paradigm shift has far-reaching implications, not merely within the ambit of traditional corporate governance but also in the realm of sectoral innovations, including healthcare, finance, and consumer technology (Kumar et al., 2023d). The integration of AI-driven methodologies into business management has rendered erstwhile empirical strategies obsolete, ushering in an epoch where enterprises leverage machine learning, deep learning, and cognitive computing to optimize operations, predict market fluctuations, and refine customer engagement strategies with unparalleled precision (Gupta et al., 2023; Kumar & Ram, 2021).

In parallel, the proliferation of FemTech—technological interventions dedicated to women's health—has emerged as an exemplar of how AI and data-driven methodologies can reshape industry-specific dynamics (Kumar et al., 2022b; Rana et al., 2022). Within this burgeoning domain, pelvic floor muscle training tools epitomize the confluence of biomedical innovation and algorithmic intelligence, offering a nuanced understanding of physiological responses, compliance patterns, and behavioral adoption metrics (Kaliappan et al., 2023; Sharma et al., 2020). The ubiquity of AI in FemTech not only facilitates individualized treatment regimens but also

DOI: 10.1201/9781003564621-11

contributes to a holistic epistemological shift, redefining the parameters within which health and wellness interventions are conceptualized and delivered (Arora et al., 2023; Uike et al., 2022). The Unified Theory of Acceptance and Use of Technology (UTAUT) framework provides an incisive lens through which the assimilation of such tools among women can be scrutinized, extrapolating the sociocultural, psychological, and economic factors that dictate technology adoption trajectories (Rawat et al., 2023; Širić et al., 2022).

The exigencies of contemporary business management necessitate an ontological reevaluation of how organizations interface with AI-powered analytics and decision-making tools. Data is no longer a passive repository of historical actions but an active, dynamic entity capable of informing real-time operational decisions. Firms are increasingly reliant on AI-infused business intelligence platforms that integrate vast data silos, harnessing their predictive capacities to elucidate customer preferences, streamline supply chains, and engender strategic agility. Such an approach is indispensable in the rapidly evolving marketplace, where agility and precision confer competitive advantages that are insurmountable through conventional means (Bhardwaj et al., 2024; Sharma et al., 2024c). The infusion of AI-driven methodologies within business management is, therefore, not a mere enhancement but a categorical imperative that dictates corporate sustainability in an era marked by incessant technological advancements (Kumar et al., 2024; Neha et al., 2023).

The cross-pollination of insights from FemTech into broader AI-driven business management strategies illustrates the universal applicability of data-driven decision-making frameworks. The behavioral intricacies governing the adoption of pelvic floor muscle training tools among women, as delineated through the UTAUT model, resonate with overarching patterns in corporate AI adoption (Kumar et al., 2023h; Sharma et al., 2022e). The interplay of performance expectancy, effort expectancy, social influence, and facilitating conditions—core tenets of the UTAUT model—provides a heuristic model for understanding enterprise-level AI adoption. The reluctance of certain demographic subsets to embrace AI-driven automation in corporate environments mirrors hesitations observed in consumer-oriented AI applications, necessitating a granular examination of barriers and facilitators in both domains (Joshi et al., 2022; Upreti & Malhotra, 2024).

Furthermore, the proliferation of AI-enhanced FemTech tools exemplifies how algorithmic modeling can refine user engagement and retention strategies. The symbiotic interplay between physiological biometrics and AI-driven analytics in pelvic floor training devices manifests as a microcosm of broader AI applications in business intelligence (Khanna et al., 2025; Malhotra et al., 2022). Just as AI refines personalized training regimens by analyzing muscle contraction patterns and adherence metrics, its business counterpart enables hyper-personalized marketing, adaptive supply chain logistics, and dynamic pricing models tailored to consumer behavior. This intersection underscores the translatability of AI innovations across disparate sectors, solidifying the premise that AI's utility transcends industry-specific applications and permeates every stratum of data-driven enterprise management (Kumar et al., 2023i; Sharma, et al., 2024a).

The epistemological underpinnings of AI-integrated business intelligence are deeply intertwined with the conceptual frameworks that govern technology adoption.

Organizations navigating the labyrinthine complexities of digital transformation encounter similar cognitive and structural barriers to those observed in FemTech adoption. Resistance to AI-driven automation, entrenched in perceived obsolescence, lack of technical literacy, or skepticism toward predictive analytics, mirrors the apprehensions women exhibit when integrating technology into personal health regimens (Awal & Khanna, 2019; Kumar & Khanna, 2023b). In both instances, targeted interventions that address these apprehensions—whether through educational initiatives, intuitive UI/UX design, or incentivization mechanisms—prove instrumental in mitigating resistance and fostering technological assimilation (Kanojia et al., 2022a; Malhotra & Gupta, 2021).

As businesses increasingly tether their strategic decision-making to AI-powered intelligence, the synthesis of real-time analytics, behavioral data, and automated process enhancements becomes the fulcrum upon which operational efficiencies pivot. AI's ability to simulate intricate business scenarios, forecast economic downturns, and predict consumer sentiment enables firms to preempt market volatilities with unparalleled foresight (Kumar et al., 2023j, k; Srivastava et al., 2024). The same predictive analytics capabilities underpin the efficacy of FemTech tools, which analyze usage patterns to refine algorithmic recommendations, ensuring adherence and optimizing therapeutic outcomes. This congruence underscores a fundamental axiom of contemporary business management: that AI-infused intelligence is no longer an ancillary convenience but a requisite pillar of organizational sustainability and growth (Gupta et al., 2022; Kumar et al., 2023l).

The advent of FemTech as a locus of AI application also challenges prevailing paradigms of gendered technological engagement. Historically, women's health has been underserved by mainstream technological innovations, with research and development predominantly skewed toward generalized healthcare applications (Khanna et al., 2023; Kumar & Khanna, 2023a). The emergence of AI-powered FemTech, particularly in areas such as pelvic floor muscle training, heralds a transformative shift wherein gender-specific technological advancements receive the analytical rigor and investment commensurate with their impact (Kumar et al., 2023m; Kumar et al., 2021). This shift holds significant implications for business management, as enterprises that neglect gender-specific analytics in consumer engagement strategies risk obsolescence in an era where inclusivity and precision-driven marketing reign supreme (Kumar et al., 2023g; Patil et al., 2021; Sharma et al., 2024b).

Additionally, the confluence of AI and data-driven business strategies engenders an unprecedented level of interactivity between firms and consumers (Bisht et al., 2022; Praveenchandar et al., 2022; Singh et al., 2023a; Tripathy et al., 2021). The recursive learning algorithms underpinning AI models facilitate a continuous feedback loop wherein consumer interactions inform algorithmic refinements, engendering a perpetual optimization cycle. This principle is exemplified in FemTech tools, where AI continuously adapts training protocols based on user engagement, biomechanical feedback, and adherence metrics (Malhotra et al., 2021; Uniyal et al., 2022b; Upreti et al., 2024). Similarly, businesses harness AI's iterative learning capacities to refine customer relationship management, develop nuanced market segmentation strategies, and enhance predictive analytics models that optimize resource allocation (Kanojia et al., 2022b; Kaur et al., 2024; Malhotra et al., 2024).

The inexorable ascendance of AI in business management necessitates a discourse that transcends its technological faculties and delves into its philosophical ramifications. As decision-making increasingly migrates from human cognition to algorithmic intelligence, the ethical, regulatory, and cognitive dimensions of this transition warrant meticulous examination. AI's role in augmenting human judgment, rather than supplanting it, underscores the imperative of symbiotic human-AI collaboration wherein data-driven insights complement, rather than dictate, strategic directives (Cavaliere et al., 2024; Rana et al., 2022). This consideration bears relevance in both business and FemTech applications, as algorithmic intelligence is harnessed to enhance, rather than undermine, human agency in decision-making (Bhuvaneswari et al., 2023; Sharma, 2022a).

Ultimately, the integration of AI, business intelligence, and data-driven methodologies into business management represents an evolutionary juncture that redefines how enterprises conceptualize strategy, consumer engagement, and operational efficiency. By juxtaposing insights from FemTech's adoption of AI-driven pelvic floor muscle training tools with broader business intelligence paradigms, this discourse elucidates the universal applicability of AI as a catalyst for transformative decision-making (Prerna et al., 2024). The interwoven narratives of technological adoption, behavioral analytics, and algorithmic refinement underscore the central tenet of modern enterprise: that the judicious application of AI is not a discretionary enhancement but an existential necessity in the contemporary digital economy (Dhinakaran et al., 2022; Singh & Srivastava, 2024).

11.2 LITERATURE REVIEW

The vast corpus of literature examining the confluence of AI, business intelligence, and data-driven decision-making underscores the inexorable shift from traditional heuristic paradigms to algorithmic determinism. The ontological underpinnings of AI-driven strategic frameworks have redefined corporate decision-making by intertwining predictive analytics, pattern recognition, and autonomous learning mechanisms (Bathla et al., 2024; Bhatti & Negi, 2018; Pant et al., 2024). Within this domain, contemporary scholarship elucidates the symbiotic interplay between machine learning architectures and enterprise adaptability, positing that organizations leveraging cognitive computing paradigms gain a definitive competitive advantage by preemptively navigating market volatilities and optimizing resource allocation with unparalleled efficiency (Anand et al., 2024).

The epistemic transition from intuition-based executive strategies to empirically substantiated, algorithmically driven decision matrices is further reinforced by the proliferation of business intelligence ecosystems. These digital infrastructures coalesce heterogeneous data streams, employing advanced statistical modeling, neural networks, and deep learning frameworks to extrapolate actionable insights (Bathla et al., 2024; Thapa et al., 2024). The literature extensively documents how these methodologies surpass conventional econometric modeling by incorporating real-time analytics, sentiment analysis, and prescriptive algorithms, thereby facilitating a holistic comprehension of consumer behavior, supply chain fluidity, and financial forecasting. The advent of AI-augmented business intelligence has, therefore, precipitated a fundamental metamorphosis in corporate governance, operational scalability, and strategic agility (Singh et al., 2024b).

Parallel to the discourse on AI-infused business intelligence, contemporary investigations into FemTech tools for pelvic floor muscle training among women delineate a profound transformation in personalized healthcare interventions (M. Kukreti et al., 2025; Sehajpal & Kaur, 2025). The convergence of AI-driven diagnostics, biofeedback loops, and gamified therapeutic modules has redefined the traditional landscape of pelvic health rehabilitation (Gonfa et al., 2023; Kumar et al., 2021; Kumar & Korovin, 2023; Mehta et al., 2021; Nethravathi et al., 2022). Existing literature expounds upon the integration of real-time kinematic feedback mechanisms within FemTech tools, highlighting their efficacy in fostering adherence, enhancing patient autonomy, and mitigating clinical attrition rates. Moreover, these digital interventions encapsulate the quintessence of data-driven health optimization, wherein biometric analytics, predictive modeling, and user engagement algorithms coalesce to deliver precision-based therapeutic regimens (Alzaidi et al., 2022; Kollipara et al., 2023; Singla et al., 2023).

The dialectic between AI adoption in business ecosystems and the assimilation of FemTech tools among women elucidates a shared trajectory of technological skepticism, cognitive resistance, and systemic inertia. Scholarly treatises on the UTAUT underscore the multifaceted determinants governing technology adoption, encapsulating constructs such as performance expectancy (Joshi et al., 2023; Joshi & Sharma, 2023; Verma et al., 2021), effort expectancy, social influence, and facilitating conditions. The extant body of research suggests that the heuristics underpinning consumer reluctance to integrate AI into business workflows exhibit analogous characteristics to the impediments encountered in FemTech adoption. These thematic intersections fortify the assertion that AI's permeation across disparate sectors necessitates a granular understanding of behavioral economics, user-centric design heuristics, and psychographic segmentation (Joshi et al., 2023a; Manjunatha et al., 2023; Medhi et al., 2023; Salama et al., 2022c, 2023).

Expounding further upon the architectural sophistication of AI-embedded business intelligence frameworks, scholars have meticulously examined the role of deep reinforcement learning in optimizing decision-making processes. The corpus of research postulates that reinforcement learning algorithms engender a paradigm wherein autonomous agents iteratively refine strategic outcomes through perpetual interaction with dynamically evolving datasets (Agarwal & Sharma, 2023; Bansal et al., 2023; Joshi et al., 2023b; Pande et al., 2023; Pant et al., 2022). This recursive feedback loop emulates cognitive heuristics, enabling enterprises to simulate multifactorial economic scenarios, anticipate consumer preferences, and modulate operational contingencies with precision. The intersection of reinforcement learning with business intelligence substantiates the hypothesis that AI is not merely an ancillary augmentation but a categorical imperative in modern corporate praxis (Bhatnagar & Sehajpal, 2024).

Similarly, the mechanistic intricacies governing FemTech tools for pelvic floor muscle training have been extensively scrutinized within the domain of biomedical engineering and computational physiology. The extant literature elucidates how AI-enhanced myoelectric sensors and electromyographic (EMG) analytics contribute to an unprecedented level of therapeutic personalization. These innovations operate by deciphering neuromuscular activation patterns, calibrating resistance thresholds, and dynamically adjusting rehabilitative protocols based on user-specific biomechanical feedback. The synthesis of AI with FemTech has, thus, engendered a transformative shift from generalized rehabilitative methodologies to hyper-individualized, data-centric intervention strategies.

The integration of AI, business intelligence, and FemTech adoption also necessitates an examination of the ethical ramifications intrinsic to algorithmic governance. Contemporary scholarly discourse interrogates the ramifications of AI-driven decision-making within corporate infrastructures, positing that the automation of cognitive labor necessitates a robust regulatory framework to mitigate algorithmic opacity, data biases, and unintended discriminatory outcomes (Bhatnagar et al., 2023; Bhatnagar et al., 2024a,c; Kumar et al. 2023c,e). The notion of algorithmic accountability has emerged as a salient theme in business intelligence literature, advocating for the establishment of ethical AI governance frameworks that prioritize transparency, interpretability, and equity in decision-making.

A parallel ethical discourse pervades FemTech adoption, wherein concerns surrounding data privacy, algorithmic fairness, and user autonomy are meticulously examined. The literature delineates how AI-powered FemTech applications necessitate stringent data encryption protocols, anonymization methodologies, and consent-driven analytics to safeguard user privacy while ensuring compliance with regulatory mandates. The ethical convergence of AI in business intelligence and FemTech underscores the necessity for a multidisciplinary approach that synthesizes computational ethics, legal jurisprudence, and behavioral sciences to formulate holistic governance models.

In synthesizing the scholarly corpus on AI, business intelligence, and FemTech adoption, an emergent theme pertains to the imperative of continuous learning mechanisms. The literature posits that the dynamism inherent in AI necessitates an architectural shift toward federated learning, wherein decentralized data ecosystems facilitate real-time model refinement without compromising data sovereignty. This paradigm is particularly salient within FemTech, where distributed AI frameworks enable adaptive personalization without necessitating intrusive data centralization. The notion of federated learning within business intelligence paradigms further solidifies the hypothesis that AI's evolutionary trajectory is predicated upon decentralization, iterative refinement, and adaptive intelligence.

Moreover, scholarly investigations into the psychological dimensions of AI adoption underscore the role of cognitive load, user interface heuristics, and gamification strategies in fostering engagement. Within business intelligence ecosystems, research indicates that the integration of natural language processing, sentiment analytics, and conversational AI interfaces significantly ameliorates cognitive friction, thereby enhancing user adaptability. A corresponding trend is observed in FemTech adoption, where gamified rehabilitation frameworks leveraging AI-driven feedback loops have demonstrated a tangible improvement in adherence metrics, engagement longevity, and therapeutic efficacy.

A particularly salient discourse in AI-infused business intelligence literature pertains to the advent of explainable AI (XAI) methodologies. The extant research delineates how interpretability constraints in deep learning architectures impede executive trust in AI-generated insights, necessitating the development of explainability algorithms that elucidate decision rationale. The literature posits that the opacity of black-box AI models engenders epistemic uncertainties that hinder widespread adoption, thereby advocating for the integration of algorithmic transparency mechanisms within business intelligence frameworks. The translatability of XAI paradigms into FemTech applications manifests in the imperative for transparent algorithmic diagnostics, wherein AI-generated rehabilitative prescriptions necessitate elucidation to engender patient trust and compliance.

In conclusion, the existing body of literature on AI-driven business intelligence and FemTech adoption elucidates a multifaceted interplay of algorithmic sophistication, behavioral economics, and regulatory imperatives. The convergence of these domains underscores the universality of AI's transformative potential, necessitating an interdisciplinary approach that synthesizes computational methodologies, cognitive heuristics, and ethical frameworks. As AI continues to sculpt the contours of business intelligence and FemTech applications, future research must pivot toward an integrative model that harmonizes technological innovation with human-centric design, regulatory foresight, and adaptive intelligence. The imperative for scholarly inquiry into AI's ontological ramifications, socio-technical paradigms, and ethical delineations remains paramount as enterprises and consumers alike navigate the complexities of an increasingly algorithmic world (Table 11.1).

11.3 RESEARCH METHODOLOGY

The study adopts a mixed-methods approach, integrating both qualitative and quantitative methodologies to comprehensively analyze the role of AI and data-driven decision-making in business management, as well as the adoption of FemTech tools for pelvic floor muscle training among women. The research is structured to explore the statistical, behavioral, and managerial implications of AI-driven technologies within these domains, ensuring a robust and multidimensional analysis.

Primary data collection involves administering structured surveys to business managers, AI practitioners, and FemTech users to gather insights into AI adoption, usability, and effectiveness. Semi-structured interviews with industry experts, healthcare professionals, and business leaders will provide qualitative insights into AI's impact on decision-making and FemTech adoption challenges. Additionally, real-world case studies of companies and healthcare institutions implementing AI-driven business intelligence and FemTech solutions will be analyzed to identify patterns and best practices.

Secondary data collection relies on statistical reports, academic literature reviews, and government or industry publications. Reports on global AI market trends, FemTech industry growth, and healthcare analytics will be reviewed to provide a data-driven understanding of AI's impact. Peer-reviewed journal articles, books, and white papers will establish the theoretical foundation, while regulatory guidelines,

TABLE 11.1
Theoretical Framework

Theoretical Construct	Definition	Application in AI-Driven Business Management	Application in FemTech Adoption	Source
Unified Theory of Acceptance and Use of Technology (UTAUT)	A framework explaining technology adoption based on performance expectancy, effort expectancy, social influence, and facilitating conditions.	Used to analyze organizational adoption of AI-driven decision-making and business intelligence tools, identifying barriers and facilitators.	Applied to assess the adoption of FemTech tools for pelvic floor muscle training, understanding women's willingness to integrate AI-based solutions into their health routines.	Bhatt et al. (2024)
Technology-Organization-Environment (TOE) Framework	A model explaining technology adoption within firms by examining technological, organizational, and environmental contexts.	Helps businesses evaluate internal capabilities, external market pressures, and technological advancements to integrate AI into strategic management.	Explains how healthcare institutions and individuals assess the technological feasibility, organizational readiness, and regulatory environment of FemTech tools.	Maaze et al. (2024)
Diffusion of Innovations Theory (DOI)	Describes how new technologies spread through society, emphasizing relative advantage, compatibility, complexity, trialability, and observability.	Used to understand how AI-powered business intelligence tools gain traction within industries and how firms move through adoption stages.	Analyzes how FemTech solutions for pelvic floor training gain acceptance, focusing on consumer perception of benefits and usability.	Verma et al. (2024)
Human-Computer Interaction (HCI) Theory	Examines the design and use of technology interfaces, emphasizing usability, accessibility, and user experience.	Guides the development of AI-powered business intelligence dashboards and automation tools to ensure seamless user interaction.	Ensures that AI-enhanced FemTech applications provide intuitive and user-friendly interfaces to promote engagement and adherence.	Kumar et al. (2023a)
Resource-Based View (RBV) Theory	Suggests that competitive advantage stems from unique resources and capabilities, including technological assets.	Helps organizations identify AI as a strategic resource that enhances business intelligence, data analytics, and competitive advantage.	Examines how FemTech companies can leverage AI-driven data analytics to differentiate their solutions and gain a competitive edge in healthcare technology.	Rajora et al. (2024)

(*Continued*)

TABLE 11.1 (*Continued*)
Theoretical Framework

Theoretical Construct	Definition	Application in AI-Driven Business Management	Application in FemTech Adoption	Source
Self-Determination Theory (SDT)	Focuses on intrinsic and extrinsic motivation in technology adoption, highlighting autonomy, competence, and relatedness.	Explains how AI-driven automation impacts employee motivation, balancing efficiency with human decision-making.	Assesses how motivational factors influence women's continued use of AI-powered pelvic floor training tools, ensuring engagement and long-term adherence.	Goyal et al. (2024)
Actor-Network Theory (ANT)	Analyzes the relationships between technology, human actors, and institutions in shaping adoption and implementation.	Explores the role of AI-driven decision-making within corporate structures, highlighting interdependencies between human managers and AI systems.	Examines how different stakeholders (users, healthcare providers, developers) influence FemTech adoption and technological evolution.	Zaheer et al. (2024)
Cognitive Load Theory (CLT)	Examines how cognitive resources are allocated during technology interaction, emphasizing ease of use and learning efficiency.	Optimizes AI-driven business intelligence tools by minimizing cognitive overload and ensuring effective information processing.	Improves the design of FemTech applications to ensure women can easily understand and use AI-based pelvic floor training tools without mental strain.	Bachheti et al. (2023)
Institutional Theory	Explains how regulatory, normative, and cognitive factors influence technology adoption.	Helps businesses understand compliance, industry standards, and social expectations regarding AI integration.	Examines the role of medical regulations, cultural norms, and healthcare guidelines in FemTech adoption.	Singh et al. (2024)
Theory of Planned Behavior (TPB)	Describes how behavioral intentions and attitudes shape the adoption of new technologies.	Predicts business leaders' willingness to integrate AI into decision-making based on perceived usefulness and ease of implementation.	Explains women's behavioral intentions in adopting FemTech tools, considering perceived benefits, ease of use, and external influences.	Nainwal et al. (2022)

healthcare policies, and AI governance frameworks will be examined to assess their influence on business and FemTech applications.

The target population includes business executives, AI professionals, FemTech users, and healthcare practitioners. The research employs a combination of purposive sampling for qualitative insights and stratified random sampling for quantitative analysis, ensuring representation across different demographics and industries. The sample size consists of at least 300 respondents for surveys and 20 in-depth interviews with key stakeholders, ensuring a comprehensive dataset for analysis.

The study applies both quantitative and qualitative analytical techniques. Quantitative analysis includes descriptive statistics to summarize AI adoption and FemTech usage trends, inferential statistical techniques such as regression analysis and correlation studies to examine the impact of AI-driven decision-making, and predictive modeling to analyze market trends and consumer behavior. Qualitative analysis involves thematic analysis of interviews to identify recurring themes related to AI implementation challenges, managerial perspectives, and FemTech usability concerns. Additionally, content analysis of case studies will extract strategic insights into AI-driven business intelligence and digital health adoption.

Ethical considerations are paramount in this study. Informed consent will be obtained from all participants involved in primary data collection, and data confidentiality will be strictly maintained in compliance with data protection regulations. Bias mitigation strategies, including the triangulation of multiple data sources, will be employed to ensure the research's validity and reliability. Measures will be taken to avoid conflicts of interest and ensure that findings remain objective and representative of actual industry trends.

Despite its rigorous approach, the study acknowledges certain limitations. Potential biases in self-reported survey responses may influence findings, and the availability of longitudinal data on AI and FemTech adoption trends may be restricted. Additionally, gaining access to proprietary AI algorithms and business intelligence models could be challenging due to confidentiality concerns. However, these limitations will be mitigated through a multi-source approach to data collection and analysis, ensuring that the research remains robust and its conclusions actionable.

This methodology ensures a rigorous, data-driven approach to understanding AI's transformative role in business management and FemTech adoption, providing actionable insights for industry practitioners and policymakers. By integrating empirical analysis, theoretical inquiry, and ethical research practices, this study aims to offer a comprehensive understanding of AI's potential to reshape business intelligence and healthcare technology.

11.4 DATA ANALYSIS

See Tables 11.2–11.4.

11.5 MANAGERIAL IMPLICATIONS

The integration of AI and data-driven decision-making into business management has profound implications for strategic operations, organizational efficiency, and market positioning. AI-driven business intelligence tools provide enterprises with

TABLE 11.2
Recent Issues in Pelvic Disorder

Factor	Statistical Data	Relevance to Pelvic Floor Muscle Training	Source
Prevalence of Pelvic Floor Disorders (PFDs) in India	34% of women experience at least one form of PFD, including urinary incontinence and pelvic organ prolapse.	High prevalence necessitates widespread awareness and access to training tools.	Joshi et al. (2024)
Urinary Incontinence Among Indian Women	21.8% of Indian women suffer from urinary incontinence, with prevalence increasing to 40% among postmenopausal women.	Strengthening pelvic floor muscles through training significantly reduces incontinence risk.	Semwal et al. (2024)
Pelvic Organ Prolapse (POP) Cases	Affects 15-20% of women over 50 years, with 3-5% requiring surgical intervention.	Non-surgical management, such as pelvic floor muscle training, can delay or prevent severe cases.	Dhinakaran et al. (2022)
Obstetric Trauma as a Risk Factor	65% of vaginal deliveries result in some level of perineal trauma, contributing to weakened pelvic floor muscles.	Training programs help postpartum recovery, improving muscle strength and reducing postpartum complications.	Bohra et al. (2024)
C-Section vs. Vaginal Delivery Impact	India's C-section rate is 21.5%, but vaginal births still account for 78.5% of deliveries, increasing the risk of pelvic floor dysfunction.	Training is essential for postpartum rehabilitation regardless of delivery type.	Sahithya et al. (2024)
Menopausal Women at Risk	51 million Indian women are postmenopausal, with estrogen decline leading to pelvic muscle atrophy.	Regular pelvic floor exercises mitigate the impact of hormonal changes on muscle tone.	Prasad et al. (2023)
Awareness and Access to Training	Only 12% of Indian women are aware of pelvic floor muscle training techniques.	Increasing awareness through FemTech solutions can enhance preventive care.	Mittal et al. (2024)
Impact of Obesity on Pelvic Health	30.7% of urban Indian women are overweight or obese, which increases intra-abdominal pressure and pelvic dysfunction risk.	Weight management combined with pelvic floor exercises can prevent complications.	Shabbiruddin et al. (2023)
Economic Burden of Pelvic Floor Disorders	The estimated cost of surgical interventions for PFDs in India exceeds ₹ 2,500 crore annually.	Preventive training programs can significantly reduce healthcare costs and improve quality of life.	Pandya et al. (2023)
Effectiveness of Pelvic Floor Muscle Training (PFMT)	Studies indicate that PFMT reduces urinary incontinence symptoms by 60-70% and improves quality of life by 80%.	Strengthening exercises provide non-invasive, cost-effective, and long-term solutions.	Kumar et al. (2023f)

TABLE 11.3
Relevance of FemTech

Factor	Latest Statistics and Current Affairs (2024-2025)	Relevance to FemTech in India	Source
Rising Women's Health Issues	67% of Indian women suffer from reproductive health issues, with PCOS affecting 1 in 5 women.	FemTech solutions can provide digital tracking, AI-based diagnostics, and personalized treatment plans.	Joshi and Sharma (2022)
Maternal Mortality Rate (MMR) Concerns	India's MMR stands at 97 per 100,000 live births, with rural areas experiencing higher fatalities.	AI-powered FemTech apps can offer prenatal monitoring, virtual consultations, and risk assessments.	Sharahiley and Kandpal (2023)
Menstrual Hygiene and Awareness	23% of girls drop out of school due to lack of menstrual hygiene facilities, and only 36% of women use sanitary products.	FemTech can promote awareness, provide digital education, and facilitate product accessibility.	Raman et al. (2023)
Government Initiatives in Digital Health	The Ayushman Bharat Digital Mission (ABDM) aims to digitize health records, and NITI Aayog has emphasized AI in healthcare.	FemTech startups can integrate with government initiatives for digital health tracking and AI-based diagnostics.	Vekariya et al. (2023)
Increasing Infertility Rates	Infertility affects 27.5 million couples in India, with demand for fertility treatments rising by 20% annually.	AI-driven FemTech fertility apps can provide ovulation tracking, hormonal monitoring, and telehealth consultations.	Ramesh et al. (2023)
Women's Digital Adoption Surge	76% of urban women and 45% of rural women now use smartphones, boosting digital healthcare accessibility.	FemTech mobile applications can bridge the gap in healthcare access through telemedicine and digital solutions.	Kukreti et al. (2023)
AI in Women's Health Research	India has seen a 30% rise in AI-driven research projects on women's health since 2023.	AI-powered FemTech solutions can enhance diagnostics, early disease detection, and personalized treatment.	Sharma and Bhadula (2023)
Rising Demand for Pelvic Health Solutions	34% of Indian women experience pelvic floor disorders, with postpartum-related issues being highly prevalent.	FemTech tools for pelvic muscle training can improve women's health outcomes through digital physiotherapy and biofeedback mechanisms.	Sharun et al. (2022) and Uniyal et al. (2022)
FemTech Startup Growth	The Indian FemTech industry is expected to grow at a CAGR of 15% and reach $1.3 billion by 2027.	Investment in FemTech can drive innovation in digital reproductive health, fertility, and wellness solutions.	Badhotiya et al. (2022) and Husen et al. (2021)
Funding and Investment Trends	Over $250 million has been invested in Indian FemTech startups since 2022, with growing interest from global investors.	Increased funding will lead to advanced AI-powered solutions in women's healthcare.	Kumar et al. (2023b)

TABLE 11.4
Relevance to AI, Business Management, and FemTech

Category	Statistical Value	Relevance to AI, Business Management, and FemTech	Source
Global AI Market Growth	$1.8 trillion projected market size by 2030	AI-driven decision-making is becoming integral to business management, improving efficiency and strategic planning.	Gupta et al. (2022)
AI Adoption in Business Intelligence	80% of enterprises use AI-powered analytics	Businesses are leveraging AI for data-driven insights, enhancing productivity and profitability.	Komkowski et al. (2023)
FemTech Market Valuation	Expected to reach $75 billion by 2025	AI-powered health solutions for women are gaining traction, revolutionizing personalized healthcare.	Saxena et al. (2022)
Prevalence of Pelvic Floor Disorders	34% of Indian women suffer from pelvic floor dysfunction	AI-driven FemTech solutions can provide personalized muscle training and improve health outcomes.	Jindal et al. (2022)
Digital Health Market in India	Valued at $16 billion in 2024, growing at 27% CAGR	AI-driven healthcare solutions, including FemTech, are accelerating in adoption due to digital health advancements.	Dhiman and Nagar (2022)
Women's Health Awareness in India	Only 12% of Indian women are aware of pelvic floor exercises	AI-powered FemTech tools can bridge the gap through digital education and engagement.	Singh et al. (2023)
AI in Predictive Analytics for Business	AI improves decision-making accuracy by 90%	AI-driven business intelligence enhances predictive capabilities, optimizing market strategies.	Singh et al. (2022)
Economic Burden of Pelvic Floor Disorders	Annual cost exceeds ₹ 2,500 crore in India	AI-driven preventive solutions can reduce economic strain on healthcare by promoting non-surgical interventions.	Sharma et al. (2022b)
Wearable and Smart FemTech Devices	Market expected to grow at 18% CAGR globally	AI-integrated wearables enhance pelvic muscle training through biofeedback and real-time analytics.	Juyal and Sharma (2020)
User Retention in AI-Driven Health Apps	75% retention rate for AI-powered health tracking apps	FemTech leveraging AI can improve adherence to pelvic muscle training regimens through personalized feedback.	Josphineleela et al. (2023)

an unparalleled ability to analyze vast amounts of structured and unstructured data, facilitating predictive analytics, automated decision-making, and enhanced operational agility (Pandey et al., 2020; Raja et al., 2022). The transition from intuition-based management to algorithmic precision is not merely an enhancement but a necessity in the contemporary business landscape. Organizations that fail to leverage AI for strategic decision-making risk obsolescence as competitors increasingly capitalize on real-time analytics, cognitive computing, and machine learning to optimize supply chains, customer engagement, and financial forecasting (Chandna et al., 2022; Kumar et al., 2022; Sharma et al., 2023).

From a managerial perspective, the adoption of AI necessitates a paradigm shift in leadership styles, skill sets, and corporate governance. Executives must cultivate a data-centric culture, ensuring that decision-making processes are grounded in empirical evidence rather than subjective judgment. This requires investment in AI literacy programs, upskilling employees to interact effectively with AI-driven insights, and fostering an organizational ecosystem that prioritizes data integrity, ethical AI deployment, and continuous technological adaptation (Chauhan et al., 2022; MITRA et al., 2022). Moreover, AI-driven decision-making demands an agile organizational structure capable of responding to rapidly evolving data streams and market conditions. Traditional hierarchical models may prove inadequate, necessitating flatter, more dynamic structures where cross-functional teams collaborate to harness AI-generated intelligence effectively (Jindal et al., 2023; Raju et al., 2022).

The managerial implications extend into the realm of FemTech, particularly in the adoption and scaling of AI-powered pelvic floor muscle training tools. The increasing prevalence of pelvic floor disorders among women, coupled with rising digital health adoption, presents a significant market opportunity for businesses operating in the FemTech sector (Singh et al., 2023b). However, widespread adoption of these solutions requires targeted strategies that address barriers such as technological skepticism, limited awareness, and accessibility challenges. Business leaders must focus on consumer education, leveraging AI-powered engagement strategies to enhance user adoption and retention. Personalized health interventions, real-time feedback loops, and gamified training modules can significantly improve adherence rates, ensuring that AI-powered FemTech solutions translate into tangible health outcomes (Goswami & Sharma, 2022; Uniyal et al., 2020).

Furthermore, AI's role in healthcare and business intelligence must be governed by ethical considerations and regulatory compliance. Data privacy concerns, algorithmic biases, and transparency in AI decision-making are critical managerial challenges that require proactive governance (Sharma et al., 2023b; Singh, et al., 2023b). Organizations developing AI-powered FemTech solutions must integrate robust data protection mechanisms, ensuring compliance with national and international health regulations while maintaining user trust. Ethical AI implementation, particularly in women's health technology, necessitates collaboration with medical professionals, policymakers, and technologists to establish frameworks that balance innovation with ethical responsibility (Bhargava et al., 2022; Rawat et al., 2023).

From an economic standpoint, AI-driven business intelligence and FemTech innovations present significant cost-saving opportunities. In business management, predictive analytics optimize resource allocation, minimize operational inefficiencies, and enhance profitability (Akana et al., 2023; Sathyapriya et al., 2022). In FemTech, AI-powered pelvic floor training tools reduce healthcare costs by mitigating the need for invasive treatments, hospitalizations, and long-term medication dependencies. Managers must therefore advocate for AI-driven investments as a long-term strategy for cost reduction and sustainable business growth, ensuring that AI initiatives align with broader corporate objectives and industry-specific demands (Gopal et al., 2023; Sharma et al., 2023a).

Additionally, the success of AI-powered business intelligence and FemTech applications hinges on strategic partnerships and ecosystem collaboration. Businesses must establish alliances with AI research institutions, healthcare providers, and regulatory bodies to accelerate innovation and ensure market penetration (Hajoary et al., 2023). Collaborative ecosystems, wherein AI developers, medical practitioners, and business leaders co-create solutions, foster innovation while addressing real-world challenges. Managers must adopt an ecosystem mindset, recognizing that AI's potential is maximized when integrated with complementary technologies, expert insights, and user-centric design principles (Bansal et al., 2023; Diddi et al., 2022; Kamra et al., 2023).

Ultimately, the managerial implications of AI-driven business intelligence and FemTech adoption underscore the necessity of a holistic, interdisciplinary approach to innovation. Businesses must balance technological advancement with ethical governance, consumer engagement with regulatory compliance, and financial investments with long-term sustainability (Bisht et al., 2023; Eswar et al., 2023; Singamaneni et al., 2022). AI's transformative impact on business management and women's health technology is undeniable, but its successful implementation requires foresight, adaptability, and strategic leadership. As AI continues to redefine industry paradigms, managers who proactively embrace data-driven strategies, foster AI literacy, and prioritize ethical deployment will be best positioned to drive organizational success in the evolving digital economy (Godbole et al., 2023; Sati et al., 2022; Sharma et al., 2022d).

11.6 CONCLUSION

The integration of AI and data-driven decision-making in business management has ushered in a transformative era, redefining how organizations operate, strategize, and innovate. AI-driven business intelligence has become an indispensable tool for enterprises seeking to optimize efficiency, predict market trends, and enhance decision-making precision. As businesses increasingly rely on machine learning, deep learning, and cognitive computing, the shift from intuition-based strategies to algorithmic rigor has become not just a competitive advantage but a necessity for survival in dynamic market landscapes.

Simultaneously, the rise of FemTech, particularly in the domain of pelvic floor muscle training tools for women, exemplifies the profound impact of AI in personalized healthcare. With a significant proportion of Indian women experiencing pelvic health disorders, AI-powered FemTech solutions offer an innovative approach to rehabilitation, preventive care, and digital health engagement. The intersection of AI and FemTech demonstrates how data analytics, predictive modeling, and biofeedback mechanisms can enhance user adoption, treatment adherence, and long-term health outcomes. The UTAUT provides valuable insights into the behavioral patterns influencing technology adoption, both in business intelligence and FemTech, underscoring the importance of strategic interventions in overcoming resistance and fostering widespread adoption.

From a managerial perspective, the implementation of AI in business intelligence and FemTech requires a nuanced understanding of ethical considerations, regulatory compliance, and consumer engagement. The success of AI-powered decision-making is contingent upon fostering a data-centric culture within organizations, promoting AI literacy among employees, and ensuring transparent, accountable AI systems. In FemTech, ensuring data privacy, algorithmic fairness, and accessibility is paramount to gaining consumer trust and regulatory approval. Strategic collaborations between AI developers, healthcare professionals, and business leaders will be instrumental in driving innovation, scaling adoption, and addressing critical gaps in women's health technology.

The economic implications of AI integration extend beyond efficiency gains, influencing cost reduction, resource allocation, and market sustainability. Predictive analytics in business intelligence enables organizations to anticipate disruptions, optimize supply chains, and personalize customer interactions, while AI-driven FemTech tools offer cost-effective alternatives to surgical and pharmacological interventions. As the global AI market continues its exponential growth, enterprises that harness AI-driven decision-making and leverage FemTech's potential will be well-positioned to lead in the digital economy.

Ultimately, the convergence of AI, business intelligence, and FemTech marks a pivotal shift in industry paradigms, emphasizing the role of technology in fostering efficiency, inclusivity, and innovation. Businesses must approach AI adoption with strategic foresight, ethical responsibility, and a commitment to data-driven excellence. The future of AI-powered decision-making lies in its ability to empower organizations, optimize healthcare, and drive technological progress, ensuring that enterprises remain agile, competitive, and future-ready in an era of digital transformation.

REFERENCES

Agarwal, P., & Sharma, S. (2023). Smart urban traffic management system using energy efficient optimized path discovery. *Proceedings of the 3rd International Conference on Artificial Intelligence and Smart Energy, ICAIS 2023*, 858–863. https://doi.org/10.1109/ICAIS56108.2023.10073840

Akana, C. M. V. S., Kumar, A., Tiwari, M., Yunus, A. Z., Vijayakumar, E., & Singh, M. (2023). An optimized DDOS attack detection using deep convolutional generative adversarial networks. *Proceedings of the 5th International Conference on Inventive Research in Computing Applications, ICIRCA 2023*, 668–673. https://doi.org/10.1109/ICIRCA57980.2023.10220745

Alzaidi, M. S., Subbalakshmi, C., Roshini, T. V, Shukla, P. K., Shukla, S. K., Dutta, P., & Alhassan, M. (2022). 5G-telecommunication allocation network using IoT enabled improved machine learning technique. *Wireless Communications and Mobile Computing, 2022*. https://doi.org/10.1155/2022/6229356

Anand, A., Singh, M. P., Sah, S., Verma, A., Khatib, M. N., Pant, M., Zahiruddin, Q. S., & Rustagi, S. (2024). H5N1 avian influenza in USA: a call for vigilance in one health surveillance. *Clinical Infection in Practice, 23*. https://doi.org/10.1016/j.clinpr.2024.100369

Arora, S., Pargaien, S., Khan, F., Misra, A., Gambhir, A., & Verma, D. (2023). Smart parking allocation using raspberry Pi based IoT system. *Proceedings of the 5th International Conference on Inventive Research in Computing Applications, ICIRCA 2023*, 1457–1461. https://doi.org/10.1109/ICIRCA57980.2023.10220619

Awal, G., & Khanna, R. (2019). Determinants of millennial online consumer behavior and prospective purchase decisions. *International Journal of Advanced Science and Technology, 28*(18), 366–378.

Bachheti, A., Deepti, Bachheti, R. K., Singh, A., Zebeaman, M., Hunde, Y., & Husen, A. (2023). Bioactive constituents and health promoting compounds of underutilized fruits of the northern Himalayas of India: a review. *Food Production, Processing and Nutrition, 5*(1). https://doi.org/10.1186/s43014-023-00140-5

Badhotiya, G. K., Avikal, S., Soni, G., & Sengar, N. (2022). Analyzing barriers for the adoption of circular economy in the manufacturing sector. *International Journal of Productivity and Performance Management, 71*(3), 912–931. https://doi.org/10.1108/IJPPM-01-2021-0021

Bansal, S., Kumar, V., Kumari, A., & Kuzmin, E. (2023). Understanding the role of digital technologies in supply chain management of SMEs. In V. Kumar, G. L. Kyriakopoulos, V. Akberdina, & E. Kuzmin (Eds.), *Lecture Notes in Information Systems and Organisation: Vol. 61 LNISO,* pp. 195–205. Springer Science and Business Media Deutschland GmbH. https://doi.org/10.1007/978-3-031-30351-7_16

Bathla, G., Kumar, S., Garg, H., & Saini, D. (2024). *Artificial Intelligence in Healthcare: Emphasis on Diabetes, Hypertension, and Depression Management.* CRC Press. https://doi.org/10.1201/9781003522096

Bhardwaj, M., Prakash, N., & Malhotra, R. K. (2024). Energy sector and carbon trading in the future: a reference to India. *2024 International Conference on Smart Devices, ICSD 2024*. https://doi.org/10.1109/ICSD60021.2024.10751388

Bhargava, A., Bhargava, D., Kumar, P. N., Sajja, G. S., & Ray, S. (2022). Industrial IoT and AI implementation in vehicular logistics and supply chain management for vehicle mediated transportation systems. *International Journal of System Assurance Engineering and Management, 13*, 673–680. https://doi.org/10.1007/s13198-021-01581-2

Bhatt, S., Dani, R., & Singh, A. K. (2024). Exploring cutting-edge approaches to sustainable tourism infrastructure and design a case studies of regenerative accommodation and facilities. In R. Rajendra, S. Azad, T. Sanjay, O. Ercan, & D. J. Paulo (Eds.), *Dimensions of Regenerative Practices in Tourism and Hospitality*. IGI Global. https://doi.org/10.4018/979-8-3693-4042-4.ch003

Bhatti, K. K., & Negi, A. (2018). Green store attributes and infrastructure: a connecting link between retailers and consumers to bring sustainability. *International Journal of Civil Engineering and Technology, 9*(11), 190–198.

Bhuvaneswari, C. A., Garg, N., Supreeth, B. R., Patjoshi, P. K., Ushapriya, C., & Tiwari, M. (2023). Optimistic boosting internet connection throughput by incorporating blockchain filters into network infrastructure. *Proceedings of the 2023 International Conference on Innovative Computing, Intelligent Communication and Smart Electrical Systems, ICSES* 2023. https://doi.org/10.1109/ICSES60034.2023.10465436

Bisht, B., Gururani, P., Aman, J., Vlaskin, M. S., Anna I, K., Irina A, A., Joshi, S., Kumar, S., & Kumar, V. (2023). A review on holistic approaches for fruits and vegetables biowastes valorization. *Materials Today: Proceedings, 73*, 54–63. https://doi.org/10.1016/j.matpr.2022.09.168

Bisht, R. K., Bisht, I. P., & Joshi, B. C. (2022). Growth of micro, small and medium enterprises (MSMES) in Uttarakhand (India). In D. Nair, & C. Andrews (Eds.), *AIP Conference Proceedings* (Vol. 2481). American Institute of Physics Inc. https://doi.org/10.1063/5.0103881

Bohra, N. S., Bansal, S. S., Johri, A., & Kathari, S. (2024). Evaluating infrastructure fund performance in India: a study of thematic investing. In R. Rajendra, S. Azad, T. Sanjay, O. Ercan, & D. J. Paulo (Eds.), *Issues of Sustainability in AI and New-Age Thematic Investing*. IGI Global. https://doi.org/10.4018/979-8-3693-3282-5.ch002

Cavaliere, L. P. L., Byloppilly, R., Khan, S. D., Othman, B. A., Muda, I., & Malhotra, R. K. (2024). Acceptance and effectiveness of industry 4.0 internal and external organisational initiatives in Malaysian firms. *International Journal of Management and Enterprise Development, 23*(1), 1–25. https://doi.org/10.1504/IJMED.2024.138422

Chandna, R., Saini, S., & Kumar, S. (2022). Selecting the most agile manufacturing system with respect to agile attribute-technology: fuzzy AHP approach. In D. Nair, & C. Andrews (Eds.), *AIP Conference Proceedings* (Vol. 2481). American Institute of Physics Inc. https://doi.org/10.1063/5.0103804

Chauhan, A., Sharma, N. K., Tayal, S., Kumar, V., & Kumar, M. (2022). A sustainable production model for waste management with uncertain scrap and recycled material. *Journal of Material Cycles and Waste Management, 24*(5), 1797–1817. https://doi.org/10.1007/s10163-022-01435-4

Dhiman, G., & Nagar, A. K. (2022). Editorial: blockchain-based 6G and industrial internet of things systems for industry 4.0/5.0. *Expert Systems, 39*(10). https://doi.org/10.1111/exsy.13162

Dhinakaran, M., Priya, P. K., Alanya-Beltran, J., Gandhi, V., Jaiswal, S., & Singh, D. P. (2022). An innovative internet of things (IoT) computing-based health monitoring system with the aid of machine learning approach. *Proceedings of 5th International Conference on Contemporary Computing and Informatics, IC3I 2022*, 292–297. https://doi.org/10.1109/IC3I56241.2022.10072528

Diddi, P. K., Sharma, P. K., Srivastava, A., Madduru, S. R. C., & Reddy, E. S. (2022). Sustainable fast setting early strength self compacting concrete(FSESSCC) using metakaolin. *IOP Conference Series: Earth and Environmental Science, 1077*(1). https://doi.org/10.1088/1755-1315/1077/1/012009

Eswar, K. N. D. V. S., Doss, M. A. N., Vishnuram, P., Selim, A., Bajaj, M., Kotb, H., & Kamel, S. (2023). Comprehensive study on reduced dc source count: multilevel inverters and its design topologies. *Energies, 16*(1). https://doi.org/10.3390/en16010018

Godbole, V., Kukrety, S., Gautam, P., Bisht, M., & Pal, M. K. (2023). Bioleaching for heavy metal extraction from e-waste: a sustainable approach. In A. Kumar, P. Gupta, & S. K. Sharma (Eds.), *Microbial Technology for Sustainable E-waste Management*. Springer International Publishing. https://doi.org/10.1007/978-3-031-25678-3_4

Gonfa, Y. H., Tessema, F. B., Tadesse, M. G., Bachheti, A., & Bachheti, R. K. (2023). Medicinally important plant roots and their role in nanoparticles synthesis and applications. In M. M. A. Abdel-Daim, A. B. Wali, & R. K. Bachheti (Eds.), *Secondary Metabolites from Medicinal Plants: Nanoparticles Synthesis and their Applications*. CRC Press. https://doi.org/10.1201/9781003213727-11

Gopal, S., Gupta, P., & Minocha, A. (2023). Advancements in fin-tech and security challenges of banking industry. *4th International Conference on Intelligent Engineering and Management, ICIEM 2023*. https://doi.org/10.1109/ICIEM59379.2023.10165876

Goswami, S., & Sharma, S. (2022). Industry 4.0 enabled molecular imaging using artificial intelligence technique. *2022 1st International Conference on Computational Science and Technology, ICCST 2022 – Proceedings*, 455–460. https://doi.org/10.1109/ICCST55948.2022.10040406

Goyal, S., Kaur, A., Sharma, V., & Batra, N. (2024). IMHS: an IoT-based machine learning healthcare system. *2024 2nd International Conference on Disruptive Technologies, ICDT 2024*, 916–919. https://doi.org/10.1109/ICDT61202.2024.10489127

Gupta, A., Dixit, A. K., Kumar, K. S., Lavanya, C., Chakravarthi, M. K., & Gangodkar, D. (2022a). Analyzing robotics and computer integrated manufacturing of key areas using cloud computing. *Proceedings of 5th International Conference on Contemporary Computing and Informatics, IC3I 2022*, 194–199. https://doi.org/10.1109/IC3I56241.2022.10072581

Gupta, C., Jindal, P., & Malhotra, R. K. (2022b). A study of increasing adoption trends of digital technologies – an evidence from Indian banking. In D. Nair, & C. Andrews (Eds.), *AIP Conference Proceedings* (Vol. 2481). American Institute of Physics Inc. https://doi.org/10.1063/5.0104572

Gupta, P., Gopal, S., Sharma, M., Joshi, S., Sahani, C., & Ahalawat, K. (2023). Agriculture informatics and communication: paradigm of e-governance and drone technology for crop monitoring. *9th International Conference on Smart Computing and Communications: Intelligent Technologies and Applications, ICSCC 2023*, 113–118. https://doi.org/10.1109/ICSCC59169.2023.10335058

Hajoary, P. K., Balachandra, P., & Garza-Reyes, J. A. (2023). Industry 4.0 maturity and readiness assessment: an empirical validation using confirmatory composite analysis. *Production Planning and Control*. https://doi.org/10.1080/09537287.2023.2210545

Husen, A., Bachheti, R. K., & Bachheti, A. (2021). *Non-Timber Forest Products: Food, Healthcare and Industrial Applications*. Springer International Publishing. https://doi.org/10.1007/978-3-030-73077-2

Jindal, G., Tiwari, V., Mahomad, R., Gehlot, A., Jindal, M., & Bordoloi, D. (2023). Predictive design for quality assessment employing cloud computing and machine learning. *2023 3rd International Conference on Advance Computing and Innovative Technologies in Engineering, ICACITE 2023*, 461–465. https://doi.org/10.1109/ICACITE57410.2023.10182915

Jindal, T., Sheoliha, N., Kishore, K., Uike, D., Khurana, S., & Verma, D. (2022). A conceptual analysis on the impact of internet of things (IoT) towards on digital marketing transformation. *2022 2nd International Conference on Advance Computing and Innovative Technologies in Engineering, ICACITE 2022*, 1943–1947. https://doi.org/10.1109/ICACITE53722.2022.9823714

Joshi, K., Patil, S., Gupta, S., & Khanna, R. (2022). Role of Pranayma in emotional maturity for improving health. *Journal of Medical Pharmaceutical and Allied Sciences*, *11*(2), 4569–4573. https://doi.org/10.55522/jmpas.V11I2.2033

Joshi, K., Sharma, R., Singh, N., & Sharma, B. (2023a). Digital world of cloud computing and wireless networking: challenges and risks. In R. Rajendra, S. Azad, T. Sanjay, O. Ercan, & D. J. Paulo (Eds.), *Applications of Artificial Intelligence in Wireless Communication Systems*. IGI Global. https://doi.org/10.4018/978-1-6684-7348-1.ch003

Joshi, S., & Sharma, M. (2022). Sustainable performance through digital supply chains in industry 4.0 era: amidst the pandemic experience. *Sustainability (Switzerland)*, *14*(24). https://doi.org/10.3390/su142416726

Joshi, S., & Sharma, M. (2023). Strategic challenges of deploying LARG approach for sustainable manufacturing: research implications from Indian SMEs. *International Journal of Internet Manufacturing and Services*, *9*(2–3), 373–397. https://doi.org/10.1504/IJIMS.2023.132791

Joshi, S., Bisht, B., Kumar, V., Singh, N., Jameel Pasha, S. B., Singh, N., & Kumar, S. (2024). Artificial intelligence assisted food science and nutrition perspective for smart nutrition research and healthcare. *Systems Microbiology and Biomanufacturing*, *4*(1), 86–101. https://doi.org/10.1007/s43393-023-00200-4

Joshi, S., Gangola, S., Bhandari, G., Bhandari, N. S., Nainwal, D., Rani, A., Malik, S., & Slama, P. (2023b). Rhizospheric bacteria: the key to sustainable heavy metal detoxification strategies. *Frontiers in Microbiology*, *14*. https://doi.org/10.3389/fmicb.2023.1229828

Joshi, S., Sharma, M., & Barve, A. (2023c). Implementation challenges of blockchain technology in closed-loop supply chain: a waste electrical and electronic equipment (WEEE) management perspective in developing countries. *Supply Chain Forum, 24*(1), 59–80. https://doi.org/10.1080/16258312.2022.2135972

Josphineleela, R., Siva Reddy, K. V., Reddy, M. V. S. S., & Rawat, R. S. (2023). Design and development of a smart sprinkler device for IoT-integrated plants irrigation. *Proceedings of the 2023 2nd International Conference on Electronics and Renewable Systems, ICEARS 2023*, 498–501. https://doi.org/10.1109/ICEARS56392.2023.10084960

Juyal, P., & Sharma, S. (2020). Estimation of tree volume using mask R-CNN based deep learning. *2020 11th International Conference on Computing, Communication and Networking Technologies, ICCCNT 2020*. https://doi.org/10.1109/ICCCNT49239.2020.9225509

Kaliappan, S., Natrayan, L., & Garg, N. (2023). Checking and supervisory system for calculation of industrial constraints using embedded system. *Proceedings of the 4th International Conference on Smart Electronics and Communication, ICOSEC 2023*, 87–90. https://doi.org/10.1109/ICOSEC58147.2023.10275952

Kamra, J., Mani, A. P., & Tripathi, V. M. (2023). Decarbonization trajectory in cement industry. In P. Šolić, S. Nižetić, J. J. P. C. Rodrigues, D. L.-de-I. G.-de-Artaza, T. Perković, L. Catarinucci, & L. Patrono (Eds.), *2023 8th International Conference on Smart and Sustainable Technologies, SpliTech 2023*. Institute of Electrical and Electronics Engineers Inc. https://doi.org/10.23919/SpliTech58164.2023.10193682

Kanojia, P., Malhotra, R. K., & Uniyal, A. K. (2022a). Impact of organizational commitment components on the teachers of higher education in Uttarakhand: an empirical analysis. *Proceedings - 2022 International Conference on Recent Trends in Microelectronics, Automation, Computing and Communications Systems, ICMACC 2022*, 360–364. https://doi.org/10.1109/ICMACC54824.2022.10093606

Kanojia, P., Malhotra, R. K., & Uniyal, A. K. (2022b). Organizational commitment and the academic staff in HEI's in North West India. *Proceedings - 2022 International Conference on Recent Trends in Microelectronics, Automation, Computing and Communications Systems, ICMACC* 2022, 365–370. https://doi.org/10.1109/ICMACC54824.2022.10093347

Kaur, J., Khanna, R., Kumar, R., & Sunil, G. (2024). Role of blockchain technologies in goods and services tax. *Proceedings - 2024 3rd International Conference on Sentiment Analysis and Deep Learning, ICSADL 2024*, 607–612. https://doi.org/10.1109/ICSADL61749.2024.00104

Khanna, R., Jindal, P., & Noja, G. G. (2023). Blockchain technologies, a catalyst for insurance sector. In K. Sood, S. Grima, G. Sharma, & B. Balusamy (Eds.), *The Application of Emerging Technology and Blockchain in the Insurance Industry*. (pp. 285–300). River Publishers.

Khanna, R., Sharma, S., Czeh, T., & Adhana, K. (2025). Challenges, interventions, and strategies for higher education transformation: vision 2047. *Lecture Notes in Networks and Systems, 1075 LNNS*, 637–655. https://doi.org/10.1007/978-981-97-6106-7_38

Kollipara, V. N. H., Kalakota, S. K., Chamarthi, S., Ramani, S., Malik, P., & Karuppiah, M. (2023). Timestamp based OTP and enhanced RSA key exchange scheme with SIT encryption to secure IoT devices. *Journal of Cyber Security and Mobility, 12*(1), 77–102. https://doi.org/10.13052/jcsm2245-1439.1214

Komkowski, T., Antony, J., Garza-Reyes, J. A., Tortorella, G. L., & Pongboonchai-Empl, T. (2023). A systematic review of the integration of Industry 4.0 with quality-related operational excellence methodologies. *Quality Management Journal, 30*(1), 3–15. https://doi.org/10.1080/10686967.2022.2144783

Kukreti, A., Shriyal, A., Sharma, S., & Bhadula, S. (2023). Internet-of-things enabled smart and portable terrace garden protection shed. *2023 4th IEEE Global Conference for Advancement in Technology, GCAT 2023*. https://doi.org/10.1109/GCAT59970.2023.10353281

Kukreti, M., Sehajpal, S., Tiwari, R., & Sood, K. (2025). *Driving Global Health and Sustainable Development Goals with Smart Technology* (1st ed.). IGI Global Scientific Publishing. https://doi.org/10.4018/979-8-3373-0240-9

Kumar, A., & Ram, M. (2021). *Systems Reliability Engineering: Modeling and Performance Improvement.* De Gruyter. https://doi.org/10.1515/9783110617375

Kumar, I., Rawat, J., Mohd, N., & Husain, S. (2021). Opportunities of artificial intelligence and machine learning in the food industry. *Journal of Food Quality, 2021.* https://doi.org/10.1155/2021/4535567

Kumar, K., Singh, V., Mishra, G., Ravindra Babu, B., Tripathi, N., & Kumar, P. (2022a). Power-efficient secured hardware design of AES algorithm on high performance FPGA. *Proceedings of 5th International Conference on Contemporary Computing and Informatics, IC3I 2022*, 1634–1637. https://doi.org/10.1109/IC3I56241.2022.10073148

Kumar, M. A., Valli Mayil, V., Jeyalakshmi, T. R., Ghanshala, A., & Gupta, A. (2023a). Wireless body area network (WBAN) centric healthcare system for continuous monitoring. *2023 9th International Conference on Advanced Computing and Communication Systems, ICACCS 2023*, 214–219. https://doi.org/10.1109/ICACCS57279.2023.10112922

Kumar, N. M., Islam, S., Podder, A. K., Selim, A., Bajaj, M., & Kamel, S. (2023b). Lifecycle-based feasibility indicators for floating solar photovoltaic plants along with implementable energy enhancement strategies and framework-driven assessment approaches leading to advancements in the simulation tool. *Frontiers in Energy Research, 11.* https://doi.org/10.3389/fenrg.2023.1075384

Kumar, P., Obaidat, M. S., Pandey, P., Wazid, M., Das, A. K., & Singh, D. P. (2023d). Design of a secure machine learning-based malware detection and analysis scheme. In M. S. Obaidat, Z. Ning, K.F. Hsiao, P. Nicopolitidis, & Y. Guo (Eds.), *Proceedings of the 2023 IEEE International Conference on Communications, Computing, Cybersecurity and Informatics, CCCI 2023.* Institute of Electrical and Electronics Engineers Inc. https://doi.org/10.1109/CCCI58712.2023.10290761

Kumar, R., Goel, R., Singh, T., Mohanty, S. M., Gupta, D., Alkhayyat, A., & Khanna, R. (2024). Sustainable finance factors in Indian economy: analysis on policy of climate change and energy sector. *Fluctuation and Noise Letters, 23*(2). https://doi.org/10.1142/S0219477524400042

Kumar, R., Kandpal, B., & Ahmad, V. (2023f). Industrial IoT (IIoT): security threats and countermeasures. *International Conference on Innovative Data Communication Technologies and Application, ICIDCA 2023 - Proceedings*, 829–833. https://doi.org/10.1109/ICIDCA56705.2023.10100145

Kumar, R., Kathuria, S., Malholtra, R. K., Kumar, A., Gehlot, A., & Joshi, K. (2023g). Role of cloud computing in goods and services tax(GST) and future application. *2nd International Conference on Sustainable Computing and Data Communication Systems, ICSCDS 2023 - Proceedings*, 1443–1447. https://doi.org/10.1109/ICSCDS56580.2023.10104597

Kumar, R., & Khanna, R. (2023a). Role of artificial intelligence in digital currency and future applications. *Proceedings of the 2023 2nd International Conference on Augmented Intelligence and Sustainable Systems, ICAISS* 2023, 42–46. https://doi.org/10.1109/ICAISS58487.2023.10250480

Kumar, R., & Khanna, R. (2023b). RPA (robotic process automation) in finance & accounting and future scope. *Proceedings of the 2023 2nd International Conference on Augmented Intelligence and Sustainable Systems, ICAISS* 2023, 1640–1645. https://doi.org/10.1109/ICAISS58487.2023.10250496

Kumar, R., Khannna Malholtra, R., & Grover, C. A. N. (2023h). Review on artificial intelligence role in implementation of goods and services tax(GST) and future scope. *2023 International Conference on Artificial Intelligence and Smart Communication, AISC* 2023, 348–351. https://doi.org/10.1109/AISC56616.2023.10085030

Kumar, R., Lande, A., Kumar, D., Malhotra, R. K., & Sharma, A. (2023i). Technology bridging -in entrepreneurs and consumers in product development. *ISED 2023 - International Conference on Intelligent Systems and Embedded Design*. https://doi.org/10.1109/ISED59382.2023.10444599

Kumar, R., Malhotra, R. K., & Grover, N. (2023k). Data mining in credit scoring and future application. *International Conference on Innovative Data Communication Technologies and Application, ICIDCA 2023 - Proceedings*, 1096–1100. https://doi.org/10.1109/ICIDCA56705.2023.10100032

Kumar, R., Malhotra, R. K., Pandey, S., Gehlot, A., Gautam, I., & Chamola, S. (2023l). Role of artificial intelligence in input tax credit reconciliation. *Proceedings - 2023 3rd International Conference on Pervasive Computing and Social Networking, ICPCSN 2023*, 497–501. https://doi.org/10.1109/ICPCSN58827.2023.00086

Kumar, R., Malholtra, R. K., Singh, R., Kathuria, S., Balyan, R., & Pal, P. (2023j). Artificial intelligence role in electronic invoice under goods and services tax. *2023 International Conference on Computational Intelligence, Communication Technology and Networking, CICTN 2023*, 140–143. https://doi.org/10.1109/CICTN57981.2023.10140870

Kumar, R., Singh, T., Mohanty, S. N., Goel, R., Gupta, D., Alharbi, M., & Khanna, R. (2023m). Study on online payments and e-commerce with SOR model. *International Journal of Retail and Distribution Management*. https://doi.org/10.1108/IJRDM-03-2023-0137

Kumar, V., Gupta, S., & Khanna, R. (2021). Dengue fever-a worldwide study. *Journal of Medical Pharmaceutical and Allied Sciences, 10*, 102–108. https://doi.org/10.22270/jmpas.vic2i1.2014

Kumar, V., & Korovin, G. (2023). A comparison of digital transformation of industry in the Russian federation with the European Union. In V. Kumar, G. L. Kyriakopoulos, V. Akberdina, & E. Kuzmin (Eds.), *Lecture Notes in Information Systems and Organisation, Vol. 61 LNISO* (pp. 45–57). Springer Science and Business Media Deutschland GmbH. https://doi.org/10.1007/978-3-031-30351-7_5

Kumar, V., Pant, B., Elkady, G., Kaur, C., Suhashini, J., & Hassen, S. M. (2022b). Examining the role of block chain to secure identity in IoT for industry 4.0. *Proceedings of 5th International Conference on Contemporary Computing and Informatics, IC3I 2022*, 256–259. https://doi.org/10.1109/IC3I56241.2022.10072516

Maaze, M. R., Garg, N., Das, S. K., & Shrivastava, S. (2024). Industrial and C&D waste as precursor material in geopolymers: a state-of-the-art review. *Innovative Infrastructure Solutions, 9*(6). https://doi.org/10.1007/s41062-024-01546-x

Malhotra, R. K., & Gupta, S. (2021). Tele health in the digital era during COVID-19 a case study of Uttarakhand. *Journal of Medical Pharmaceutical and Allied Sciences, 10*, 109–112. https://doi.org/10.22270/jmpas.vic2i1.2016

Malhotra, R. K., Gupta, C., & Jindal, P. (2022). Blockchain and smart contracts for insurance industry. In S. Misra, A. Mukherjee, S. K. Sahay, & S. Obaid (Eds.), *Blockchain Technology in Corporate Governance: Transforming Business and Industries*. https://doi.org/10.1002/9781119865247.ch11

Malhotra, R. K., Ojha, M. K., & Gupta, S. (2021). A study of assessment of knowledge, perception and attitude of using tele health services among college going students of Uttarakhand. *Journal of Medical Pharmaceutical and Allied Sciences, 10*, 113–116. https://doi.org/10.22270/jmpas.vic2i1.2020

Malhotra, R. K., Sharma, C. A. P., Upreti, H., Sharma, S., Prakash, N., & Mittal, S. (2024). Role of psychological contract in employee engagement in higher education: empirical study on Uttarakhand. In P. K. Prabhakar (Ed.), *E3S Web of Conferences* (Vol. 556). EDP Sciences. https://doi.org/10.1051/e3sconf/202455601030

Manjunatha, B. N., Chandan, M., Kottu, S., Rappai, S., Hema, P. K., Singh Rawat, K., & Sarkar, S. (2023). A successful spam detection technique for industrial IoT devices based on machine learning techniques. *Proceedings of the 2nd International Conference on Applied Artificial Intelligence and Computing, ICAAIC 2023*, 363–369. https://doi.org/10.1109/ICAAIC56838.2023.10141275

Medhi, M. K., Ambust, S., Kumar, R., & Das, A. J. (2023). Characterization and purification of biosurfactants. In *Advancements in Biosurfactants Research*. Springer International Publishing. https://doi.org/10.1007/978-3-031-21682-4_4

Mehta, K., Sharma, S., & Mishra, D. (2021). Internet-of-things enabled forest fire detection system. *Proceedings of the 5th International Conference on I-SMAC (IoT in Social, Mobile, Analytics and Cloud), I-SMAC* 2021, 20–23. https://doi.org/10.1109/I-SMAC52330.2021.9640900

Mitra, D., Mondal, R., Khoshru, B., Senapati, A., Radha, T. K., Mahakur, B., Uniyal, N., Myo, E. M., Boutaj, H., Sierra, B. E. G., Panneerselvam, P., Ganeshamurthy, A. N., Elković, S. A., Vasić, T., Rani, A., Dutta, S., & Mohapatra, P. K. D. A. S. (2022). Actinobacteria-enhanced plant growth, nutrient acquisition, and crop protection: advances in soil, plant, and microbial multifactorial interactions. *Pedosphere*, *32*(1), 149–170. Https://doi.org/10.1016/s1002-0160(21)60042-5

Mittal, A., Pandey, H., Arsh, M., Soni, M., Goyal, H. R., & Sharma, S. (2024). Advancements and applications of artificial intelligence in various sectors: from internet of vehicles to 6g and healthcare. *Proceedings - 2024 4th International Conference on Pervasive Computing and Social Networking, ICPCSN 2024*, 107–112. https://doi.org/10.1109/ICPCSN62568.2024.00026

Nainwal, P., Lall, S., Bhatt, J., Dobhal, N., & Nawaz, A. (2022). Public behavioural design for corona health care professionals during covid-19. *Journal of Medical Pharmaceutical and Allied Sciences, Int. Confe* (2), 232–234. https://doi.org/10.22270/jmpas.VIC2I2.1835

Neha, Mohanty, S., Alfurhood, B. S., Bakhare, R., Poongavanam, S., & Khanna, R. (2023). The role and impact of artificial intelligence on retail business and its developments. *2023 International Conference on Artificial Intelligence and Smart Communication, AISC 2023*, 1098–1101. https://doi.org/10.1109/AISC56616.2023.10085624

Nethravathi, K., Tiwari, A., Uike, D., Jaiswal, R., & Pant, K. (2022). Applications of artificial intelligence and blockchain technology in improved supply chain financial risk management. *Proceedings of 5th International Conference on Contemporary Computing and Informatics,* IC3I 2022, 242–246. https://doi.org/10.1109/IC3I56241.2022.10072787

Pande, S. D., Bhatt, A., Chamoli, S., Saini, D. K. J. B., Kute, U. T., & Ahammad, S. H. (2023). Design of atmel PLC and its application as automation of coal handling plant. *2023 International Conference on Sustainable Emerging Innovations in Engineering and Technology, ICSEIET 2023*, 178–183. https://doi.org/10.1109/ICSEIET58677.2023.10303627

Pandey, K., Joshi, H., Paliwal, S., Pawar, S., & Kumar, N. (2020). Technology transfer: an overview of process transfer from development to commercialization. *International Journal of Current Research and Review*, *12*(19), 188–192. https://doi.org/10.31782/IJCRR.2020.121913

Pandya, D. J., Kumar, Y., Singh, D. P., Vairavel, D. K., Deepak, A., Rao, A. K., & Rana, A. (2023). Automatic power factor compensation for industrial use to minimize penalty. *Proceedings of International Conference on Contemporary Computing and Informatics, IC3I 2023*, 2499–2504. https://doi.org/10.1109/IC3I59117.2023.10398095

Pant, A., Joshi, R. C., Sharma, S., & Pant, K. (2024). Air quality and public health risk assessment: a case of an industrial area in Haridwar City, Uttarakhand (India). *Indian Journal of Public Health*, *68*(2), 222–226. https://doi.org/10.4103/ijph.ijph_279_23

Pant, R., Gupta, A., Pant, G., Chaubey, K. K., Kumar, G., & Patrick, N. (2022). Second-generation biofuels: facts and future. In D. P. Singh, A. Manchanda, & R. K. Upadhyay (Eds.), *Relationship between Microbes and the Environment for Sustainable Ecosystem Services: Microbial Tools for Sustainable Ecosystem Services: Volume 3* (Vol. 3). Elsevier. https://doi.org/10.1016/B978-0-323-89936-9.00011-4

Patil, S. P., Singh, B., Bisht, J., Gupta, S., & Khanna, R. (2021). Yoga for holistic treatment of polycystic ovarian syndrome. *Journal of Medical Pharmaceutical and Allied Sciences, 10*, 120–125. https://doi.org/10.22270/jmpas.2021.V10S2.2035

Prasad, M. S. G., More, P., Christa, S., & Kumar, M. A. (2023). A chronic disease diagnosis model for smart healthcare systems enabled by artificial intelligence and the internet of things. In J. P. Patra & Y. K. Rathore (Eds.), *Applications of Artificial Intelligence in the Healthcare Sector*. Nova Science Publishers, Inc.

Praveenchandar, J., Vetrithangam, D., Kaliappan, S., Karthick, M., Pegada, N. K., Patil, P. P., Rao, S. G., & Umar, S. (2022). IoT-based harmful toxic gases monitoring and fault detection on the sensor dataset using deep learning techniques. *Scientific Programming, 2022*. https://doi.org/10.1155/2022/7516328

Prerna, Singh, P., & Singh, D. P. (2024). Role of disruptive technologies in the smart healthcare sector of society 5.0. In A. Saxena, A. Roy, & S. C. Satapathy (Eds.), *Artificial Intelligence and Society 5.0: Issues, Opportunities, and Challenges*. CRC Press. https://doi.org/10.1201/9781003397052-13

Raja, S., Agrawal, A. P., P Patil, P., Thimothy, P., Capangpangan, R. Y., Singhal, P., & Wotango, M. T. (2022). Optimization of 3D printing process parameters of polylactic acid filament based on the mechanical test. *International Journal of Chemical Engineering, 2022*. https://doi.org/10.1155/2022/5830869

Rajora, R., Gupta, H., Malhotra, S., Devliyal, S., & Sunil, G. (2024). Advancing renal health: unleashing the power of CNNS in multi-class classification for precise kidney condition diagnosis. *2024 IEEE 9th International Conference for Convergence in Technology, I2CT 2024*. https://doi.org/10.1109/I2CT61223.2024.10544018

Raju, K., Balakrishnan, M., Prasad, D. V. S. S. S. V., Nagalakshmi, V., Patil, P. P., Kaliappan, S., Arulmurugan, B., Radhakrishnan, K., Velusamy, B., Paramasivam, P., & El-Denglawey, A. (2022). Optimization of WEDM Process Parameters in Al2024-Li-Si_{3N4}MMC. *Journal of Nanomaterials, 2022*. https://doi.org/10.1155/2022/2903385

Raman, R., Kumar, R., Ghai, S., Gehlot, A., Raju, A. M., & Barve, A. (2023). A new method of optical spectrum analysis for advanced wireless communications. *2023 3rd International Conference on Advance Computing and Innovative Technologies in Engineering, ICACITE 2023*, 1719–1723. https://doi.org/10.1109/ICACITE57410.2023.10182414

Ramesh, S. M., Rajeshkannan, S., Pundir, S., Dhaliwal, N., Mishra, S., & Saravana, B. S. (2023). Design and development of embedded controller with wireless sensor for power monitoring through smart interface design models. *Proceedings of the 2023 2nd International Conference on Augmented Intelligence and Sustainable Systems, ICAISS* 2023, 1817–1821. https://doi.org/10.1109/ICAISS58487.2023.10250506

Rana, M. S., Cavaliere, L. P. L., Mishra, A. B., Padhye, P., Singh, R. R., & Khanna, R. (2022). Internet of things (IoT) based assessment for effective monitoring data against malicious attacks on financial collectors. *2022 2nd International Conference on Advance Computing and Innovative Technologies in Engineering, ICACITE 2022*, 177–181. https://doi.org/10.1109/ICACITE53722.2022.9823612

Rawat, R., Sharma, S., & Goyal, H. R. (2023). Intelligent digital financial inclusion system architectures for industry 5.0 enabled digital society. *Winter Summit on Smart Computing and Networks, WiSSCoN 2023*. https://doi.org/10.1109/WiSSCoN56857.2023.10133858

Sahithya, B., Guru Prasad, M. S., Sahithi, B., Madhu Krishna, K., Devarlla, A. C., & Yashavanth, T. R. (2024). Empowering healthcare with AI: advancements in medical image analysis, electronic health records analysis, and AI-driven chatbots. *2024 3rd International Conference for Innovation in Technology, INOCON 2024.* https://doi.org/10.1109/INOCON60754.2024.10511753

Salama, R., Al-Turjman, F., Bordoloi, D., & Yadav, S. P. (2023). Wireless sensor networks and green networking for 6g communication- an overview. *2023 International Conference on Computational Intelligence, Communication Technology and Networking, CICTN 2023*, 830–834. https://doi.org/10.1109/CICTN57981.2023.10141262

Sathyapriya, G., Natarajan, U., Sureshkumar, B., Navaneethakrishnan, G., Palanisamy, R., Bajaj, M., Sharma, N. K., & Kitmo. (2022). Quality and tool stability improvement in turning operation using plastic compliant damper. *Journal of Nanomaterials, 2022.* https://doi.org/10.1155/2022/8654603

Sati, P., Sharma, E., Soni, R., Dhyani, P., Solanki, A. C., Solanki, M. K., Rai, S., & Malviya, M. K. (2022). Bacterial endophytes as bioinoculant: microbial functions and applications toward sustainable farming. In S. K. Singh, R. Prasad, A. Varma, & L. Tripathi (Eds.), *Microbial Endophytes and Plant Growth: Beneficial Interactions and Applications.* Elsevier. https://doi.org/10.1016/B978-0-323-90620-3.00008-8

Saxena, A., Pant, B., Alanya-Beltran, J., Akram, S. V., Bhaskar, B., & Bansal, R. (2022). A detailed review of implementation of deep learning approaches for industrial internet of things with the different opportunities and challenges. *Proceedings of 5th International Conference on Contemporary Computing and Informatics, IC3I 2022*, 1370–1375. https://doi.org/10.1109/IC3I56241.2022.10072499

Sehajpal, S., & Kaur, N. (2025). Neuroscience, nursing, and financial innovation : a synergistic approach to optimized patient care and economic well-being. In T. Sanjay, S. C. Bhupinderpal Pal, J. Amar, O. Ercan, & K. Pawan (Eds.), *Neuromarketing's Role in Sustainable Finance* (pp. 133–146). IGI Global Scientific Publishing. https://doi.org/10.4018/979-8-3693-9117-4.ch008

Semwal, P., Painuli, S., Shabaaz Begum, J. P., Jamloki, A., Rauf, A., Olatunde, A., Mominur Rahman, M., Mukerjee, N., Khalil, A. A., Aljohani, A. S. M., Al Abdulmonem, W., & Simal-Gandara, J. (2024). Corrigendum to "Exploring the nutritional and health benefits of pulses from the Indian Himalayan region: a glimpse into the region's rich agricultural heritage" [Food Chemistry 2023, 422, 136259]. *Food Chemistry, 445,* 138830. https://doi.org/10.1016/j.fo odchem.2024.138830

Shabbiruddin, Kanwar, N., Jadoun, V. K., Jayalakshmi, N. S. J., Afthanorhan, A., Fatema, N., Malik, H., & Hossaini, M. A. (2023). Industry - challenge to pro-environmental manufacturing of goods replacing single-use plastic by Indian industry: a study toward failing ban on Single-Use Plastic Access. *IEEE Access, 11*, 77336–77346. https://doi.org/10.1109/ACCESS.2023.3296097

Sharahiley, S. M., & Kandpal, V. (2023). The impact of monetary and non-monetary reward systems upon creativity: how rational are Saudi professional employees? *International Journal of Work Organisation and Emotion, 14*(4), 339–358. https://doi.org/10.1504/IJWOE.2023.136599

Sharma, A., Sharma, A., Juneja, P. K., & Jain, V. (2020). Spectral features based speech recognition for speech interfacing to control pc windows. In S. Semwal & P. Dhuliya (Eds.), *Proceedings - 2020 International Conference on Advances in Computing, Communication and Materials, ICACCM 2020* (pp. 341–345). Institute of Electrical and Electronics Engineers Inc. https://doi.org/10.1109/ICACCM50413.2020.9212827

Sharma, A., Sharma, R. K., Nanda, S., & Misra, A. (2022a). Emerging real time analytics based health start-ups: opportunities during covid-19. *Journal of Medical Pharmaceutical and Allied Sciences, 11*(4), 5017–5025. https://doi.org/10.55522/jmpas.V11I4.2360

Sharma, E., Rana, S., Sharma, I., Sati, P., & Dhyani, P. (2023a). Organic polymers for CO_2 capture and conversion. In A. A. Arof, M. F. Z. Kadir, & R. Othman (Eds.), *CO2-Philic Polymers, Nanocomposites and Solvents: Capture, Conversion and Industrial Products*. Elsevier. https://doi.org/10.1016/B978-0-323-85777-2.00002-0

Sharma, M., Hagar, A. A., Krishna Murthy, G. R., Beyane, K., Gawali, B. W., & Pant, B. (2022b). A study on recognising the application of multiple big data technologies and its related issues, difficulties and opportunities. *2022 2nd International Conference on Advance Computing and Innovative Technologies in Engineering, ICACITE 2022*, 341–344. https://doi.org/10.1109/ICACITE53722.2022.9823623

Sharma, M., Luthra, S., Joshi, S., & Joshi, H. (2022c). Challenges to agile project management during COVID-19 pandemic: an emerging economy perspective. *Operations Management Research*, *15*(1–2), 461–474. https://doi.org/10.1007/s12063-021-00249-1

Sharma, M., Luthra, S., Joshi, S., & Kumar, A. (2022d). Analysing the impact of sustainable human resource management practices and industry 4.0 technologies adoption on employability skills. *International Journal of Manpower*, *43*(2), 463–485. https://doi.org/10.1108/IJM-02-2021-0085

Sharma, P., Malhotra, R. K., Altybekova, D., & Singh, C. P. (2024b). AI and machine learning as catalysts for educational equity: bridging the urban–rural divide. *Lecture Notes in Networks and Systems*, *1074 LNNS*, 41–55. https://doi.org/10.1007/978-981-97-6103-6_4

Sharma, P., Malhotra, R. K., & Kumar, R. (2024a). How is digitalization reshaping the landscape of rural banking in India? *2024 International Conference on Smart Devices, ICSD 2024*. https://doi.org/10.1109/ICSD60021.2024.10750981

Sharma, P., Malhotra, R. K., Ojha, M. K., & Gupta, S. (2022e). Impact of meditation on mental & physical health and thereby on academic performance of students: a study of higher educational institutions of Uttarakhand. *Journal of Medical Pharmaceutical and Allied Sciences*, *11*(2), 4641–4644. https://doi.org/10.55522/jmpas.V11I2.2309

Sharma, S., & Bhadula, S. (2023). Secure federated learning for intelligent industry 4.0 IoT enabled self skin care application system. *Proceedings of the 2nd International Conference on Applied Artificial Intelligence and Computing, ICAAIC 2023*, 1164–1170. https://doi.org/10.1109/ICAAIC56838.2023.10141028

Sharma, S., Gupta, A., & Tyagi, R. (2023b). Sustainable natural resources utilization decision system for better society using Vedic scripture, cloud computing, and IoT. In R. C. Biradar, K. M. Sudharshan, & M. Devanathan (Eds.), *Proceedings of IEEE 2023 5th International Conference on Advances in Electronics, Computers and Communications, ICAECC 2023*. Institute of Electrical and Electronics Engineers Inc. https://doi.org/10.1109/ICAECC59324.2023.10560335

Sharma, S., Gupta, C., Khanna Malhotra, R., & Upreti, H. (2024c). Sustainable banking practices: impact, challenges and opportunities. In P. K. Prabhakar (Ed.), *E3S Web of Conferences* (Vol. 556). EDP Sciences. https://doi.org/10.1051/e3sconf/202455601031

Sharun, V., Rajasekaran, M., Kumar, S. S., Tripathi, V., Sharma, R., Puthilibai, G., Sudhakar, M., & Negash, K. (2022). Study on developments in protection coating techniques for steel. *Advances in Materials Science and Engineering*, *2022*. https://doi.org/10.1155/2022/2843043

Singamaneni, K. K., Dhiman, G., Juneja, S., Muhammad, G., AlQahtani, S. A., & Zaki, J. (2022). A novel QKD approach to enhance IIOT privacy and computational knacks. *Sensors*, *22*(18). https://doi.org/10.3390/s22186741

Singh, A., & Srivastava, A. (2024). Innovation needed to improve road construction and road infrastructure in the future in India. In G. Anand, & V. R. Jain (Eds.), *IOP Conference Series: Earth and Environmental Science* (Vol. 1326, Issue 1). Institute of Physics. https://doi.org/10.1088/1755-1315/1326/1/012097

Singh, A. R., Vishnuram, P., Alagarsamy, S., Bajaj, M., Blazek, V., Damaj, I., Rathore, R. S., Al-Wesabi, F. N., & Othman, K. M. (2024a). Electric vehicle charging technologies, infrastructure expansion, grid integration strategies, and their role in promoting sustainable e-mobility. *Alexandria Engineering Journal, 105*, 300–330. https://doi.org/10.1016/j.aej.2024.06.093

Singh, K. D., Singh, P., Chhabra, R., Kaur, G., Bansal, A., & Tripathi, V. (2023a). Cyber-physical systems for smart city applications: a comparative study. In R. Kumar, R. Kumar, M. Gupta, M. Gupta, R. Srivastava, & R. Srivastava (Eds.), *2023 International Conference on Advancement in Computation and Computer Technologies, InCACCT 2023* (pp. 871–876). Institute of Electrical and Electronics Engineers Inc. https://doi.org/10.1109/InCACCT57535.2023.10141719

Singh, K. D., Singh, P., Kaur, G., Khullar, V., Chhabra, R., & Tripathi, V. (2023b). Education 4.0: exploring the potential of disruptive technologies in transforming learning. *Proceedings of International Conference on Computational Intelligence and Sustainable Engineering Solution, CISES* 2023, 586–591. https://doi.org/10.1109/CISES58720.2023.10183547

Singh, N. K., Singh, Y., Rahim, E. A., Senthil Siva Subramanian, T., & Sharma, A. (2023a). Electric discharge machining of hybrid composite with bio-dielectrics for sustainable developments. *Australian Journal of Mechanical Engineering*. https://doi.org/10.1080/14484846.2023.2249577

Singh, S., Singh, A., Mohan, A., Batola, M., & Kumar, N. (2024b). Health expenditures and health outcomes in Central Europe and the Baltic region. *Baltic Region, 16*(2), 126–143. https://doi.org/10.5922/2079-8555-2024-2-7

Singh, Y., Rahim, E. A., Singh, N. K., Sharma, A., Singla, A., & Palamanit, A. (2022). Friction and wear characteristics of chemically modified mahua (madhuca indica) oil based lubricant with SiO2 nanoparticles as additives. *Wear, 508–509*. https://doi.org/10.1016/j.wear.2022.204463

Singla, R. K., De, R., Efferth, T., Mezzetti, B., Sahab Uddin, M., Sanusi, Ntie-Kang, F., Wang, D., Schultz, F., Kharat, K. R., Devkota, H. P., Battino, M., Sur, D., Lordan, R., Patnaik, S. S., Tsagkaris, C., Sai, C. S., Tripathi, S. K., Găman, M.-A., … Shen, B. (2023). The International Natural Product Sciences Taskforce (INPST) and the power of Twitter networking exemplified through #INPST hashtag analysis. *Phytomedicine, 108*. https://doi.org/10.1016/j.phymed.2022.154520

Širić, I., Eid, E. M., Taher, M. A., El-Morsy, M. H. E., Osman, H. E. M., Kumar, P., Adelodun, B., Abou Fayssal, S., Mioč, B., Andabaka, Ž., Goala, M., Kumari, S., Bachheti, A., Choi, K. S., & Kumar, V. (2022). Combined Use of Spent Mushroom Substrate Biochar and PGPR Improves Growth, Yield, and Biochemical Response of Cauliflower (Brassica oleracea var. botrytis): a Preliminary Study on Greenhouse Cultivation. *Horticulturae, 8*(9). https://doi.org/10.3390/horticulturae8090830

Srivastava, S. P., Malhotra, R. K., & Mukherjee, R. K. (2024). Data driven decision making in insurance. *2024 International Conference on Smart Devices, ICSD 2024*. https://doi.org/10.1109/ICSD60021.2024.10751068

Thapa, D., Kumar, V., Naik, B., Kumar, V., Gupta, A. K., Mohanta, Y. K., Mishra, B., & Rustagi, S. (2024). Harnessing probiotic foods: managing cancer through gut health. *Food Science and Biotechnology, 33*(9), 2141–2160. https://doi.org/10.1007/s10068-024-01638-5

Tripathy, S., Verma, D. K., Thakur, M., Patel, A. R., Srivastav, P. P., Singh, S., Chávez-González, M. L., & Aguilar, C. N. (2021). Encapsulated Food Products as a Strategy to Strengthen Immunity Against COVID-19. *Frontiers in Nutrition, 8*. https://doi.org/10.3389/fnut.2021.673174

Uike, D., Agarwalla, S., Bansal, V., Chakravarthi, M. K., Singh, R., & Singh, P. (2022). Investigating the role of block chain to secure identity in IoT for industrial automation. In R. K. Dwivedi, A. Kr. Saxena, G. Khan, & S. Bhardwaj (Eds.), *Proceedings of the*

2022 11th International Conference on System Modeling and Advancement in Research Trends, SMART 2022 (pp. 837–841). Institute of Electrical and Electronics Engineers Inc. https://doi.org/10.1109/SMART55829.2022.10047385

Uniyal, A. K., Kanojia, P., Khanna, R., & Dixit, A. K. (2022a). Quantitative analysis of the impact of demography and job profile on the organizational commitment of the faculty members in the HEI'S of Uttarakhand. *Communications in Computer and Information Science, 1742 CCIS*, 24–35. https://doi.org/10.1007/978-3-031-23647-1_3

Uniyal, S., Mangla, S. K., & Patil, P. (2020). When practices count: implementation of sustainable consumption and production in automotive supply chains. *Management of Environmental Quality: An International Journal, 31*(5), 1207–1222. https://doi.org/10.1108/MEQ-03-2019-0075

Uniyal, S., Sarma, P. R. S., Kumar Mangla, S., Tseng, M.-L., & Patil, P. (2022b). Ict as "knowledge management" for assessing sustainable consumption and production in supply chains. In Information Resources Management Association (Ed.), *Research Anthology on Measuring and Achieving Sustainable Development Goals* (Vol. 3). IGI Global. https://doi.org/10.4018/978-1-6684-3885-5.ch048

Upreti, H., & Malhotra, R. K. (2024). Bridging the urban-rural education gap in India through CSR (corporate social responsibility) initiatives: a conceptual study with special reference to sustainable development goal 4 (quality education). In P. K. Prabhakar (Ed.), *E3S Web of Conferences* (Vol. 556). EDP Sciences. https://doi.org/10.1051/e3sconf/202455601032

Upreti, H., Malhotra, R. K., Kumar, R., & Karki, B. (2024). Leveraging smart devices for corporate social responsibility in India: advancements, trends, and concerns towards sustainability. *2024 International Conference on Smart Devices, ICSD 2024*. https://doi.org/10.1109/ICSD60021.2024.10751578

Vekariya, D., Rastogi, A., Priyadarshini, R., Patil, M., Kumar, M. S., & Pant, B. (2023). Mengers authentication for efficient security system using blockchain technology for industrial IoT(IIoT) systems. *2023 3rd International Conference on Advance Computing and Innovative Technologies in Engineering, ICACITE 2023*, 894–896. https://doi.org/10.1109/ICACITE57410.2023.10182454

Verma, A., Sharma, D., Pant, M., Khatib, M. N., Singh, M. P., Zahiruddin, Q. S., Rustagi, S., Gaidhane, S., & Gaidhane, A. M. (2024). First sighting of human H5N1 in Australia: A detailed account and public health implications. *New Microbes and New Infections, 60–61*. https://doi.org/10.1016/j.nmni.2024.101447

Verma, M., Sharma, S., Kumar, A., Kumar, V., Kim, M., Hong, Y., Lee, I., & Kim, H. (2021). Application of green nanomaterials in catalysis industry. In S. Thomas, A. N. Grace, & D. Ponnamma (Eds.), *Green Nanomaterials for Industrial Applications*. Elsevier. https://doi.org/10.1016/B978-0-12-823296-5.00013-7

Zaheer, A., Zehra Hussain, D., Verma, A., Sah, S., Pant, M., Nazli Khatib, M., Pratap Singh, M., Syed Zahiruddin, Q., & Rustagi, S. (2024). Marburg virus: an emerging threat to global health. *Clinical Infection in Practice, 23*. https://doi.org/10.1016/j.clinpr.2024.100371

12 Digitalization in the Banking Sector
The Case of Turkey

Cem Sabutay Turkdonmez, Adalet Hazar, and Şenol Babuşcu

12.1 INTRODUCTION

Henry Ford, the founder of the Ford Motor Company, broke new ground in the industry when he began mass production in the late 1800s. Ford, who had introduced hitherto unknown concepts of mass production and assembly lines, also triggered a paradigm shift: Ford's new order for middle-class American citizens to own automobiles was revolutionary in the industry.

More than a century has passed since Ford, and in that time, the world has changed dramatically. Nowadays, artificial intelligence-based programs and their future are being discussed. However, the idea of programmable computers was put forward by Charles Babbage a little more than two centuries ago in 1791. Almost 100 years later, in 1981, the first personal computers were made by IBM. Less than 50 years after the use of the first personal computer, first the desktop, then the laptop, then the tablet and finally the mobile devices have become an integral part of our lives (Collier and MacLachlan, 1998).

In parallel with the development of digital-oriented technologies that are rapidly being used in every field, the establishment of digitalization in every field of human life is growing exponentially. Digitalization is also increasing in the field of banking, which is one of the most dynamic sectors. It is possible to use mobile banking applications from anywhere in the world via a phone and a functioning internet network. At this point, it should not be wrong to say that a smartphone with the mobile application of any bank functions as a bank branch.

In the following sections, after briefly talking about how digitalization and automation applications have developed in terms of the banking sector in the world, the brief history of digitalization in the banking sector in Turkey and the latest situation we have reached in Turkey with figures, artificial intelligence applications, open banking issues will be emphasized.

12.2 DIGITALIZATION IN THE BANKING SECTOR

Although the concept of digitalization in the banking sector brings banking applications on mobile phones to the minds of most people at the first moment, digitalization in banking is not limited to this. In the 1990s, the increase in the use of desktop

DOI: 10.1201/9781003564621-12

computers and the widespread use of the internet around the world were simultaneous. The process that started with the spread of the internet has led banks to offer alternative channels other than the branch to the service of the customer.

These channels can be expressed as follows:

- Internet Banking: With the internet banking branches that have been implemented with the increase in the use of desktop computers, banks have started to serve their customers, and some banking transactions, from inquiring the account balance to paying bills, have begun to be carried out by the customer themselves from desktop devices with a self-service approach.
- Mobile Banking: Customers can use their mobile phones as a bank branch to perform many transactions from learning the balance, making money transfers, paying and questioning credit card debts to loading money to Global System for Mobile Communications (GSM) lines.
- Automatic Teller Machines (ATMs): ATMs, which are now widely used, function in many areas such as transferring money to customers, depositing and withdrawing money. With these machines, some of the more complex transactions can also be carried out, for example, deposit passbook processing, etc.
- Plastic Cards: Plastic cards, credit cards or debit cards, issued by the bank to its customers allow customers to spend, shop and withdraw money through a payment point (POS).

We can list the abovementioned issues as electronic funds transfer and mass payment systems as new and useful tools of the digital age (Sardana and Singhania, 2018).

Although the concept of online banking and digital banking are often used synonymously, it is necessary to evaluate the concept of online banking in a narrower sense. Online banking primarily focuses on online issues such as money transfer, remote account opening, etc. For the concept of online banking, concepts such as e-banking, virtual banking and internet banking are also used from time to time. Digital banking, on the other hand, includes all activities carried out by financial institutions and customers, including digitalization (Harchekar, 2018).

In their study, Chen and Lam found that banks operating in large parts of India and Asia use three types of strategies to position themselves digitally. The first of these is to provide low-level services to the customer through direct channels to support the branch-centered product-oriented structure, the second is to use direct channels together with the physical branch channel, which can also be called customer-oriented multi-channel, and the third and last is the completely digital-centric model. Banks have to choose one of these three models according to their customer base and business models. In addition, banks that want to adapt to this changing business model need to focus on the IT infrastructure quickly (Chen and Lam, 2014).

Almost all over the world, banks are institutions with very well-organized, relatively trained staff. Since the banking sector operates with relatively little capital and high leverage, the element of trust in this sector is more important than in any other sector. This leads to banks being supervised by supervisory institutions. In this new period that has emerged with digitalization, banks are subject to supervision by regulatory institutions. The digital age is one of the main reasons for the emergence of financial technology companies, FinTechs, along with banks. FinTechs are alternative digital finance solutions to banks that aim to meet the banking needs of small businesses and retail customers.

In their study, Karsh and Abufara showed that FinTechs offer the same banking products as banks, while FinTech companies grow faster in countries where digital technologies and mobile phones are widely used (Karsh and Abufara, 2020).

Sustainability has been an important issue for all sectors today, and the banking sector is not exempt from this trend. One of the important elements of banking in the 21st century has been digitalization and this has become directly linked to sustainability. Digital banking represents an important step toward sustainability in many aspects, such as reducing the use of energy and paper, serving a wider and more diverse customer base and offering a wider range of financial services (Anis et al., 2023).

Digital banking enables banking services to become more efficient by automating transaction processes and minimizing the use of physical resources. This helps banks reduce the use of energy and other resources, which improves both operational efficiency and environmental sustainability.

In addition, digital banking enables the expansion of access to banking services. This means the potential to serve communities that previously had little or no access to financial services, especially those living in rural areas and people with low incomes. This is important for social sustainability (Fong, 2017).

In summary, digital banking's relationship with sustainability is multidimensional and supports environmental, social and economic sustainability. Therefore, digitalization plays a critical role in achieving the sustainability goals of the banking sector.

In addition to the many advantages that digitalization brings to the banking sector, there are also possible negative effects on the environment and climate. First of all, the digitalization of the banking sector requires technological infrastructures such as data centers and cloud services. The use and operation of these systems consume a significant amount of energy and lead to carbon emissions (Malmodin and Lundén, 2018).

Moreover, the implementation of digital banking solutions can lead to the generation of large amounts of electronic waste. Banks regularly provide hardware updates for mobile banking apps, ATMs and other digital devices. In this process, aging devices often become waste, which can exacerbate the problem of electronic waste.

Finally, digital banking may increase the use of technologies such as energy-intensive cryptocurrencies. Mining cryptocurrencies such as Bitcoin consumes a large amount of energy, which contributes to carbon emissions (Krause & Tolaymat, 2018).

In summary, the environmental and climatic impacts of digital banking are not limited to the direct impacts of the application and use of technologies alone. It is also important to consider the side effects and indirect effects of using these technologies.

12.3 TURKISH BANKING SYSTEM AND DIGITALIZATION

In this section, we will briefly touch on the crises that have occurred in the history of Turkish banking. The regulatory institutions that emerged as a result of the experiences gained as a result of the crises resulted in the efficient and effective operation of the banking sector in the country compared to many other sectors. The 1994 banking and financial crisis became widespread due to the CBRT's (Central Bank of Republic of Turkey) lack of reserves to intervene in the situation in a timely manner and to the necessary extent, and became a threat to the entire banking sector and the economy. The main reason why the banking sector was seriously affected by the 1994 crisis was the decrease in profit rates with the end of the low-exchange-rate and high-interest-rate policies followed in the 1989–1993 period. The 1994 financial crisis and the mini-financial crises of the following years showed that the banking sector should be regulated at the beginning of the structural measures to be taken in order to ensure and maintain economic stability in Turkey.

Turkey entered the year 2000 in an environment in which very important economic decisions were taken. In the negotiations held with the IMF (International Money Fund) in June 1999, the transformation of the Close Monitoring Agreement into a programmatic and financial finance-supported agreement was adopted and the framework of the macroeconomic policies to be implemented in the period of 2000–2002 was drawn. Following its Letter of Intent to the IMF of 9 December 1999, which was accepted, the government implemented a program of reducing inflation from 1 January 2000, covering a three-year economic period, including structural changes as well as fiscal, monetary, exchange rate and income policies. Although some positive developments were observed in the economy in a very short time with the implementation of this program, a crisis occurred in November 2000 caused by the rapid increase in foreign exchange demand caused by the liquidity crunch in the Turkish financial markets, the deterioration in international markets and the negative effects experienced at home. This crisis could only be prevented with an IMF loan, but the program to reduce inflation was badly damaged.

In February 2001, the fragile structure of confidence in financial markets once again led to a financial crisis, and as a result, the monetary and exchange rate policies foreseen in the 2000 Inflation Reduction Program were abandoned.

On 22 February 2001, the floating exchange rate system was introduced, thus ending the Inflation Reduction Program. In the aftermath of the November 2000 and February 2001 crises, excessively rising interest rates increased the funding losses of banks with maturity mismatches and led to the depreciation of the securities they held in their portfolios. At the same time, with the introduction of the floating exchange rate in February 2001, banks operating with high open positions due to the rapid depreciation of the TL against foreign currencies also faced significant foreign exchange losses.

Both financial crises in the Turkish economy in the 2000s adversely affected all sectors and economic life, especially the banking sector.

As a result of these negativities, the stand-by agreement with the IMF came to the agenda and the Banking Regulation and Supervision Agency (BRSA) was established and put into operation for the effective supervision of the banking sector. This was a period when the number of banks in Turkey was rapidly decreasing and banks with foreign capital started to operate in Turkey.

With the tight monetary policy implemented as a result of the stand-by agreement with the IMF and a series of measures taken in the economy, the Turkish economy started to recover rapidly. A number of issues such as the survival of only banks with strong capital, the injection of capital into public banks and the fact that banks with foreign capital already had strong capital raised the capital adequacy ratio of the Turkish Banking Sector above the standards. The stabilization of interest rates and inflation and the fact that fluctuations were almost never experienced until 2008 led to the strengthening of banks and an increase in their profitability.

The actors of the Turkish Banking System, which is highly dynamic and has gained a healthy structure with the regulations, have also kept pace with the developments in the world. Digitalization is an area where the Turkish Banking Sector is rapidly adapting. In this part of the study, information about digitalization in Turkey is included.

12.3.1 Turkish Banking Sector Bank, Branch, Employee Development and Current Situation

The developments in the number of banks, branches and employees in the last 20 years of the Turkish Banking Sector are as follows:

As of March 2023, in the Turkish Banking Sector, there are 35 deposits, 6 participation and 16 development and investment banks. According to the ownership groups, a total of 57 banks are operating, including 13 public, 16 domestic private and 28 foreign. The total number of employees increased by 3,114 people compared to the previous quarter and amounted to 2,09,367 people. Again, compared to the previous quarter, the number of branches increased by 26 and the number of branches reached 11,060. Table 12.1 shows the number of employees of these banks, the number of branches and the changes that have occurred in them compared to the previous quarter.

As of May 2023, the total asset size of banks in Turkey was 16.8 billion TL and the sector's asset size increased by approximately 2.5 billion TL compared to the end of 2022. As of March 2023, the number of employees in the sector is 1,91,209 people. As of the same date, the number of branches is 9,667 and the number of branches per bank in deposit banks is 274. The number of branches per 1,00,000 people is 11.3 and the number of personnel per 1,00,000 people is 224.

Information on digitalization in the sector is included in the later stages of our study. The statistics consist of data from 28 banks that are members of the Banks

TABLE 12.1
Banks in Turkey by March 2023

Number of Banks	35 Deposit Banks	6 Participation Banks	16 Development and Investment Banks
Bank ownership	6 State, 8 Domestic Private, 21 Foreign	3 State, 3 Foreign	4 State, 8 Domestic private, 5 Foreign
Personnel number	185.350 People	18.413 People	5.603 People
Branch number	9.530 Branches	1.402 Branches	68 Branches

Source: BRSA (https://www.bddk.org.tr/Veri/EkGetir/8?ekId=140).

Association of Turkey and provide internet banking services and 22 banks that provide mobile banking services and are as of March 2023. (www.bddk.org.tr)

12.3.2 The Digitalization Process in the Turkish Banking Sector

In the January–March 2023 period, the total number of active digital banking customers (individual and corporate) reached 99 million 519 thousand people (while the data of the banks were aggregated, the number of customers among the banks was not deduplicated). Of this number, 2 million 114 thousand people made "internet banking only" transactions, while 87 million 362 thousand people made "mobile banking only" transactions. The number of users who make both internet and mobile banking transactions is 10 million 44 thousand people. The total number of active digital banking customers (retail and corporate) increased by 18 million 892 thousand people compared to the previous year. The graph below shows the development in this area.

Within the framework of the latest data, active digital banking customer information as of March 2023 is included in Figure 12.1.

As can be seen from Figure 12.1, the number of digital banking customers is increasing rapidly. The number has increased every year from 35 million in December 2017 and has exceeded 99 million by the end of March 2023. The graph also shows the number of people using not only mobile banking. The highest increase was on the mobile banking side, with the number of users reaching 87 million in March 2023 from 22 million in December 2017. Another noteworthy point is that the number of internet banking users, which was 5 million as of December 2017, decreased to 2 million as of March 2023. This rapid search for mobile banking follows a parallel course with the spread of both mobile phones and access to the internet.

The distribution according to age groups and gender is shown in the charts.

As can be seen from Chart 12.1, in March 2023, approximately 63 million users, or 66% of active retail digital banking customers, are men, and approximately 32 million are women, or 34%.

As can be seen from Chart 12.2, in the January–March 2023 period, the 36–55 age group takes the first place in the distribution of active retail digital banking customers on the basis of age groups with 37 million 678 thousand people. This is followed by the 26–35 age group with 25 million 831 thousand people and the 18–25 age group with 18 million 126 thousand people.

Especially the situation in the gender distribution is striking. Considering that mostly men alone work in households, especially in rural areas, this seems to be normal, but it is seen that many women with mobile phones still do not benefit from digital banking channels. In this respect, it would be correct to say that there is a great potential for the banking sector. On the other hand, when we look at the age groups, it is seen that 64% of the users are in the 26–55 age group. Although the rate of digital banking customers between the ages of 0 and 17, who are not yet involved in working life, is at very low levels such as 1% of the total number of digital banking customers, it is known that many young people in this age group have smartphones.

According to the results of the study on the use of information technology in children conducted by Turkish Statistical Institute (TURKSTAT) in 2021, smartphone

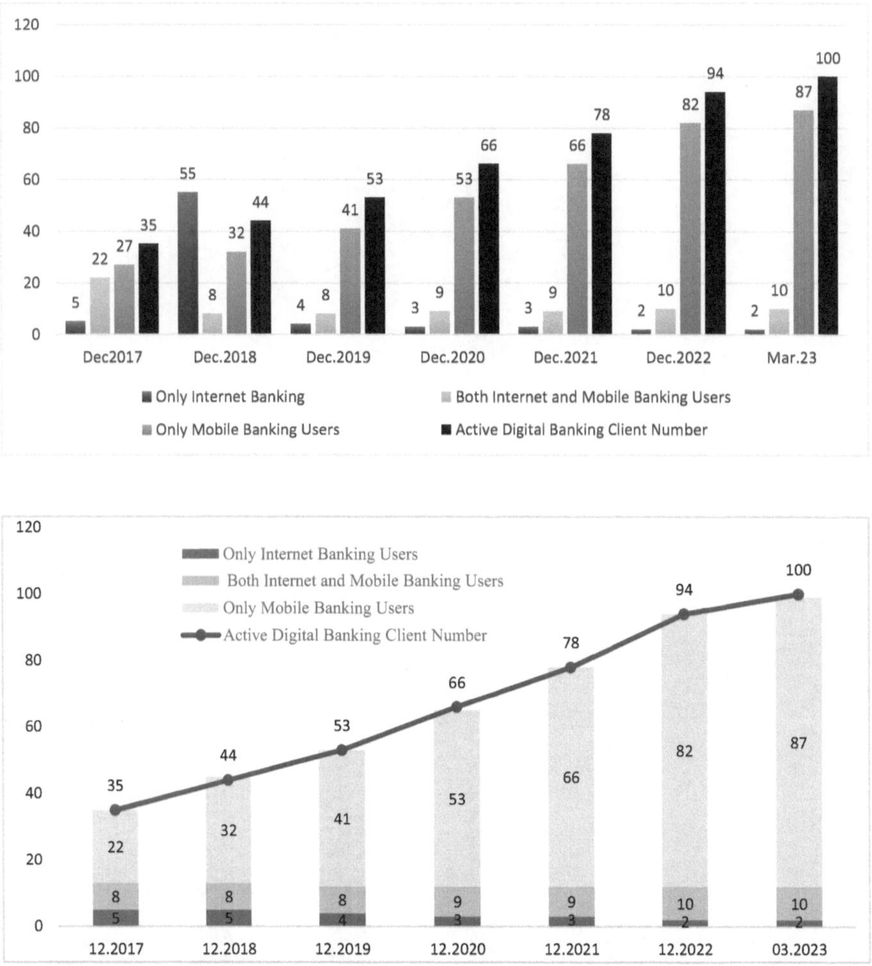

FIGURE 12.1 Active digital banking customers (million people). (*Source*: Turkish Bank Association, March 2023.)

ownership of children and young people between the ages of 11 and 15 increased from 24% in 2013 to 55.5% in 2021 when compared to the number of all children and young people in this group. On the other hand, the ratio of the 18–25 age group to the total number of customers using retail digital banking is 18%, and considering that the people in this group are either university students or people who have just started working life, it can be said that there is a high potential in this group.

As of March 2023, the number of retail customers who are registered in the system to conduct internet banking and have logged in at least once is 92 million 88 thousand people. The number of individual customers who have logged in at least once in the last year is 25 million 834 thousand people (Table 12.2).

As of January–March 2023, the total number of financial transactions made using the internet banking service is 126 million and the amount is approximately

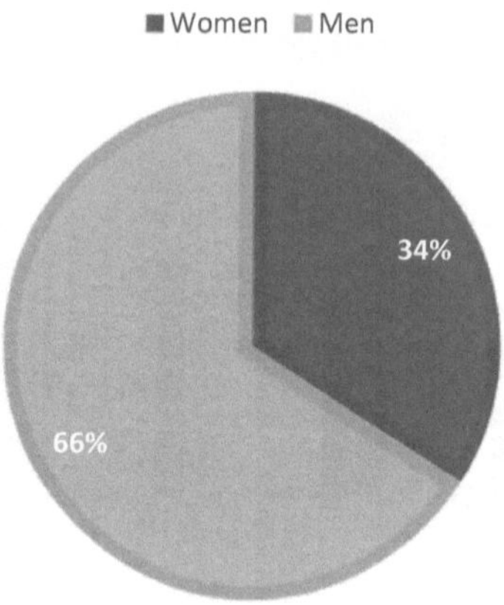

CHART 12.1 Active retail digital banking customers by gender. (*Source*: Turkish bank association, March 2023.)

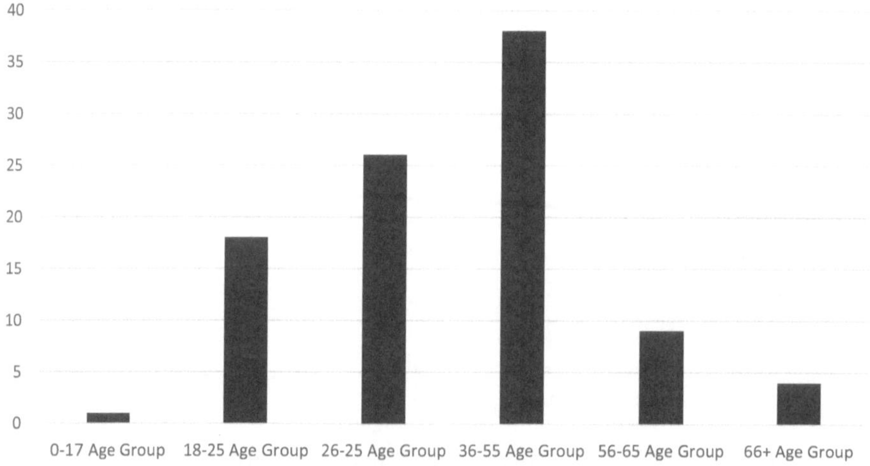

CHART 12.2 Active retail digital banking customers by age groups. (*Source*: Turkish bank association, March 2023.)

6 trillion TL. Money transfer transactions, including EFT, remittance and foreign exchange transfer transactions, had the largest share among financial transactions (70% of financial transaction volume) with a transaction volume of 4 trillion 195 billion TL and 70 million transactions. In second place are investment transactions with a transaction volume of 1 trillion 283 billion TL and 24 million transactions **(Turkish Bank Association, March 2023)**.

The total number of financial transactions made using the internet banking service increased by 7 million compared to the previous period and by 12 million compared to the previous year. Total transaction volume increased by 239 billion TL compared to the previous period and by 2 trillion 945 billion TL compared to the previous year (Table 12.3).

TABLE 12.2
Number of Customers Using Internet Banking

	January–March 2022	October–December 2022	January–March 2023
Number of Individual Customers (Thousand People)			
Active (A) (*logged once in the last 3 months*)	9.937	10.545	10.564
Registered (B) (*logged at least once*)	84.030	91.137	92.088
Registered (C) (*logged at least once in the last 1 year*)	23.969	25.147	25.834
Active (A)/registered (B) customer ratio (%)	**12**	**12**	**11**
Number of Corporate Customers (Thousand People)			
Active (A) (*logged once in the last 3 months*)	1.497	1.541	1.593
Registered (B) (*logged at least once*)	5.444	6.069	6.172
Registered (C) (*logged at least once in the last 1 year*)	2.076	2.199	2.284
Active (A)/registered (B) customer ratio (%)	**28**	**25**	**26**
Total Number of Customers (Thousand People)			
Active (A) (*logged once in the last 3 months*)	11.434	12.086	12.158
Registered (B) (*logged at least once*)	89.474	97.206	98.260
Registered (C) (*logged at least once in the last 1 year*)	26.045	27.346	28.118
Active (A)/registered (B) customer ratio (%)	**13**	**12**	**12**

Source: Turkish Bank Association, March 2023.

As of January–March 2023, the first place in the volume of investment transactions carried out was taken by stock transactions with 477 billion TL and 19.32 million transactions. This is followed by mutual fund transactions with 421 billion TL and approximately 2 million transactions. The highest average transaction volumes were realized in VIOP transactions with 264 thousand TL. This was followed by mutual fund transactions with an average transaction volume of TL 245 thousand. In the January–March 2023 period, the number of instant loans extended through internet banking is 125 thousand and the volume is 6 billion TL (**Turkish Bank Association, March 2023).** The number of insurance sales realized in the same period is 60 thousand. As can be seen, especially the number and volume of investment transactions are increasing rapidly (Table 12.4).

The total number of customers (individual and corporate) who are registered in the system to conduct mobile banking and who have logged in at least once is 151 million 528 thousand people as of March 2023. Of these, 97 million 405

TABLE 12.3
Financial Transactions Using Internet Banking

	Number of Transactions (Millions)			Transaction Volume (Billion TL)		
	01-03/2022	01-03/2023	Worth. %	01-03/2022	01-03/2023	Worth %
Money transfers	65	70	7.7	2.256	4.195	85.9
Payments	26	22	−15.4	127	263	107.1
Trading operations	14	24	71.4	529	1.283	142.5
Credit card processing	6	5	−16.7	40	74	85.0
Other financial transactions	4	4	0.0	99	182	83.8
Sum	**114**	**126**	10.5	**3.051**	**5.996**	96.5

Source: Turkish Bank Association, March 2023.

TABLE 12.4
Investment Transactions in Internet Banking

	Number of Transactions		Throughput	
	(000)	Pay %	(Billion TL)	Pay %
Mutual funds	1.719	7.1	421	32.8
Foreign exchange transactions	1.557	6.5	165	12.9
Term accounts	630	2.6	88	6.9
Stock	19.292	80.1	477	37.2
Repo transactions	53	0.2	8	0.6
Bonds and bills	101	0.4	14	1.1
Gold	342	1.4	9	0.7
VİOP	380	1.6	100	7.8
Sum	24.074	100	1.282	100

Source: Turkish Bank Association, March 2023.

thousand people (64%) made at least one entry during the period from January to March 2023. The total number of active customers (individual and corporate) increased by 5 million 316 thousand people compared to the previous period and by 19 million 145 thousand people compared to the previous year. As of January–March 2023, the total number of financial transactions made using the mobile banking service was 1 billion 856 million and the amount was 13 trillion 193 billion TL. Money transfers, which include EFT, remittance and foreign exchange transactions, accounted for 62% of the volume of financial transactions and 25%

TABLE 12.5
Number of Customers Using Mobile Banking

	October–December 2022	January–March 2023	Worth %
Number of Individual Customers (Thousand People)			
Active (A) (*logged once in the last 3 months*)	88.647	93.733	5.7
Registered (B) (*logged at least once*)	136.095	144.837	6.4
Registered (C) (*logged at least once in the last 1 year*)	103.182	110.427	7.0
Active (A)/registered (B) customer ratio (%)	**65**	**65**	
Number of Corporate Customers (Thousand People)			
Active (A) (*logged once in the last 3 months*)	3.443	3.672	
Registered (B) (*logged at least once*)	6.534	6.692	
Registered (C) (*logged at least once in the last 1 year*)	4.152	4.456	
Active (A)/registered (B) customer ratio (%)	**53**	**55**	
Total Number of Customers (Thousand People)			
Active (A) (*logged once in the last 3 months*)	92.090	97.405	5.8
Registered (B) (*logged at least once*)	142.629	151.528	6.2
Registered (C) (*logged at least once in the last 1 year*)	107.334	114.883	7.0
Active (A)/registered (B) customer ratio (%)	**65**	**64**	

Source: Turkish Bank Association, March 2023.

of investment transactions. From January to March 2023, the number of instant loans extended through mobile banking is 5 million 187 thousand, and the volume is approximately 178 billion TL. Insurance sales realized in the same period amounted to 3 million 904 thousand units (Turkish Bank Association, March 2023) (Table 12.5).

12.4 OPEN BANKING IN TURKEY

In the field of financial services, one of the most important digital transformations of the 2020s is the concept of open banking. The concept of open banking first came to the agenda in the United Kingdom in 2016, and then the legal infrastructure was prepared with the regulation published by the BRSA in March 2020 in our country.

Open banking is a model that basically means that the private financial data of customers in banks are shared with third parties, namely TPPs, with the permission[1] of the customers. Thanks to this model, the financial data belonging to the customers and not even shared among the banks for today are no longer private to the banks and are put on a common platform with the request and consent of the customer. Within the framework of the European Union legislation, open banking practices were first paved with the Payment Services Directive (PSD) in 2007, and then in 2015, the PSD 2 and[2] the regulations related to these applications were made comprehensive and placed on a legal basis.

Payment services in Turkey were first regulated by Law No. 6493 on Payment and Securities Settlement Systems, Payment Services and Electronic Money Institutions. As the legal regulation that forms the basis of open banking, the Law No. 7192 on Payment and Securities Settlement Systems, Payment Services and Electronic Money Institutions and the Law on the Amendment of Some Laws were adopted in 2019.

The definition of open banking was first included in the Regulation on Information Systems and Electronic Banking Services of Banks published by the BRSA on 15 March 2020 in Turkey. The data are made available to FinTechs through this platform. In the open banking model, all financial data, from monthly credit card expenditures to savings habits in investment accounts, are shared with all banks and FinTechs for data analysis purposes. Thanks to the open banking model, both banks and FinTechs can realize custom products and interfaces according to the needs of customers. In this way, they can design systems that allow customers to access all their accounts and financial data from a single interface. In other words, customers can now manage all their banking transactions or investments through a single app or website. The potential of open banking is both an advantage and a cause for concern. Its advantage is the ease of use on the customer side and the ability to access every product and price with a single interface. However, since the service provider that will implement this interface may become a monopoly, it may not be stingy in transaction costs and may not always consider the customer side in the products it will offer to customers when the competitive environment is eliminated.

Currently, open banking applications are used by many banks in the market. However, transactions made through a bank's mobile application are still limited. In the coming period, the use of all other functions such as mutual funds and stocks, opening term deposits and withdrawing loans in open banking applications will increase the transitivity among bank customers. The inclusion of digital banks in the system will change the game even more.

Almost all of the conventional banks that conduct traditional banking in the market have a mobile banking application. This situation shows that these banks, which are engaged in traditional branch banking, also perform digital banking transactions with the field of mobile channels and a kind of hybrid structure has emerged. In hybrid vehicles, just as the engine works with the power it receives from the electric battery according to the traditional gasoline method based on its location, this new model hybrid banking also works in this way. Mobile banking applications will be an important factor that will reduce operational costs such as branch personnel etc. included in non-interest expense, which is the most important expense item of banks. This situation has also revealed a new model that will make the branch banking in the system.

12.5 DIGITAL BANKING IN TURKEY

With the Digital Banking Regulation published by the BRSA on 1 January 2022, digital banks in Turkey have been established. The regulation also sets out the necessary infrastructure and conditions for the operation of digital banks, which will operate through digital channels without any branch network. Accordingly, digital banks, whose capital will be determined as 1 billion TL, will not have branches other than the general directorate and its affiliated units. At the same time, according to the regulation, these banks will only be able to serve consumers and customers of small medium enterprises (SME) quality, and they will only be able to provide foreign currency loans for companies exceeding the size of SMEs. One of the important issues is that if these traditional banks want to operate in the field of digital banking, they will be able to do only digital banking provided that they obtain permission from the BRSA and close their physical branches within the specified periods.

As of March 2023, five organizations have received digital banking licenses.

The fact that remote customer acquisition has started allows customers to become customers of a bank through mobile channels without going to any bank branch. As of March 2023, the number of banks providing remote customer acceptance services in Turkey is 17.

As of March 2023, the number of remote customer applications was approximately 1 million 399 thousand. Remote customer acquisition concluded with a customer representative is approximately 617 thousand. As of March 2023, the number of customers acquired through collective acquisition, online application and branch channels was 1 million 178 thousand (UMT remote and branch customer acquisition statistics).

12.6 RESULT

In the last century following the Industrial Revolution, everything has changed in a dizzying way. The spread of the internet has accelerated access to information. The rapid proliferation of smartphones has allowed many elements of daily use to be gathered on a single device. With the digital transformation, banks have also begun to change their traditional roles and have rapidly adapted to digital channels. Banks have perhaps the highest customer data among all sectors. Digital technologies have led to a faster and healthier assessment of customer trends.

On the other hand, the influence of "culture" on the adoption of electronic banking applications has also begun to be questioned. In their study, Zhang and others conducted a meta-analysis of 62 samples from 27 countries, and the results showed the effects of each of Eric Hofstede's national cultural dimensions. When using electronic banking, people living in countries with high power distances give more importance to social interaction and trust, while in countries where individualism is strong, performance expectation and risk perception are more prominent. In country cultures where avoidance of uncertainty is more dominant, performance expectation and confidence are more prominent (Zhang et al., 2018).

Mobile banking, the use of which has become widespread rapidly, developing FinTechs and the widespread use of open banking applications have led to the evolution of traditional branch banking to a hybrid model. Digital banks, which have emerged recently and are considered certain to become widespread, will provide services only through mobile channels in a branchless manner. While performing all the transactions that a traditional bank can do, digital banks, whose operational expenses will be incomparably lower than traditional banks, will also change the rules of the game in the near future.

NOTES

1 Third Pary Provider (TPP) is a common term used specifically in the field of technology and finance and refers to a third-party provider or service provider. It is the service or product that a company or organization does not offer within its own structure from another company or from an external source and offering it to its users or customers.
2 The Payment Systems Directives are two main regulatory documents issued to regulate and integrate payment services in the European Union and aim to regulate the activities of companies providing payment services in EU members.

REFERENCES

Anis, I., Gani, L., Fauzi, H., Hermawan, A.A., Adhariani, D. (2023). "The sustainability awareness of banking institutions in Indonesia, its implication on profitability by the mediating role of operational efficiency". *Asian Journal of Accounting Research*, 8(4): 356–372. https://doi.org/10.1108/AJAR-06-2022-0179

Banks Association of Turkey March 2023 Digital, Internet, Mobile Banking Statistics. Access https://www.tbb.org.tr/tr/bankacilik/banka-ve-sektor-bilgileri/istatistiki-raporlar/59

Chen, J., Lam. K. (2014). "How to prepare for Asia's digital banking boom." McKinsey and Company. Retrieved from. https://asia.udp.cl/Informes/2014/Asias_digital

Cocuklarda Bilisim Teknolojileri Kullanim Arastirmasi (2021). https://data.tuik.gov.tr/Bulten/Index?p=Cocuklarda-Bilisim-Teknolojileri-Kullanim-Arastirmasi-2021-41132 (Date of access 13.04.2023).

Collier and MacLachlan (1998). *Charles Babages and the Engines of Perfection*, (Chapter 7, pp. 104–111), Oxford University Press.

Dijital Internet Mobil Bankacilik Istatistikleri-Aralik (2022). https://www.tbb.org.tr/Content/Upload/istatistikiraporlar/ekler/3999/Dijital-Internet-Mobil_Bankacilik_Istatistikleri-Aralik_2022.pdf (Date of Access 13.04.2023).

Fong, M. W. L. (2017). "Digital divide between urban and rural regions in China". *The Electronic Journal of Information Systems in Developing Countries*, 36(1): 1–31.

Harchekar, J. S. (2018). "Digitalization in Banking Sector" international journal of trend in scientific research and development (IJTSRD) International Open Access Journal | www.ijtsrd.com ISSN No: 2456–6470. *Conference Issue – ICDEBI-2018 International Conference on Digital Economy and Its Impact on Business and Industry*. Maharashtra, India.

Karsh, A. S., Abufara, Y. (2020). "The new era of financial technology in banking industry". *Journal of Southwest Jiaotong* Üniversity, 55(4): 1–13.

Sardana, V., Singhania, S. (2018). "Digital technology in the realm of banking: A review of literature". *International Journal of Research in Finance and Management*, 1(2): 28–32.

TurkStat. https://data.tuik.gov.tr/Bulten/Index?p=Cocuklarda-Bilisim-Teknolojileri-Kullanim-Arastirmasi-2021-41132

Uzaktan ve Subeden Musteri Edinimi Istatistikleri Mart (2023). https://www.tbb.org.tr/Content/Upload/istatistikiraporlar/ekler/4031/Uzaktan_ve_Subeden_Musteri_Edinimi_Istatistikleri-Mart_2023.pdf (Date of Access 14.04.2023).

Zhang, Y., Weng, Q., Zhu, N. (2018). "The relationships between electronic banking adoption and its antecedents: A meta-analytic study of the role of national culture". *International Journal of Information Management,* 40: 76–87.

13 The Impact of Artificial Intelligence and Business Intelligence on Banking Industry for Enhanced Data-Driven Decision-Making

Shantharuby Buvanendra, Harshi Garg, Mohammad Kashif, and Priyank Sharma

13.1 INTRODUCTION

Artificial intelligence (AI) is a general, umbrella word. The objective of this chapter is to spread knowledge among individuals who use AI and business intelligence (BI) to make choices in the banking field. In validating the idea, the writers offer many examples in the chapter. This section discusses the use of AI in business decision-making. Overall, each choice depends on likelihood; however, with the support of AI choices will be dependent on assurance. AI relies on brain networks and fuzzy reasoning. Advertising choices require several kinds of issues. Machine learning models can address the division and satisfy customers' requirements. The use of AI helps some decision-making processes while also reducing time and effort. A system with AI can acquire data, predict, and evaluate trends. The technology is capable of calculating an individual's life cycle (Zohuri, 2020).

Figure 13.1 represents the historical development of AI in the financial sector, which illustrates significant transitions from algorithms to powerful AI tools. This figure indicates how each stage eventually increased banking procedures, resulting in improved reliability and rational decisions.

AI decreases the likelihood of bounces in an algorithm and is frequently referred to as information mining or sentiment mining. Opinion analysis enables internet searches for thoughts and feelings. This is how companies can discover more about their particular services and intended demographics. AI provides numerous search platforms, along with particular internet pages. This chapter aims to present strategies to enhance the banking sector through AI and BI. The financial sector's difficulties can be overcome through the use of AI (Usman et al., 2024).

DOI: 10.1201/9781003564621-13

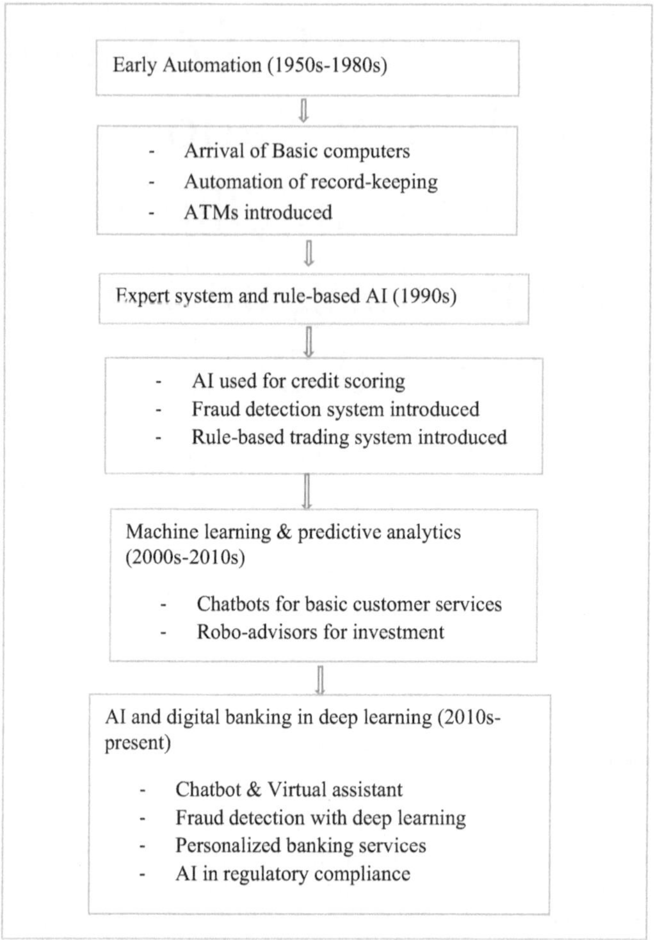

FIGURE 13.1 History of artificial intelligence in the banking sector. (Source: Authors' compilation.)

This research examines the scientific innovations needed in the banking field and offers predicted novel approaches that match the requirements of interested parties and help them create significant profits from their offerings. The philosophical foundations of contemporary scholarship in artificial awareness had been evolving for two or three centuries when Mary Shelley definitively connected modern science to the genesis myth. The section emphasizes more on AI educational history. Aristotle's influence provides the basis for such a history that includes AI. Aristotle linked the creativity, attractions, and worries of the earliest Greek tradition with the careful evaluation and ready conviction that would set the norm for contemporary science (Sun et al., 2018).

Aristotle proved that variation is a particularly appealing element in creation. He described 'the act of thinking' as 'the study of things that have transformed in his metaphysics'. He differentiated between the problem and a particular kind of item: an image created from metal that looks like an individual. When the metal is formed into a unique framework, it transforms. The problem of copyright offers an ethical reason for integrating new concepts, such as transfer calculation and analysis. We are managing models that are various kinds of magnetic stuff, with the movements of this material resolving various aspects of the program of action strategy (Kashif et al., 2023b).

13.1.1 AI in the Banking Services

To increase production, people have consistently developed stronger equipment. The arrival of motorcycles and, afterward, vehicles completely altered the way we travel. These gadgets were developed with engines that burned fuel and wheels, each of which was a universal invention. AI, the latest broad-purpose innovation, has been applied to change the banking sector and corporate, similar to what the personal computer and web did before it. The emergence of AI has created considerable chaos in the field of financial services because more banks are interested in employing AI-powered methods to enhance current operations. AI is transforming our relationships with electronics and moving the role of the intellectual workload from people to robots. You are likely to ask Google, or Siri now, when we have to understand where we are going and what to perform to accomplish something (Sun et al., 2018).

The world of finance has grown more volatile and intense as a consequence of internationalization and growing openness to trade. Users nowadays assume unusual service when they employ the goods or services of a business, or, to phrase it differently, there is a greater focus on client fulfillment. As a consequence, independent management of information that doesn't need human involvement might prove extremely beneficial to institutions to improve efficiency and precision. Four different groups can be employed to group numerous possible uses of AI in the financial industry. There are original front-office apps designed for customers and administration programs at businesses. Another issue is the rules and laws affecting investing and managing a portfolio. Most banks remain in the testing phase, but certain ones have completely incorporated current approaches throughout their business processes. Digital money laundering has been investigated as a possible use for AI. Financial sectors are working with AI to present information from accounts and statutory files, like financial statements, and derive the appropriate sections (Dhamija & Bag, 2020).

Software with AI can create algorithms by examining information and performing retrospective testing to gain insight from previous errors. Numerous existing banking technology instruments have evolved into accurate AI algorithms as time passed. Digital financial savings software assists individuals in making better investing portfolio decisions, robot counselors enable 100% automation for certain investment operations, and a new PWC study estimates that AI could add over $16 trillion to the world's economy by 2030. Over $6 billion has reportedly been expended worldwide on AI programs. It is predicted that the financial services firm will have earned $1.5 trillion as an outcome of AI execution, primarily through the demise of offices. In the past few years, advances in technology have allowed AI to provide professional

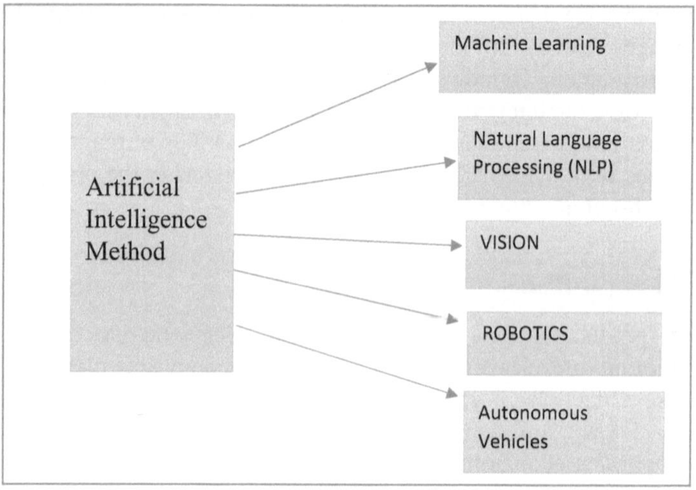

FIGURE 13.2 Artificial intelligence method. (Source: Authors' compilation.)

cognitive technology that involves introducing analytics into applications to assist with organizational duties (Abusweilem & Abualoush, 2019; Garg et al., 2024).

This involves enhancing the efficacy of knowledge, creating more precise and reliable information results, and enabling workers to carry out a high degree of activities. Over the past decade, AI-based solutions have proven to be efficient and helpful. However, numerous business executives continue to remain unsure about how to properly integrate AI into their companies. Even though 85% of enterprise leaders noticed AI as an essential resource to offer companies an ongoing competitive edge, an investigation revealed that only 39% of business executives had a plan of action for the implementation of AI while they were uncertain how to use it throughout their companies (Kashif et al., 2024b).

Figure 13.2 describes several types of AI approaches, especially NLP (natural language processing), Robotics, and Vision, that are commonly used in the banking sector. Banks may oversee massive amounts of data, eliminate repetitive tasks, simplify relationships with clients, and increase safety procedures due to these AI algorithms. Every tool has an individual purpose. For example, machine learning allows customized products and predictions. NLP supports chatbots and consumer service; Robotics simplifies daily operations; and Vision can detect theft and fraud. Collectively, these tools boost company precision, efficiency, and client satisfaction.

13.1.2 Overview of the BI

BI is crucial for helping managers in obtaining knowledge to enhance efficiency or make better choices. BI could improve and help the efficacy of operational regulation and its effect on corporate-level choices, supervisory structure, and financial management, offering better strategies. BI pertains to the internet, devices, platforms, and programs that are employed to create, analyze, interact, and present company data to make the best company choices. This approach will provide them with limitless support in acquiring, studying, and managing their information to make more informed choices while developing legal procedures and guidelines. BI is often characterized as a computerized method for generating predictions and conclusions from initial information obtained from different sources and organized in a structured way to enhance company processes and methods. In business BI layouts, the most efficient approach is to distinguish the collecting information and statistical management processes related to the background infrastructure from the analysis of information and presentation to consumers through the user interface (Kashif et al., 2024a; Zhou et al., 2023).

In BI, executed transactions provide information that is stored in functional databases referred to as Online Transaction processing (OLTP). Throughout the collection and processing activities, OLTP information is kept in an organized data store referred to as the data warehouse. With a database, numerous query optimization strategies can be employed to expedite information and perform statistics inquiries. To accomplish this speed, data stores produce subgroups of the data store known as data marts. Conventional business analytical systems additionally incorporate reporting capabilities for accessing information about transactions kept in a data repository. Hence, studying this information about transactions can assist us in discovering trends and identifying business developments (Zhou et al., 2023).

BI information sources are now going outside standard transaction information to include contemporary information streams like cell phones and data from sensors, along with internet communications produced by corporate intranets and worker and client profiles. The vast majority of current information is informal like comments put on social media sites and data received through different devices. As a consequence, the primary challenge is to maintain these current sources of data as comparable to traditional database structures, while also maintaining query speed. Regarding data analysis, additional information corresponds to a greater opportunity for finding fresh perspectives (Shajar, et al., 2024b).

Figure 13.3 highlights the history of BI in the financial sector, which indicates significant improvements from the initial implementation of statistical reporting instruments to the advanced BI systems of today. This milestone reveals how organizations can now acquire, analyze, and employ information to enhance their business advantage, tactical understanding, and decisions through BI. From basic information collection to advanced analysis, BI systems have changed in recent years, providing valuable insights into risk control, consumer habits, and business efficacy. This comprehensive overview reveals how important BI is to boosting the utilization of data and promoting development in the field of banking.

In recent years, throughout the latest commercial valuation, an immense quantity of information has been produced by computing devices like GPS, detectors, internet or app infrastructure, or by individuals using social networks; each minute, storage devices include enormous quantities of information produced by businesses. This huge amount of information derives from internet pages, networks of people, evaluations, apps for the Internet of Things, gauges, and electronic articles. Technological developments in information and communication gadgets are making it easier to gather tremendous quantities of diverse information from different places. This material contains both organized and

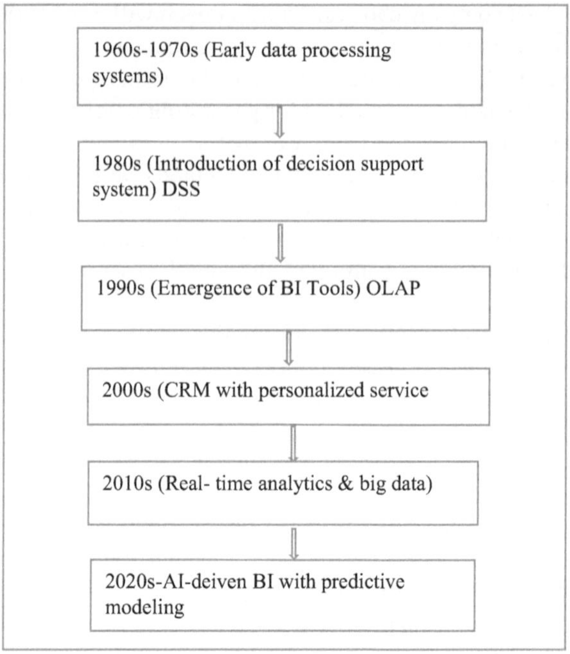

FIGURE 13.3 History of business intelligence in the banking sector. (Source: Authors' compilation.)

unorganized data, along with complex and basic data. Nowadays, companies receive up to 80% of their earnings from the examination of raw information. As a consequence of this examination of unorganized data including helpful data, the company can enhance its company's efficiency cycle. Likewise, it has significant consequences for safety, education, banking, and business. AI, big data analytics, and administration can assist in achieving business knowledge. Big data challenges represent an essential problem from an analysis perspective. Because of the rise in the information, there are fresh possibilities within the realm of BI, which serves as more than merely an instrument to evaluate past data patterns; it may additionally combine data gathered by detectors along with additional actual time private data to derive knowledge that is not frequently accessible, which is known as contextual BI. For company operations, BI is referred to as functional BI, and it provides information in actual time regarding these operations, like getting immediate input for customer service as a consequence of their job. Analytics rules can be created using meta-information concerning the data that has been offered, leading to self-service options. As a consequence, these new BI methods must be handled effectively to guarantee that the regulatory frameworks and management are not violated (Lim et al., 2013).

13.1.3 BI and the Banking Sector

A BI platform provides an established function as an interaction and data diffusion route, preferably one that is reliable, accessible, open, clear, and durable, enabling the investigation and assessment of business outcomes while upholding the confidentiality of information. BI is a method of analyzing, gathering, and sharing data to minimize risk for each business choice. Business researchers describe BI as an enterprise administration word that describes the software and hardware employed to analyze information regarding a company to assist them make better choices for their company. The increasing significance of BI technology represents three attached trends, increasing requirements for providing key risk and achievement signals to collaborators, and serious issues in correctly dealing with the heavily linked operations (Lim et al., 2013).

Figure 13.4 shows the BI process utilized by institutions that reveals the crucial phases in changing unstructured information into valuable knowledge. The collection of data, acceptance, and presentation is often included in this method. Banks can identify trends, assess efficiency, and arrive at informed choices by carefully examining data from various sources. Banks cannot effectively differentiate between excellent and terrible users until they have a piece of complete information about their clients from the incorporated databases. Data storage software can assist in integrating numerous subsystems into a BI structure. The BI method includes the fundamental business structure and takes a patterned nature (Jourdan et al., 2008).

BI is the deliberate, systematic conversion of knowledge from every source of data into novel formats to present economically driven and focused outcomes. It often includes a combination of tools, databases, and suppliers to provide a framework that not only provides the foundational answer but also combines the ability to evolve with the company's needs. The primary categories of BI advantages can be effectively connected to the existing business planning method. The financial commitment assists the organization in realizing its business objectives and is a crucial factor in assessing if the purchase of BI is worthwhile. In a bank, BI could bring together customer data to offer significant insights for choice makers, enabling them to increase their ability while offering enhanced client service (Kashif et al., 2023a; Lim et al., 2013).

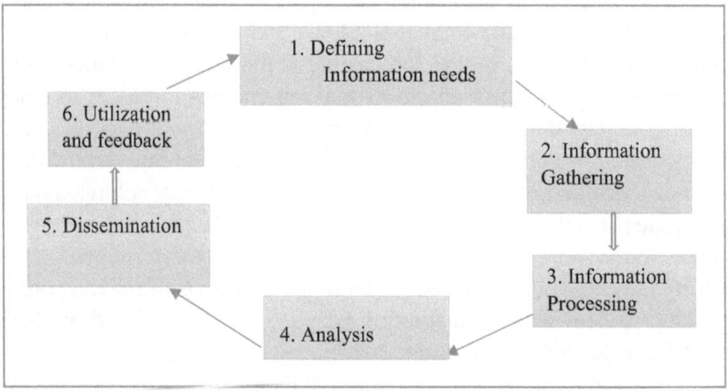

FIGURE 13.4 Business intelligence process in banking sector. (Source: Authors' compilation.)

13.2 LITERATURE REVIEW

The investigator tried to conduct an extensive assessment of current studies on BI, AI, and decision-making in the financial sector, BI and employees, and the performance of banks. This project encourages the investigator to recognize multiple research spaces, which results in the creation of particular assertions and queries for the study as a whole. The literature review is essential for study since it defines its scope and particular issues. The investigator has separated the review of research according to the topics included, such as studies linked to BI and making choices in the banking sector, and research related to BI.

13.2.1 AI, BI, and Decision-Making

Several scholars and practitioners around the country seek to concentrate on educating company executives along with other persons tasked with managing firms. Although the foundations of information systems for management have improved significantly, effective fulfillment of leadership data requirements has seen fewer advances. The development of structures to meet executive data needs has a long track record, it has been driven by a variety of specifications: some keep secure (developing, performing, and changing approach; observing personal exercise), whereas others grow or have an unstable life phase: verifying closest situation; focusing special consideration on possibilities and hazards (Skyrius et al., 2013).

The theory of a commercial exercise promotes the knowledge situation. This kind of behavior is constant and is usually supported by a management information system (MIS)'s abilities. Yet, this nature is unstable and unreliable, requiring skilled and able support; this is an aspect of a BI system and associated with programs, such as support for decisions, strategic intelligence, functional intelligence, early alerting system, and a variety of structures that help in verification, perceptions, and thinking critically (Shafi Reshi & Ahmad Khan, 2014). Actions are taken depending on the accuracy of the accessible data, instruments, internet, confidence, adaptability, and choice-making competency of the decision. BI ideas ought to motivate managers to consider their gut feeling by using rational resources that can confirm and support instinct before deploying (Ruiz-Real et al., 2021; Shajar et al., 2024a).

Administrators can also employ forecasting techniques to enhance their ability to make choices. The present scenario of making choices forces us to enjoy every advantage of company data. Collecting enormous quantities of data does not ensure successful decisions, a company's business analytics system must transform precise data into knowledge and render it accessible in actual time to policymakers as they choose a decision (Perifanis & Kitsios, 2023). Relevant data has to be provided when it is required. It uncovers patterns and trends that would otherwise go unreported by those making the decisions. An effective bank needs a basis of reliable, precise, and complete data, to deal with an institution based on sense, updated predictions, or averages (Palanivelu Head & Vasanthi, 2020).

BI can enhance the return on investment (ROI) for highly incorporated banking applications and other optional systems by revealing the vast amount of knowledge kept within these platforms by combining data from numerous sources, like a variety of clients, broad extend networks, investors at a data center, and the use of numerous tools for BI and AI (Oyekunle & Boohene, 2024). A decision support system can be described as an electronic system of information that includes data and models for addressing informal and unorganized issues throughout an extensive operational area. Many companies are turning toward decision assistance programs to enhance their capacity for making choices.

13.2.2 AI and BI in the Banking Sector

The company's information processes contain several phases: data requirement means, material collection, data processing, assessment, dissemination of information, evaluation, input, and reporting. The process design has been verified if the provided input revisits or reorganizes the requirement of information. In the company's business process, there is usually insufficient emphasis on a particular subject or issue, and the assets of a corporate intelligence platform are employed for ongoing monitoring of inside and outside company circumstances (Moro et al., 2015). These systems fulfill essential data requirements by keeping users updated on the condition of the company, often integrating a tracking ability with alarms, news accounts, and additional tools to attract attention to shifts (Marshall et al., 2004). A key characteristic of BI systems is their propensity to give a full synthesis context, which helps in the avoidance of errors. Automation is changing the financial service sector because of internet connectivity and the increasing number of cell phones and applications; current banking institutions have higher rivalry, changing customer needs, and a requirement for strict supervision and risk control in a highly volatile market (Lim et al., 2013).

Innovation has provided enormous analytical intelligence skills. The instruments that are accessible to the finance and banking sector for using information about clients offer information that may result in improved procedures and company choices (Kitsios & Kamariotou, 2021). This is an overview of some of the ways financial companies and banks are employing BI tools to improve revenue, minimize risks, make profitable and efficient choices, and achieve an advantage over their competitors. Information technology promotes the exchange of information and improves understanding (Khan & Quadri, 2012). It improves the group by offering the required assistance at all stages. The finance industry in the nation has been eager to accept BI. The level of complexity of BI increases with the size of the bank of staff, information, and so on. Transmission of BI is an expensive activity, but its outcomes are advantageous because it decreases operating expenses (Kashif et al., 2024c).

In India, they must manage individuals with different attitudes and actions. Clients and the multi-segment structure of ongoing financial services offer client protection. The integration of various parts is needed for smooth operation. BI is vitally essential. This allows one to establish an individual thing by combining and merging different components (Jourdan et al., 2008).

The finance industry has seen substantial shifts. With rising shifts, the structure should be reinforced to prevent a downturn. AI acts as a means for pushing this industry ahead. The banking system was ineffective, tedious, and unfavorable to clients. There was no real drive to figure out what the buyer was searching for. As they increased in size, operation, and agreements, physical labor became exorbitant. As time went on, they pushed up with different products and multi-segmentation to offer users one convenient solution (Herschel & Jones, 2005). This range should be selected for smooth collaboration and organizing with BI, where BI serves an essential part. Intelligence and ability are essential for handling and extracting huge quantities of information. It must be deployed with the correct equipment, culture, and methods. The real advantage of such analytical instruments can only be attained if they are utilized correctly (Fombellida et al., 2020).

13.2.3 BI and Employee Performance

Globalization has influenced the tactical environment of the company. It has led to significant relationships and difficulty, which has additionally become the root cause, surrounding almost any sector in all nations. The processes of knowledge output, dispersion, and preservation have been fundamentally altered, to a level where enormous quantities of data are generated and acquired beyond the control of any company. As the cost of computing and transmission capabilities power has decreased, it has grown cheaper for businesses and individuals to acquire and employ technology (Choi et al., 2022). To address the difficulties of internationally spread operations and resolve regional independence with worldwide oversight, companies have turned to modern technological solutions (Carter, 2018).

Modern technologies have also allowed companies to offer goods and services to their consumers. Companies have become more technologically sophisticated. Empowering staff members, process enhancement, and technological advances are a few examples of durability and long-lasting tactics employed by the banking sector to keep up with modern times (Ahmad & Mustafa, 2022). The consumers are more engaged, politically involved, updated, and envious. Banks have to fully comprehend the requirements of their customers before providing products and goods that satisfy them. To attract new clients, maintain current ones, and find improved methods to manage the company, banks have to provide their decision-makers with an in-depth overview of their function. The BI system offered the idea of information exploration, which banks quickly adopted to assist with their decision process regarding the company. AI enables a company to acquire knowledge and comprehension from its own experiences. Particular knowledge management tasks concentrate on the company organizing, storing, and applying information for problem resolution, adaptive learning, and strategic thinking (Kashif et al., 2024a).

13.3 RESEARCH METHOD

The research utilizes a combination of methods, including sentiment analysis, a comprehensive case investigation of JP Morgan Chase, and a SWOT analysis, to assess the impact of AI and BI advances on consumer attitudes and choices. The chapter additionally contains a comprehensive investigation of JP Morgan Chase, which analyzes its tactical adoption of AI and BI tools to enhance operations. This case study employs current articles, reports from the market, and public statements to explain the bank's strategy for applying technology for greater consumer experiences and cost savings.

This model exposes both the internal and external variables affecting the bank's success in an evolving scenario, leading to an in-depth comprehension of how JP Morgan Chase can maximize its upside while resolving the obstacles of implementing technological advances. NLP instrument was employed to evaluate client sentiment, especially programs such as VADER and TextBlob. Remarks collected from various pathways, like social media and customer reviews, have been processed by representing, establishing a normal, and clearing popular stop phrases. Each word gets a sentiment score determined by the present vocabulary, enabling a comprehensive assessment of overall sentiment.

The research provided beneficial data regarding how users see the financial institution's advancements in technology. Moreover, a SWOT analysis was carried out to identify the advantages, disadvantages, possibilities, and risks associated with JP Morgan Chase's utilization of these improvements.

Figure 13.5 presents the data extraction model used in this chapter. Several approaches, like sentiment analysis, a thorough case study of JP Morgan Chase, and

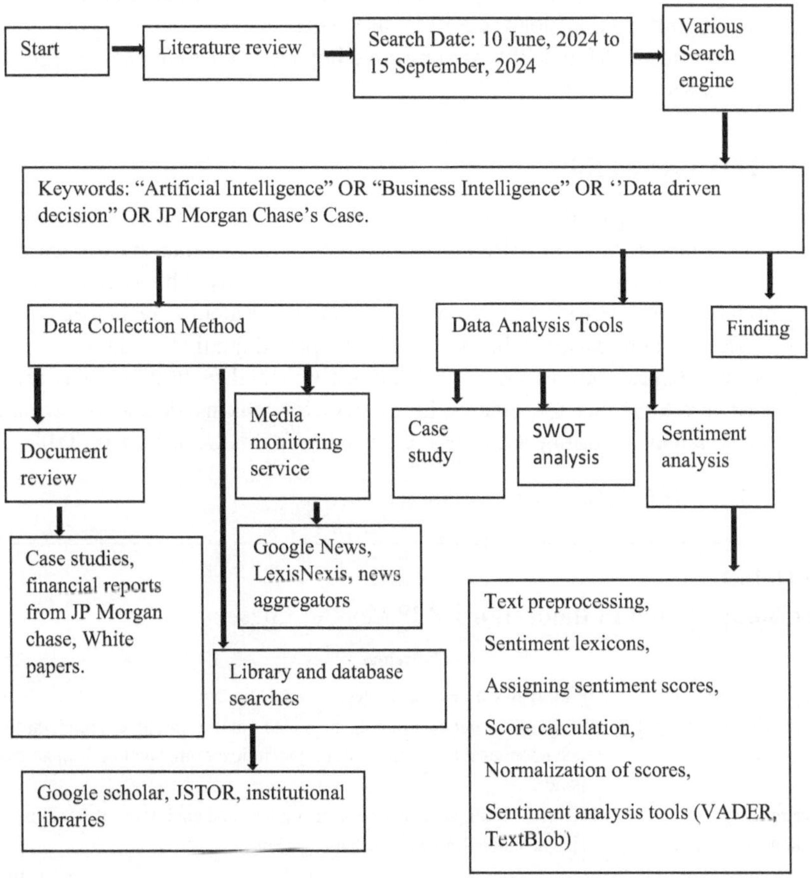

FIGURE 13.5 Data extraction model. (Source: Authors' compilation.)

SWOT analysis were used. This framework investigates how innovations in business and AI influence consumer perspectives and choices.

13.4 RESEARCH OBJECTIVES

1. To assess how AI and BI innovations affect banking and investment management decision-making processes.
2. To measure the effect of AI and BI advances on data-driven choices through SWOT analysis at the JP Morgan Chase banking industry.
3. To determine the impact of AI and BI advancements on customers through sentiment analysis on banking industry.

The hypothesis of the study:

H_1: AI and BI innovations significantly improve the banking and investment sector decision-making process.

H_2: AI and BI developments significantly increase JP Morgan Chase's accuracy in making decisions according to SWOT analysis.

$H_{3:}$ Innovations in AI and BI have a positive impact on customer sentiments in the banking industry.

13.5 DATA ANALYSIS

To effectively investigate the importance and responsiveness of the JP Morgan Chase case study, a broad range of different secondary sources of data have been examined as a component of the present investigation's data study. The assessment relied on information from news organizations, periodicals, academic evaluations, online communities, and professional discussions. It comprised qualitative data.

JP Morgan Chase, one of the world's largest financial institutions, has utilized AI and BI developments to enhance its process for making decisions in managing risks, financial planning, and interaction with clients as shown in Table 13.1.

TABLE 13.1
Case Study: AI and BI Innovations at JP Morgan Chase

	Overview
Banking organization	JP Morgan Chase is the richest universal financial services organization.
AI technology employed	JP Morgan's (contract intelligence) medium uses intelligent retrieval for document evaluation and choosing significant information from financial bonds.
BI instruments implemented	Business intelligence manifesto as Tableau and exclusive information analysis tools for actual decision-making.

(*Continued*)

TABLE 13.1 (*Continued*)
Case Study: AI and BI Innovations at JP Morgan Chase

	Overview
Purpose of revolution	To increase decision-making in loss prevention, boost investment management, and increase operational productivity using AI and BI tools.
	Area of Impact
Loss prevention	JP Morgan applies automatic technology algorithms to observe business action for loss prevention, detect fraud, and risk of malfunction, especially decreasing the event of transaction fraud.
Portfolio management	Machine learning algorithms support financial advisers to examine market needs and frame data-centric investment decisions, exceeding the bank's asset allocation.
Support service	Automation 'interactive agent' and 'digital assistant' handle client problems and queries, unburdening individual staff for complicated functions and decreasing reply time by 40%.
	Data Retrieval
Information system	BI instrument examines bulky statistics as market trade and client accounts, to give anticipating perception and improve decision-making.
AI-generated predictions	Predictive consequences are employed for trade estimation and transaction risk valuation, permitting financial managers to make more precise decisions.
Decision-making sense	Real-time information insight offered by the BI instrument helps JP Morgan respond speedily to understand trade conditions.
	Impact Process
Impact on accuracy	AI tools have decreased human faults in physical processes, by over 85%.
Impact on revenues	The use of AI and BI instruments provided a 60% depletion in runtime for complicated matters.

Source: Authors' compilation.

This financial institution processes immense quantities of information employing influential machine learning and analytical intelligence tools, leading to improved understanding and precise choices. This case study explores how JP Morgan has incorporated AI and BI into its business to enhance and oversee investments and streamline its banking procedures.

Table 13.2 describes the review of JP Morgan Chase's corporate papers, white papers, and case histories and offers an in-depth examination of their AI and BI innovations. These written materials include comprehensive details regarding the effectiveness, execution of strategies, and outcomes associated with AI and BI apps.

Table 13.3 depicts an examination of company reports, study papers, and publications that suggest that JP Morgan Chase's AI and BI applications have had a significant effect on the banking industry. Publications such as The Financial Times explore the developments and complications related to AI and BI, noting their significance for deciding the future of finance.

TABLE 13.2
Company Reports

Document Type	Description	Aspect Covered
Internal reports	Annual reports of JP Morgan Chase	Comprehensive standards on AI/BI execution, project landmark
Project reports	Reports on AI and BI execution program	Difficulties faced, progress reports, portfolio strategies using AI, BI
White papers	AI for loss prevention, financial optimization papers	Perception of AI applications, an advantage in loss prevention and fraud detection
Case studies	Case studies on AI, COin platform	Particular executions, results, and lessons learned

Source: Authors' compilation.

TABLE 13.3
Publication and Research

Source	Description
Industry reports	Gartner, McKinsey, Accenture
Journals	Articles from The Financial Times
Academic papers	IEEE Transactions, Harvard Business Review

Source: Authors' compilation.

Table 13.4 highlights the thorough spotlight on the Press release and media coverage report, which was published in popular publications and on global news outlets, and has served a significant part in highlighting the institution's technical successes and its consequences for the banking sector.

Table 13.5 presents the thematic analysis of JP Morgan Chase, emphasizing significant challenges concerning the banking industry execution of AI and BI. The main themes that appear from the study include controlling risk, data-driven choices, superior customer service, and effectiveness in operations. Each theme represents how JP Morgan Chase embraces innovation to improve its methods and capabilities. The table shows the bank's devotion to creativity and its deliberate emphasis on employing AI and BI tools.

Table 13.6 depicts the sentiment analysis carried out on JP Morgan Chase through various social media sites, such as Twitter, App Store, Financial Report, LinkedIn posts, and other blog posts. The thoughts and opinions of the general population concerning the bank's operations and AI tasks and innovations are included in this

TABLE 13.4
Press Release and Media Coverage

Region	Media Outlets Covered
United States	CNBC, Bloomberg
Asia	The Economic Times
Europe	BBC, Financial Times
Global	AI journal, TechCrunch

Source: Authors' compilation.

TABLE 13.5
Thematic Analysis of This Case Study

Theme	Findings	Impact	Implications
AI in loss prevention	Fraudulent activities and credit exposure assessment.	Fraud detection, risk analysis.	Enhance trust and security, and prevent portfolio loss.
Portfolio management	Trade examination, asset management.	Competitive returns, boost investment tactics.	Increase consumer satisfaction and competition in the market.
Client service	Digital assistant, artificial conversation entity.	60% saves time in reply.	Transforms customer involvement and support.
AI and BI processing	BI instruments for actual information insights.	Quick solution.	Keep competition in the market.
Data analysis activities	Fraud forecasting and trade tendency.	Reliable prediction and decreases losses.	Help dynamic portfolio decisions.
Resource optimization	Document review mechanization.	60% fast run time.	Lowers costs, prevents mistakes, and transforms service delivery.
Employee adaptation	Guidance on AI and BI tools.	Increased cooperation between individuals and AI.	Motivate revolution, and lessen resistance.

Source: Authors' manual thematic analysis.

TABLE 13.6
Sentiment Analysis

Platform	Comment	Sentiment	Sentiment Score
Twitter	AI chatbot is flexible for client service.	Positive	0.8
	JP Morgan's AI advancements are game changers in enhancing efficiency.	Positive	0.9
	AI is an amazing instrument to prevent fraud detection in financial sectors.	Positive	0.85
App Store	The AI bot doesn't solve my queries.	Negative	-0.6
	AI function feels robotic, it isn't able to understand my queries.	Negative	-0.5
Financial report	Potential to increase growth but the cost may influence short-term profitability.	Neutral	0.1
	AI-based investment insights are promising, but need more improvement.	Neutral	0.2
LinkedIn post	AI-based fraud detection is amazing at JP Morgan Chase.	Positive	0.7
	AI-based fraud detection systems are revolutionary and enhance the decision-making process.	Positive	0.8

Source: Authors' compilation.

assessment. The study reveals positive, negative, and neutral emotions by examining the content created by users to identify sentiment themes. The sentiment analysis provides helpful information into the opinion of customers concerning JP Morgan Chase's gadgets' success. These results empower the financial sector to further develop its business model and communicate with customers concerning feedback from the general populace.

Table 13.7 describes JP Morgan Chase's SWOT analysis for the implementation of AI and BI. Strengths entail a solid reputation and technological proficiency, whereas negative aspects are confidentiality of information, the possibility of enhanced service to clients, and difficulties from the market and laws and regulations. This brief study reveals an understanding of how the bank could benefit from its strengths while dealing with banking problems.

13.6 DISCUSSION

H_1: Testing hypothesis 1, thematic analysis was employed using a case study alongside media and company reports to examine the impact of AI and BI advances on customer views. The examination of quantitative evidence indicated significant themes

TABLE 13.7
SWOT Analysis

Strength	Weaknesses
Data-driven accuracy: JP Morgan Chase employs huge amounts of information using AI, authorizing error-free hazard evaluation and better economic prediction.	**High execution costs:** The money from combining AI and BI tools is material, influencing the interim benefits.
Operational organization: AI-generated networks simplify operations, decreasing human mistakes and boosting speed in procedures like fraud identification, product service, and support service.	**Reliance on information:** AI networks depend completely on information accuracy, and any data contravention and incorrectness could result in wrong decision-making.
Customer incorporation: BI permits JP Morgan Chase to give extremely customized finance solutions, increasing customer loyalty and client services.	**Expertise gaps:** The requirement for technical aptitude in AI and BI creates a professionalism gap, which can reduce the productivity and planning of the organization.
Industry-leading: JP Morgan Chase benefits from a competitive edge over rivals, as this organization is a creator in introducing AI and BI in the banking industry.	**Limited data:** Difficult AI functions can make it complex for relevant parties to acknowledge how decisions are made.
Advanced data processing: The bank uses AI for difficult data examination, allowing for better managerial decisions.	**Short-term benefits concern:** High financing in AI and BI devices may control short-term returns.
Operational streamlines: AI networks increase speed, lessen human mistakes, and boost client support and services.	
Strengthen customer awareness: AI and BI tools provide individualized banking support, enhancing client satisfaction and awareness.	
Power-source advantage: Early developers of AI and BI have spotted this organization as a leading firm in the high-tech industry.	
Opportunities	**Threat**
Expansion of AI-based commercial services: Introducing unique AI-based commercial products, like individualized and customized portfolio management solutions.	**Commencing data security and information privacy concern:** AI increases risks of information breaches and information technology security hazards.
Universal growth through AI-generated scheme: Deploying AI to evaluate current markets and efficient global operations.	**Extreme rivalry:** AI may disturb the market of rising FinTech organizations and digital giants financing.
Mechanization of compliance ordinance process: AI can assist with robotic difficult tasks, verifying compliance with complicated laws.	**Public judgment issue:** AI may cause trust issues among clients.
Data monetization: Using consumer and market information for monetization using collaboration or unique service offerings.	**Governmental unreliability surrounding AI method:** Modifying laws on AI in portfolio services could apply operational restrictions.

(Continued)

TABLE 13.7 (*Continued*)
SWOT Analysis

Strength	Weaknesses
Strengthening of client loyalty schemes: BI instruments can provide insights to plan more successful customer loyalty and faithfulness.	**Profitable unreliability influencing AI investment:** Universal economic variations could restrict the bank's capacity to extend financing in AI and BI tools.
Sustainability inventiveness: Employing AI to improve portfolio in sustainable finance.	**Technological disturbance:** Continual development in AI innovation may surpass JP Morgan Chase's capacity to adapt.

Source: Authors' compilation.

like improved productivity, satisfied customers, and innovations in technology. These themes emphasized the bank's intentional adoption of AI and BI to enhance its offerings and customer experience. The study provided an in-depth analysis of how customers and business experts evaluate the financial institution's advances, proving hypothesis 1 that AI and BI increase the banking service.

H_2: Testing hypothesis 2, SWOT analysis is approached to evaluate the inside and outside variables that affect JP Morgan Chase's AI and BI innovations. It complements the objective of the research by highlighting key areas of expertise that motivate making choices and reflecting on risks and possibilities for expansion in the future, thereby providing a helpful tool to examine the hypothesis, showing that hypothesis 2 is accepted.

H_3: Testing hypothesis 3, the authors used a NLP tool to acquire a clear understanding of customer sentiment regarding JP Morgan Chase's AI and BI developments. The contributors of this chapter have applied software like VADER and TextBlob to comprehend the feedback that people posted. First, the authors divided the responses and feedback into distinct phrases, making sure that every word was in smaller letters and eliminating frequently used phrases that offered little sense. Then contributors evaluated every word based on how favorable or adverse it was. By integrating these evaluations, the authors were enabled to figure out the general sentiment reflected in the remarks. This technique allows the contributors of this chapter to carefully evaluate customer input without the requirement for charts, revealing how proficient NLP is at understanding sentiments. Hypothesis 3 is accepted.

13.7 CONCLUSION

Nowadays, AI and BI are a need for organizations to enhance their skills and upgrade their services. This chapter introduced a thoughtful journey of AI and BI, especially focusing on the banking sector. The main purpose was to explore the growth of BI and AI and decision-making choices.

The purpose was achieved using a real-world study of the banking sector of JP Morgan Chase, a thorough literature review, and academic and corporate insights.

Media and published reports were also used to understand the foundation of the case study. A SWOT analysis was used to explore the strengths, weaknesses, opportunities, and threats of the JP Morgan Chase financial institution. The SWOT analysis revealed a lot of unique insights that will be advantageous to the marketers, stakeholders, and researchers in this field.

The methodology carried out for this research was a thorough literature review, a real-world case study, a SWOT analysis, and a sentiment analysis. The case study was analyzed through thematic analysis manually by the contributors of this chapter. Swot analysis was performed using newspaper headlines, media reports, published articles, literature reviews, and observation methods. Sentiment analysis was performed using software like VADER and TextBlob to understand the reviews that people posted on social media.

The outcomes of this study showed a significant shift in AI and BI advancements. The growth from conventional data processing to robotics, and individualized analytics has redefined the era of the financial sector's decision-making choice. The study described that the merger of AI and BI in financial institutions and businesses is not just a fashion but a basic movement in how banking sectors perform their services.

The role of AI and BI is pivotal to understanding that these advancements are today's requirement in the business sectors; these technologies save a lot of time, decrease human errors, improve customer satisfaction and customer loyalty, and provide a valuable service quality. The features of AI chatbots and virtual assistants have reduced a lot of manual work for employees. Future research suggests a comprehensive observation of the societal implications and ethical quandaries of AI, BI, and data analytics in the context of other business fields. As organizations continue to steer the internet era, the importance of AI and BI will play a significant role in building the success and improvement of the business sectors.

REFERENCES

Abusweilem, M. A., & Abualoush, S. (2019). The impact of knowledge management process and business intelligence on organizational performance. *Management Science Letters*, *9*(12), 2143–2156. https://doi.org/10.5267/j.msl.2019.6.020

Ahmad, H., & Mustafa, H. (2022). The impact of artificial intelligence, big data analytics and business intelligence on transforming capability and digital transformation in Jordanian telecommunication firms. *International Journal of Data and Network Science*, *6*(3), 727–732. https://doi.org/10.5267/j.ijdns.2022.3.009

Carter, D. (2018). How real is the impact of artificial intelligence? The business information survey 2018. *Business Information Review*, *35*(3), 99–115. https://doi.org/10.1177/0266382118790150

Choi, L. K., Panjaitan, A. S., & Apriliasari, D. (2022). The effectiveness of business intelligence management implementation in industry 4.0. *Startupreneur Business Digital (SABDA Journal)*, *1*(2), 115–125. https://doi.org/10.34306/sabda.v1i2.106

Dhamija, P., & Bag, S. (2020). Role of artificial intelligence in operations environment: a review and bibliometric analysis. *TQM Journal*, 32(4), 869–896.https://doi.org/10.1108/TQM-10-2019-0243

Fombellida, J., Martín-Rubio, I., Torres-Alegre, S., & Andina, D. (2020). Tackling business intelligence with bioinspired deep learning. *Neural Computing and Applications*, *32*(17), 13195–13202. https://doi.org/10.1007/s00521-018-3377-5

Garg, H., Ike, G. N., Kashif, M., & Azam, M. S. (2024). Dynamics of sustainable and ethically responsible business practices in corporate social responsibility. In K. Singh, J. Yadav, & R. Abraham (Eds.), *Technology-Driven Evolution of the Corporate Social Responsibility Ecosystem* (pp. 92–113). IGI Global. https://doi.org/10.4018/979-8-3693-3238-2.ch005

Herschel, R. T., & Jones, N. E. (2005). Knowledge management and business intelligence: The importance of integration. *Journal of Knowledge Management*, *9*(4), 45–55. https://doi.org/10.1108/13673270510610323

Jourdan, Z., Rainer, R. K., & Marshall, T. E. (2008). Business intelligence: An analysis of the literature. *Information Systems Management*, *25*(2), 121–131. https://doi.org/10.1080/10580530801941512

Kashif, M., David, A., Azam, Md. S., & Kumar, S. (2024a). Utilization of robo-advisory tools in decision support systems. In S. Gupta, S. Taneja, V. Kumar, E. Ozen, & L. Vardari (Eds.), *Robo-Advisors in Management* (pp. 76–89). IGIGlobal. https://doi.org/10.4018/979-8-3693-2849-1.ch005

Kashif, M., Garg, H., Weqar, F., & David, A. (2024b). Regulatory strategies and innovative solutions for deepfake technology. In G. Lakhera, S. Taneja, E. Ozen, M. Kukreti, & P. Kumar (Eds.), *Navigating the World of Deepfake Technology* (pp. 262–282). IGIGlobal. https://doi.org/10.4018/979-8-3693-5298-4.ch013

Kashif, M., Kumar, P., Ghai, S., & Kumar, S. (2023a). Disruptive technologies in computational finance. In A. Singh, S. Taneja, & P. Kumar (Eds.), *Algorithmic Approaches to Financial Technology: Forecasting, Trading, and Optimization* (pp. 46–60). IGIGlobal.

Kashif, M., Shajar, S. N., Singhal, N., & Kumar, P. (2023b). Achieving sustainable investment practices through green finance: Challenges and opportunities. In S. Taneja, P. Kumar, E. Ozen, & S. Grima (Eds.), *Sustainable Investments in Green Finance* (pp. 234–244). IGIGlobal.

Khan, R. A., & Quadri, S. M. K. (2012). Business intelligence: An integrated approach. *Business Intelligence Journal*, 5(1), 64–70.

Kitsios, F., & Kamariotou, M. (2021). Artificial intelligence and business strategy towards digital transformation: A research agenda. *Sustainability (Switzerland)*, *13*(4), 1–16. https://doi.org/10.3390/su13042025

Lim, E. P., Chen, H., & Chen, G. (2013). Business intelligence and analytics: Research directions. *ACM Transactions on Management Information Systems*, 3(4), 1–10. https://doi.org/10.1145/2407740.2407741

Marshall, B., McDonald, D., Chen, H., & Chung, W. (2004). EBizPort: Collecting and analyzing business intelligence information. *Journal of the American Society for Information Science and Technology*, *55*(10), 873–891. https://doi.org/10.1002/asi.20037

Moro, S., Cortez, P., & Rita, P. (2015). Business intelligence in banking: A literature analysis from 2002 to 2013 using text mining and latent Dirichlet allocation. *Expert Systems with Applications*, *42*(3), 1314–1324. https://doi.org/10.1016/j.eswa.2014.09.024

Oyekunle, D., & Boohene, D. (2024). Digital transformation potential: The role of artificial intelligence in business. *International Journal of Professional Business Review*, *9*(3), e04499. https://doi.org/10.26668/businessreview/2024.v9i3.4499

Palanivelu Head, V. R., & Vasanthi, B. (2020). Role of artificial intelligence in business transformation. *International Journal of Advanced Science and Technology*, *29*(4s), 392–400.

Perifanis, N. A., & Kitsios, F. (2023). Investigating the influence of artificial intelligence on business value in the digital Era of strategy: A literature review. *Information (Switzerland)*, *14*(2), 85. https://doi.org/10.3390/info14020085

Ruiz-Real, J. L., Uribe-Toril, J., Torres, J. A., & Pablo, J. D. E. (2021). Artificial intelligence in business and economics research: Trends and future. *Journal of Business Economics and Management*, *22*(1), 98–117. https://doi.org/10.3846/jbem.2020.13641

Shafi Reshi, Y., & Ahmad Khan, R. (2014). Information and knowledge management creating business intelligence through machine learning: an effective business decision making tool. *Information and Knowledge Management*, *4*(1). www.iiste.org

Shajar, S. N., Beg, K., Kashif, M., Khan, M., Saleem, S., & Usmani, F. (2024a). Enhancing firm performance through effective working capital management: a study of indian manufacturing firms listed at S&P BSE 500. *International Research Journal of Multidisciplinary Scope*, *5*(3), 661–669. https://doi.org/10.47857/irjms.2024.v05i03.0806

Shajar, S. N., Kashif, M., George, J., & Nasir, S. (2024b). The future of green finance: artificial intelligence-enabled solutions for a more sustainable world. In M. Kukreti, R. Tiwari, S. Sehajpal (Eds.), *Harnessing Blockchain-Digital Twin Fusion for Sustainable Investments*. (pp. 11–36). IGIGlobal. https://www.igi-global.com/book/harnessing-blockchain-digital-twin-fusion/330583

Skyrius, R., Kazakevičienė, G., & Bujauskas, V. (2013). From management information systems to business intelligence: the development of management information needs. *International Journal of Interactive Multimedia and Artificial Intelligence*, *2*(3), 31. https://doi.org/10.9781/ijimai.2013.234

Sun, X., Zhang, C., Li, G., Sun, D., Ren, F., Zomaya, A., & Ranjan, R. (2018). Detecting users' anomalous emotion using social media for business intelligence. *Journal of Computational Science*, *25*, 193–200. https://doi.org/10.1016/j.jocs.2017.05.029

Sun, Z., Sun, L., & Strang, K. (2018). Big data analytics services for enhancing business intelligence. *Journal of Computer Information Systems*, *58*(2), 162–169. https://doi.org/10.1080/08874417.2016.1220239

Usman, M., Moinuddin, M., & Khan, R. (2024). Unlocking insights: harnessing the power of business intelligence for strategic growth. *International Journal of Advanced Engineering Technologies and Innovations*, 1, 23–34.

Zhou, J., San, O. T., & Liu, Y. (2023). Design and implementation of enterprise financial decision support system based on business intelligence. *International Journal of Professional Business Review*, *8*(4), e0873. https://doi.org/10.26668/businessreview/2023.v8i4.873

Zohuri, B. (2020). From business intelligence to artificial intelligence. *Modern Approaches on Material Science*, *2*(3). https://doi.org/10.32474/mams.2020.02.000137

14 An Empirical Investigation into Analyzing the Impact of Implementing Artificial Intelligence in the Banking Industry

Amar Johri, Reepu, Sanjay Taneja, and Zelhuda Shamsuddin

14.1 INTRODUCTION

In recent decades, the application of artificial intelligence (AI) in different business areas has been considered the key to achieving sustainable growth and development. There is growing research on the application of AI in other places, which is now considered more comprehensively. There are considerable benefits that can be gained by the financial services industry and the global economy through the utilization of AI. Recent research indicates that the worldwide banking industry estimates that the investment in these new technological areas will increase to $39 trillion by 2029. The emergence of the first generation in the sector has developed through several stages throughout its history. The most current stage, financial 4.0, incorporates modern technology across various financial domains, including AI technologies. Financial institutions have begun implementing cutting-edge technologies to maintain relevant and keep ahead of their rivals (Kurni et al., 2022). During the 1960s, for instance, Barclays Bank was the first financial institution to use automated teller machines (ATMs), a famous example of Banking 2.0. In the current decade, it is identified that the new technologies have supported operational efficiencies with a focus on processing information, enabling addressing queries and cost reduction; this is now being stated as Banking 4.0.

A significant factor that contributed to the rapid transformation of banking services was the allure of AI-based financial components with lower operating expenses. According to the findings of our investigation, electronic money institutes (EMI) have grown their EMIs by a factor of three, resulting in a total of 40.8 million through the commercial licensing initiatives they have undertaken. Financial institutions have invested significantly in developing AI infrastructures and new financial

DOI: 10.1201/9781003564621-14

features to meet the overall demands (Kaur et al., 2020), even though AI technology is appealing and could have several possible effects. This situation can be mainly ascribed to the fact that clients have a significantly low level of knowledge regarding technologies that utilize AI. This component has been investigated in a general sense in previous study findings.

However, it is crucial to note that additional research is needed to fully comprehend the technical capabilities of AI. This investigation focuses on the specific understanding of AI rather than just the general knowledge of technological applications. The study has collected data from five different Asian countries, a sample size that exceeds the recommendations of other studies. The banking sector must stay abreast of how customers perceive various financial technologies, particularly the role of AI. Researchers must understand the level of consumer knowledge regarding AI technologies, especially in the banking business, to ensure the successful implementation and acceptance of AI in the industry.

Research is now being conducted to investigate the significant developments that have been made in financial technology. According to statistical research findings, by the year 2027, the banking sector will be responsible for around 13% of the estimated 450 billion USD that will be spent on AI worldwide. Suppose bank operating earnings improve by 9 to 15%. In that case, this investment has the potential to yield an annual return ranging from $200 to $340 billion. Because of the COVID-19 outbreak, the need for implementing digitization in financial services has been highlighted. Concerns have been raised regarding the existing knowledge and skill levels of the general people, as well as those of persons who provide these services, as a result of the significant increase in the consumption of digital services (Thaker et al., 2019).

According to the findings of a poll that was carried out among banking professionals, it was discovered that 77% of them think that the ability to apply AI adequately will determine whether financial institutions are successful. Based on previous research, nearly 6% of retail banks have successfully implemented an all-encompassing AI-enabled transformation plan throughout their business. However, it has been regarded that there is a higher potential for misperception in understanding the potential of AI in the current world scenario.

14.2 REVIEW OF LITERATURE

There is a growing focus among academicians and practitioners in different domains of supply chain, agriculture, and service-related enterprises. Implementing AI in banking and financial services helps reshape how banking operations are carried out and ensures a seamless flow of cash and capital. Due to this solution's adaptability, it can enhance the performance of different corporations. In addition, incorporating AI into banking operations is unavoidable, considering the intrinsic characteristics of the technology as well as the very competitive nature of the industry's working environment. Through AI, banking companies can now enhance overall productivity and efficiency, enabling better services to customers at a higher speed (Belanche et al., 2019).

The core aspect of implementing AI in banking services is to enhance the quality of services delivered to the customers, reduce operational costs, enable effective cross-selling of products, and ensure timely delivery of queries. The application of

AI in banking companies has significantly reduced overall costs, mainly associated with paperwork and long queues in banks for making payments or withdrawals. It is noted that the investment of AI in the banking sector will cross more than $400 billion in the coming years to ensure better service delivery.

One of the unique application areas of AI in banking is the implementation of a chatbot, which provides the necessary answers to queries and supports customers in using banking and related services through their smartphones without any hassles. The chatbots enable communication with customers in a more personalized manner, ensuring that queries are resolved quickly. Providing automated solutions to customer inquiries and collecting essential data that can be used to anticipate and address future concerns are two of the multiple functions that chatbot technology serves within the banking business (Suhartanto et al., 2020).

Furthermore, using sophisticated technologies leads to a better customer experience as they are not standing in long queues. The customers are satisfied with the experiences through AI, and their level of happiness is high. Over time, it has been noted that customers' preferences have undergone significant changes (Biswas et al., 2020). Customers today have a strong preference for tailored material and responses that are provided promptly. Researchers can assess and forecast consumer behavior and credit ratings by applying AI and machine learning techniques. These techniques are obtained through the utilization of a specific algorithm. Because of this, banks can build individualized plans for each of their customers. Financial institutions can utilize AI to assist them in digitizing their procedures to fulfill their customers' expectations (Garg et al., 2021).

Investigations into the analysis of sentiment: To produce and provide, the fundamental priority of every financial institution is to study and forecast customers' behavior. To better understand and cater to consumers' preferences, researchers use technology powered by AI to perform sentiment analysis. The goal of this technology is to forecast the emotions, sentiments, and responses of consumers. Based on the preferences and choices made by users, this technology gathers data to develop and show content specifically suited to those users (Maja and Letaba, 2022).

Automated work: The banking industry utilizes AI technology to minimize the need for human involvement in counting money. This is accomplished through digital computers that count money accurately and fast. This automated technology helps to reduce the stress on personnel and minimize the incidence of arithmetic errors during cash counting by facilitating the development of the daily business volume of banks. This technological advancement also helps to reduce the number of errors that occur (Patel et al., 2022).

Because of the complex nature of employment operations and the large number of financial transactions that corporations conduct, financial institutions are frequently more susceptible to becoming victims of fraud. As was said earlier, researchers in the field of AI use unsupervised learning programs to perform mathematical calculations and sophisticated algorithms to monitor the behavior of individuals and consumers (Rahman et al., 2022). When it comes to automating human operations within the banking industry, AI relies significantly on the programming method of machine learning. This helps to reduce the potential risks that could result in adverse effects on corporate performance.

14.3 OBJECTIVES

To analyze the critical aspect of implementing AI in supporting cost reduction in banking companies

To understand the impact of implementing AI in enhancing customer experience in the banking industry

To apprehend the effect of using AI in augmenting perceived usefulness among banking customers

To analyze the role of AI in increasing operational efficiencies in the banking sector

14.4 METHOD

The critical variables considered for the article are cost reduction, enhanced customer experience, perceived usefulness, and increased operational efficiency toward the impact of implementing AI in banking services. The research is conducted in India with selected banking companies. The researcher applied a descriptive research design to understand the topic more effectively. The researcher uses primary and secondary data for the article; the researcher chooses the sample population, which consists of employees working in banking companies, to source the data, and the sample respondents are selected through the convenience sampling method. For secondary data, the information was collected from different online libraries; the keywords used to filter the research articles are "Artificial Intelligence", "Banks", "Banking Sector", "AI in banking services", etc. The primary data collected were then analyzed through statistical tools; for this purpose, SPSS and Python were used. The following section provides a detailed data analysis and interpretation.

14.5 DATA ANALYSIS

This section involves presenting the statistical analysis (Table 14.1).

It is noted that Age group 26–35 years, with those under 25 coming in a close second. The gender distribution is exactly split between males and females. Statistics

TABLE 14.1
Demographic Analysis

Category	Sub-Category	Frequency	Percent
Age	Below 25 years	42	30.90
	26–35 years	48	35.30
	36–45 years	16	11.80
	46–55 years	18	13.20
	56–65 years	4	2.90
	Above 65 years	8	5.90
Gender	Male	68	50.00
	Female	68	50.00

(Continued)

TABLE 14.1 (*Continued*)
Demographic Analysis

Category	Sub-Category	Frequency	Percent
City	Metro	96	70.60
	Non-metro	40	29.40
Marital status	Single	28	20.60
	Married	106	77.90
	Others	2	1.50
Education	Graduation	78	57.40
	Postgraduation	20	14.70
	Professional	11	8.10
	Others	27	19.90
Annual income	Below 5 Lakhs	12	8.80
	6–10 Lakhs	25	18.40
	11–15 Lakhs	44	32.40
	16–20 Lakhs	24	17.60
	Above 20 Lakhs	31	22.80
Experience	Less than 2 years	20	14.70
	2–5 years	42	30.90
	5–10 years	38	27.90
	10–15 years	20	14.70
	More than 15 years	16	11.80
Implementation of AI	Yes	80	58.80
	No	56	41.20

Source: Authors.

show that city dwellers outnumber rural residents, and most city dwellers reside in metropolitan areas. Most people are married, whereas the level of education peaks at graduation. The distribution of income levels is pretty even throughout several ranges, with just a little concentration between 11 and 15 lakh. There is a noticeable concentration in the 2- to 5-year experience range. However, there are other groups as well. A slight bias exists in favor of using AI since more than 59% of respondents claim to utilize it. Additional research, possibly using visual aids, might reveal more specific insights and relationships between these aspects (Figure 14.1).

14.6 CORRELATION ANALYSIS

The provided correlation coefficients are all moderately high, suggesting strong positive connections between the variables. So, when one variable goes up, the others usually go up. Value perception, cost management, and enhanced customer experience are highly connected. This indicates a tight relationship between reducing expenses, improving the customer experience, and increasing the apparent value of AI in banking. Additionally, these three goals are likely to work together. Increasing operational efficiency and widespread use of AI in banking are also closely related to

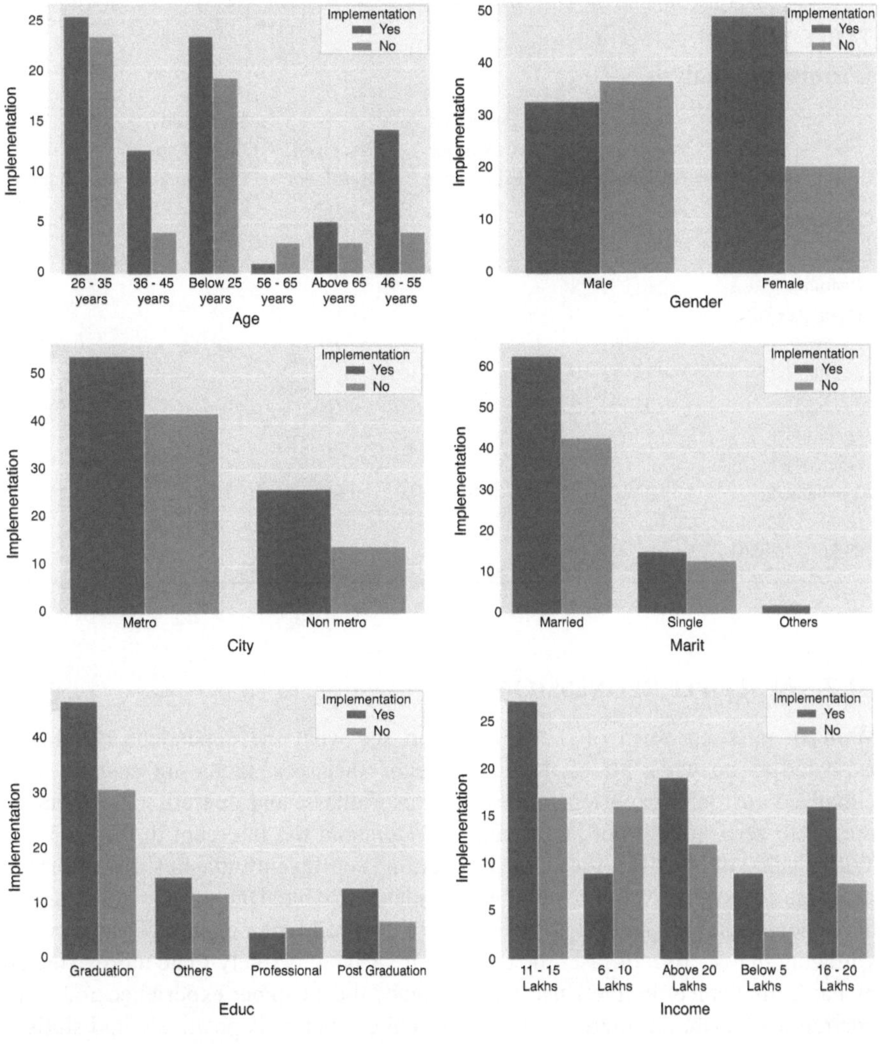

FIGURE 14.1 Exploratory data analysis between demographic variables and implementation of AI in banking.

these factors. This means that enhancing any of these parts will improve the others, which will lead to the banking industry adopting AI more generally. The correlation matrix reveals a mutually beneficial relationship between the following: cost reduction, enhanced customer experience, perceived usefulness, increased operational efficiency, and the widespread adoption of AI in banking. An all-encompassing and comprehensive AI adoption plan is likely to be produced by improving one area, as it will likely impact the others (Table 14.2).

TABLE 14.2
Correlation Analysis

Correlations	Cost Reduction	Enhanced Customer Experience	Perceived Usefulness	Increased Operational Efficiency	AI in Banking
Cost reduction	1	0.835	0.808	0.718	0.748
Enhanced customer experience	0.835	1	0.816	0.717	0.794
Perceived usefulness	0.808	0.816	1	0.713	0.822
Increased operational efficiency	0.718	0.717	0.713	1	0.764
AI in banking	0.748	0.794	0.822	0.764	1

Source: Authors

14.7 MULTIPLE REGRESSION

With an intercept term of 0.0613, we can see what the dependent variable is expected to be when all of the independent variables, including cost savings, enhanced customer experience, perceived usefulness, and operational efficiency, are set to zero. It may not be crucial to understand the intercept in this context. With a value of −0.0047, the "Cost Reduction" coefficient indicates that efforts to decrease costs negatively impact the dependent variable. Due to a lack of data, the dependent variable's statistical relevance of cost reduction cannot be determined with confidence. To rephrase, the dependent variable is likely to be more affected by banks that can effectively use AI to enhance the customer experience and demonstrate its worth. An increase in operational efficiency is positively and statistically significant ($p = 0.0$) with a coefficient of 0.2966. This indicates a positive correlation between the dependent variable and increased operational efficiency (Table 14.3).

Omnibus and Jarque-Bera tests determine if the regression model's errors, or residuals, are expected. No longer can we accept the null hypothesis of normalcy as both the Omnibus test statistic (43.424) and the Jarque–Bera test statistic (149.69) have p-values less than 0.05. This suggests that the error terms may not follow a normal distribution. The Durbin–Watson statistic (2.093) is used to identify residual autocorrelation. It is still difficult to determine whether autocorrelation exists due to the ambiguous value (between 0 and 4). Despite significant issues with error normalcy and maybe autocorrelation shown by the model fit statistics, the findings provide an exciting examination for the article. Customer experience and higher operational efficiency all appear to positively correlate with the dependent variable,

TABLE 14.3
Regression Analysis

Variables	Coeff	Std Err	t	p> |t|	0.025	0.075
Intercept	0.0613	0.134	0.458	0.648	−0.204	0.326
Cost reduction	−0.0047	0.085	−0.056	0.956	−0.172	0.163
Enhanced customer experience	0.2642	0.093	2.85	0.005	0.081	0.448
Perceived usefulness	0.3901	0.079	4.954	0	0.234	0.546
Increased operational efficiency	0.2966	0.067	4.426	0	0.164	0.429
Omnibus	43.424	Durbin–Watson	2.093			
Prob (Omnibus)	0	Jarque–Bera	149.69			
Skew	−1.128	Prob (JB)	0.03			
Kurtosis	7.618	Cord NO	14.4			

Source: Authors.

likely indicative of a successful AI deployment in the banking business. Despite the unfavorable correlation they imply, attempts to reduce costs do not have statistical significance and necessitate further research. Knowing what the dependent variable represents would allow for a more comprehensive interpretation, even though it is not explicitly stated.

14.8 STRUCTURAL EQUATION MODELING

Figure 14.2 shows the nature of the association between the variables; a corresponding latent variable represents each component; the path coefficients reveal the interplay between these components and the level toward acceptance. The model's cost-reduction variable is a latent one with a 1.71 variance in the error term. With a route coefficient of 0.26 and an error term variance of 1.47, AI in banking is associated with improved customer experience. In other words, this demonstrates a modestly positive influence, which explains the increasing usage of AI in banking by demonstrating that AI improves the client experience. Perceived usefulness possesses a more substantial effect with a path coefficient of 0.39 and an error term variance of 1.80. Improving operational efficiency also aids AI adoption, as demonstrated by a path coefficient of 0.30 and an error term variance of 1.59. According to this link, there has been an uptick in the likelihood of banks implementing AI systems to boost operational efficiency and simplify processes.

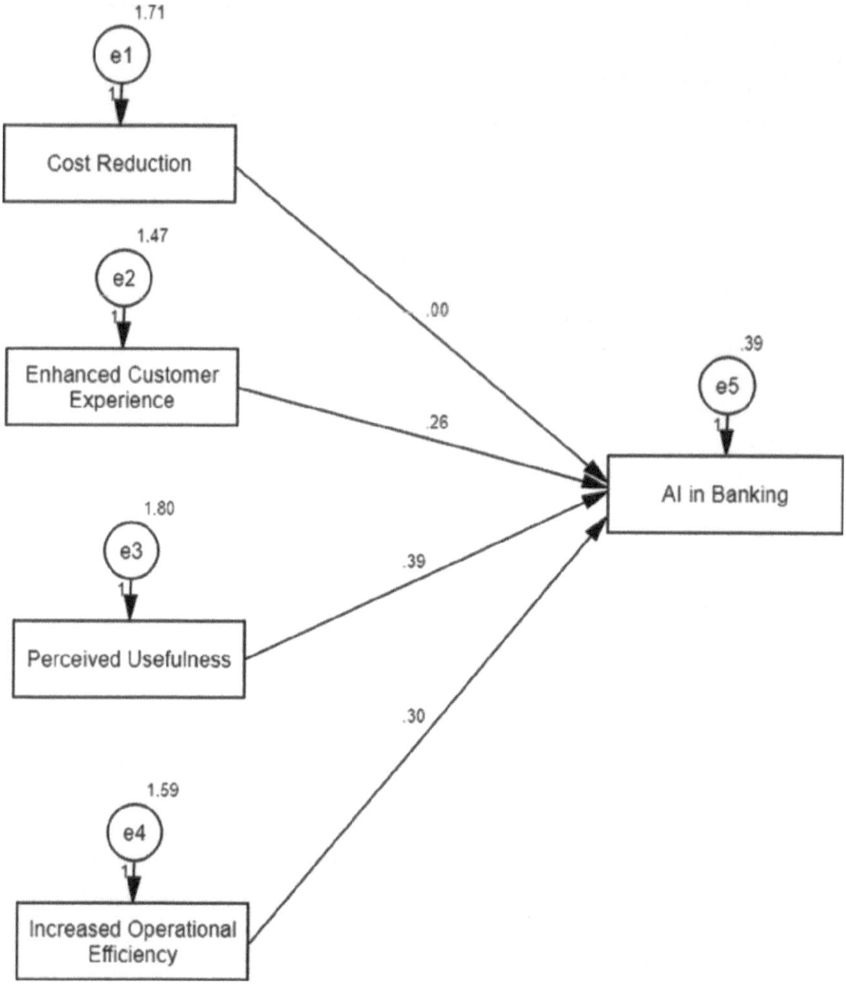

FIGURE 14.2 Overall model.

14.9 DISCUSSION

AI has emerged as a dominant factor in all aspects of contemporary life. Furthermore, it is worth noting that both developed and developing nations are actively working toward integrating AI into various other fields. AI is becoming increasingly significant for researchers and businesses, notably in the banking sector, which is characterized by the necessity for speedy transactions and the presence of potential fraud risks. Research shows that technology is advancing at a breakneck pace. The research in India aims to determine individuals' overall intent, as stated by Pitchay et al. (2019), who stated that using these tools reduces overall costs, enhances customer experience, increases perceived usefulness, and improves operational efficiency.

The first hypothesis posits that there exists a better association between the recognition of the significance of delivering exceptional customer service. The results stated that these tools also reveal that perceived usefulness and increasing operational efficiencies can be achieved sustainably. This is consistent with the research findings of Huang and Lee (2022); hence, these tools enable the organization to offer advantages effectively.

14.10 CONCLUSION

Over time, the banking industry experienced continuous upheaval. It is further noted that this transformation holds great importance in the banking industry. Beginning in 1472, the banking system followed a traditional approach before transitioning to AI-based banking in 2017. The analysis reveals that these tools help mitigate the risks associated with digital technology in transactional processes. Furthermore, this research study suggests that it is imperative to enhance security and protection policies in order to ensure improved customer service.

14.11 FUTURE DIRECTION

Further research is required to explore various aspects of this study. The scope of this research was restricted to a single country. Further research could be undertaken to explore various Asian and European countries, aiming to compare consumers' perspectives on AI. The research conducted exclusively examined the financial sector of the economy. In order to further our research, we will also consider the involvement of other financial sectors. In order to collect data through self-reporting, the methodology employed in this study was a survey. To enhance the precision of forthcoming research findings, it is advisable to employ alternative data collection methods, such as conducting tests in outdoor settings, and augment the sample size. Although mediation and moderation were not included in this study due to time constraints, future research will incorporate these variables to better understand their impact on AI. The model will be modified accordingly to accommodate these moderating and mediating factors.

REFERENCES

Belanche, D.; Casaló, L.V.; Flavián, C. (2019). Artificial intelligence in FinTech: Understanding robo-advisors adoption among customers. *Ind. Manag. Data Syst.* 2019, 119, 1411–1430.

Biswas, A.; Bhattacharjee, U.; Chakrabarti, A.K.; Tewari, D.N.; Banu, H.; Dutta, S. (2020). Emergence of novel coronavirus and COVID-19: Whether to stay or die out? *Crit. Rev. Microbiol.* 2020, 46, 182–193.

Garg, P.; Gupta, B.; Chauhan, A.K.; Sivarajah, U.; Gupta, S.; Modgil, S. (2021). Measuring the perceived benefits of implementing blockchain technology in the banking sector. *Technol. Forecast. Soc. Change* 2021, 163, 120407.

Huang, S.Y.; Lee, C.-J. (2022). Predicting continuance intention to fintech chatbot. *Comput. Hum. Behav.* 2022, 129, 107027.

Kaur, D.; Sahdev, S.L.; Sharma, D.; Siddiqui, L. (2020). Banking 4.0: 'The influence of artificial intelligence on the banking industry & how AI is changing the face of modern day banks'. *Int. J. Manage.* 2020, 11, 577–585.

Kurni, M.; Saritha, K.; Nagadevi, D.; Reddy, K.S. (2022). A forefront insight into the integration of AI and blockchain technologies. In S. K. Hafizul Islam, A. K. Pal, D. Samanta, D. Samanta, & S. Bhattacharyya (Eds.), *Blockchain Technology for Emerging Applications*; Elsevier, pp. 297–320.

Maja, M.M.; Letaba, P. (2022). Towards a data-driven technology roadmap for the bank of the future: Exploring big data analytics to support technology roadmapping. *Soc. Sci. Humanit. Open* 2022, 6, 100270.

Patel, R.; Migliavacca, M.; Oriani, M. (2022). Blockchain in Banking and Finance: Is the best yet to come? A bibliometric review. *Res. Int. Bus. Financ.* 2022, 62, 101718.

Pitchay, A.B.A.; Thaker, M.A.B.M.T.; Azhar, Z.; Mydin, A.A.; Thaker, H.B.M.T. (2019). Factors persuade individuals' behavioral intention to opt for Islamic bank services: Malaysian depositors' perspective. *J. Islam. Mark.* 2019, 11, 234–250.

Rahman, M.; Ming, T.H.; Baigh, T.A.; Sarker, M. (2022). Adoption of artificial intelligence in banking services: An empirical analysis. *Int. J. Emerg. Mark.* 2022, 18, 4270–4300.

Suhartanto, D.; Dean, D.; Ismail, T.A.T.; Sundari, R. (2020). Mobile banking adoption in Islamic banks: Integrating TAM model and religiosity-intention model. *J. Islam. Mark.* 2020, 11, 1405–1418.

Thaker, M.A.B.M.T.; Pitchay, A.B.A.; Thaker, H.B.M.T.; Amin, M.F.B. (2019). Factors influencing consumers' adoption of Islamic mobile banking services in Malaysia: An approach of partial least squares (PLS). *J. Islam. Mark.* 2019, 10, 1037–1056.

15 The Face of Deception
Unmasking Deepfake Fraud in Financial Markets through AI, BI Tools

Timilehin Olasoji Olubiyi, Pawan Kumar, Arvinder Kaur, and Ercan Ozen

15.1 INTRODUCTION

The unprecedented proliferation of deepfake technology, characterized by hyper-realistic synthetic media generated through advanced machine learning algorithms, poses an existential threat to the integrity and stability of global financial markets, as it enables nefarious actors to orchestrate sophisticated fraudulent schemes that manipulate market sentiments, erode investor confidence, and destabilize the very bedrock of economic systems through deceptive impersonations of corporate executives, dissemination of false information, and artificial inflation of stock prices, thus compelling regulatory bodies and financial institutions to adopt cutting-edge countermeasures to mitigate these multifaceted risks (Al-Khazraji et al., 2023).

The burgeoning advancement of deepfake technology in financial markets has engendered a complex and insidious landscape where nefarious actors exploit synthetic media to craft meticulously deceptive narratives. This manipulation of audiovisual content not only distorts perceptual reality but also precipitates a perilous erosion of trust within global fiscal ecosystems (Sharma et al., 2024). By engineering fraudulent representations of key financial stakeholders and institutions, these actors catalyze cascading economic destabilization, undermining regulatory frameworks and challenging the very bedrock of market integrity. The ramifications of such duplicitous machinations reverberate across the financial domain, instigating unprecedented vulnerabilities in an increasingly digitalized economy (Arora et al., 2024).

Concept of deepfake and financial market: The confluence of deepfake technology and the financial market introduces a perilous frontier where hyper-realistic, algorithmically generated fabrications of corporate executives, financial analysts, and market-moving influencers can be weaponized to orchestrate sophisticated disinformation campaigns, ultimately subverting market dynamics, eroding investor confidence, and engendering systemic risks that threaten the stability and integrity of global financial infrastructures (Sardana et al., 2024).

DOI: 10.1201/9781003564621-15

15.2 VARIABLES OF THE STUDY

15.2.1 Dependent

Market stability: It refers to the consistency and stability in the market. The market observes the fluctuations in the trading activities, and the stability of the market is affected by various factors in the market (Westerlund, 2019).

15.2.2 Independent Variables

Prevalence of deepfake incidents: This refers to the rising use of highly pragmatic artificial intelligence (AI)-generated media that maneuver visuals or audio content, frequently for misleading purposes, undermining trust in digital information (Afanasyeva & Yumasheva, 2022).

Investor's confidence: Investor confidence characterizes the psychological proclivity and confidence of stakeholders to assign capital amidst multifaceted economic uncertainties and market volatilities (Roy & Bag, 2023).

Technological sophistication tools of deepfake technology: Technological sophistication in deepfake technology includes superior algorithmic paradigms, neural system architectures, and hyper-realistic generative models, synergistically manipulating audiovisual data to fabricate indistinguishable artificial media artifacts (Lakhera et al., 2024).

Cybersecurity measures in financial institutions: Cybersecurity actions in financial institutions involve multifaceted cryptographic protocols, interruption detection systems, and pliability structures mitigating complicated cyber intimidation targeting susceptible financial ecosystems (Abbas et al., 2022).

Hypothesis 1: To know the effect of the prevalence of deepfake incidents on market stability

Hypothesis 2: To investigate the effect of investors' confidence on market stability

Hypothesis 3: To study the impact of technological sophistication tools of deepfake detention tools on market stability

Hypothesis 4: To investigate the cybersecurity measures in financial organizations on market stability

15.3 LITERATURE REVIEW

15.3.1 H1 Effect of the Prevalence of Deepfake Incidents on Market Stability

The occurrence of deepfake incidents has been widely scrutinized in literature, making clear their deleterious consequences on market stability through eroding investor trust, strengthening information asymmetry, spreading market deformation, and engendering systemic vulnerabilities, thereby requiring robust regulatory structures, adaptive technological countermeasures, and interdisciplinary model to mitigate cascading economic outcomes within global financial ecosystems (Kerner & Risse, 2021).

15.3.2 H2 Effect of Investors' Confidence on Market Stability

Existing literature highlight that investor confidence, as an essential psychological determinant, intensely influences market strength by modulating capital flows, mitigating volatility, enhancing liquidity, and nurturing equilibrium within financial ecosystems, thus underscoring its criticality in shaping macroeconomic course and engendering pliability against systemic perturbations (Matli, 2024).

15.3.3 H3: Impact of Technological Sophistication Tools of Deepfake Detention Tools on Market Stability

Literature exposes that the use of technologically sophisticated deepfake uncovering tools strengthens market strength by curtailing propaganda propagation, protecting informational integrity, restoring stakeholder trust, and mitigating destabilizing economic consequences, thereby reinforcing systemic resilience against the harmful effects of synthetic media exploitation (Zendran & Rusiecki, 2021).

15.3.4 H4: Investigate the Cybersecurity Measures in Financial Organizations on Market Stability

Literature investigates cybersecurity dealings in financial institutions as essential stabilizers, leveraging sophisticated cryptographic frameworks, irregularity detection algorithms, and buoyancy strategies to anticipate systemic disruptions, preserve data integrity, and strengthen market stability against increasing cyber threat vectors (Hasani et al., 2024).

Research Gap: The research gap lies in the paucity of empirical investigation into the multifaceted, multidimensional interrelations among deepfake incidents, investor confidence, technological complexity of deepfake detection tools, and cybersecurity trial within financial establishments, particularly regarding their combined and nuanced impact on market strength, as obtainable literature insufficiently deals with the developing risks and adaptive methods required to mitigate the systemic vulnerabilities by increasingly sophisticated AI-driven disinformation and digital safety intimidation in worldwide financial ecosystems.

15.4 RESEARCH METHODOLOGY

Table 15.1 outlines the research tools utilized to assess five pivotal constructs—prevalence of deepfake incidents, market stability, technological sophistication of deepfake detection tools, investor confidence, and the robustness of cybersecurity criteria in financial organizations—each gauged through meticulously prepared Likert scale items varying from minimal deal or effectiveness to the most elevated degree of like-mindedness or knowledge, thus delivering a subtle framework to

TABLE 15.1
Research Instruments

Serial Number	Name of the Variable	Question Statement	Scale Description
1	Market stability	How stable do you believe the financial markets are in light of current technological advancements?	1 = Strongly disagree, 7 = Strongly agree
2	Prevalence of deepfake incidents	To what extent do you perceive deepfake incidents as a growing threat to financial markets?	1 = Not a threat at all, 7 = Extremely high threat
3	Investor confidence	How confident are you in the reliability of information shared in financial markets amidst potential deepfake fraud?	1 = No confidence, 7 = Very high confidence
4	Technological sophistication of deepfake detection tools	How effective do you believe current deepfake detection tools are in safeguarding financial market integrity?	1 = Not effective, 7 = Extremely effective
5	Cybersecurity measures in financial institutions	How well do you think financial institutions are equipped with cybersecurity measures to prevent deepfake-related fraud?	1 = Not equipped at all, 7 = Very well equipped

Source: Authors.

explore perceptions of financial market stability amidst the escalating interplay of technical advancements and occurring cyber perils.

15.4.1 Data Analysis

Table 15.2 presents the distribution of respondents, and reflects a diverse asymmetrical representation, with retail investors comprising an overwhelming majority at 60.83%, juxtaposed against the relatively modest participation of financial analysis, market experts, and technology experts in cybersecurity, who together account for 23.96%, while institutional investors, corporative executives, legal experts in financial crimes, and journalists collectively contribute to a more minor albeit critical 15.2%, demonstrating a multifaceted cross-section of expertise that underscores the predominance of individual investors in this evaluative cohort of 217 participants.

Table 15.3 provides an exhaustive statistical examination of several critical variables—namely "Prevalence of Deepfake Incidents," "Investor Confidence," "Technological Sophistication of Deepfake Detection Tools," "Cybersecurity Measures in Financial Institutions," and "Market Stability"—where each variable's mean and standard deviation are meticulously adjusted for bootstrapped bias and standard errors, with confidence intervals delineating the range of plausible values, thus offering a robust depiction of the central tendencies and dispersions within the

TABLE 15.2
Respondents

Respondents	Number	%
Financial analysts and market experts	25	11.52
Institutional investors	6	2.76
Corporate executives and financial officers	5	2.30
Technology experts in cybersecurity	27	12.44
Retail investors	132	60.83
Legal experts in financial crimes	10	4.61
Journalists and media professionals	12	5.53
Total	217	100.00

Source: Authors.

TABLE 15.3
Descriptive Statistics

			Bootstrap[a]			
				95% Confidence Interval		
		Statistic Bias	Std. Error	Lower	Upper	
Prevalence_of_ deepfake_incidents	Mean	5.1429	0.0013	0.1081	4.9217	5.3456
	Std. Deviation	1.57023	−0.00626	0.06986	1.42168	1.69899
	N	217	0	0	217	217
Investor_confidence	Mean	4.5622	−0.0020	0.0874	4.3825	4.7372
	Std. Deviation	1.25722	−0.00181	0.06268	1.13454	1.37734
	N	217	0	0	217	217
Technological_ sophistication_of_ deepfake_ detection_tools	Mean	4.1751	−0.0015	0.1014	3.9770	4.3641
	Std. Deviation	1.58603	−0.00156	0.06388	1.44808	1.70374
	N	217	0	0	217	217
Cybersecurity_ measures_in_ financial_ institutions	Mean	4.5760	−0.0019	0.0866	4.4055	4.7327
	Std. Deviation	1.31056	−0.00246	0.06384	1.17801	1.42991
	N	217	0	0	217	217
Market_stability	Mean	4.0000	0.0012	0.1002	3.8112	4.2120
	Std. Deviation	1.50308	−0.00372	0.06148	1.37437	1.61437
	N	217	0	0	217	217

Source: Authors.

[a] Unless otherwise noted, bootstrap results are based on 1,000 bootstrap samples.

dataset, ensuring statistical precision and reliability, and presenting a comprehensive quantitative portrait of the variability and relationships across these pivotal constructs in the financial ecosystem.

Table 15.4 reveals a nuanced interplay among the variables, where investor confidence exhibits a significant positive correlation with market stability, technological sophistication of deepfake detection tools, and cybersecurity measures, underscoring its central role in maintaining market equilibrium; conversely, the prevalence of deepfake incidents shows an insignificant relationship with market stability, while technological sophistication and cybersecurity measures demonstrate robust positive associations with market stability, suggesting their critical influence in bolstering financial resilience against the destabilizing effects of deepfake threats and cyber vulnerabilities.

Figure 15.1 represents a multilayer perceptron (MLP) neural network architecture comprising an input layer with multiple nodes corresponding to distinct features, followed by a hidden layer where each neuron employs the hyperbolic tangent activation function, modulated by bias terms, and weighted synaptic connections (both positive and negative) to propagate transformed inputs through non-linear combinations; the output layer utilizes the softmax activation function to deliver probabilistic classifications across the output nodes, indicating the system's ability to map complex, multi-dimensional input data into distinct output categories with precision across weighted interconnections.

Table 15.5 explores the result of the case processing summary that reveals a comprehensive analysis wherein the entirety of the dataset was included for all examined pairwise interactions between market stability and the independent variables—prevalence of deepfake incidents, investor confidence, technological sophistication of deepfake detection tools, and cybersecurity measures in financial institutions—without any exclusions, ensuring that the statistical analysis remains unimpeded by missing data and thus provides a robust and complete evaluation of the interdependencies between these critical variables within the context of financial stability and technological influences.

Table 15.6 reveals that market stability fluctuates across varying levels of deepfake prevalence, with bootstrap-adjusted means and standard deviations highlighting a tendency toward higher market stability at increased prevalence levels, while bias-corrected confidence intervals ensure the robustness of these findings across extensive resampling iterations, confirming the statistical reliability and consistency of the observed relationships.

Table 15.7 presents an intricate statistical analysis of the correlation between "Market Stability" and varying degrees of "Investor Confidence," revealing a nuanced interplay where each confidence level is characterized by distinct central tendencies and dispersions, meticulously adjusted for bias and standard errors via bootstrapping, with confidence intervals offering a comprehensive understanding of variability; the analysis underscores how shifts in investor confidence exert differential impacts on market stability, with higher confidence levels associated with more stable outcomes and lower levels introducing greater volatility, all while ensuring statistical rigor and precision through robust methodological frameworks.

TABLE 15.4
Correlations

				Prevalence_of_ Deepfake_ Incidents	Investor_ Confidence	Technological_ Sophistication_of_ Deepfake_Detection_Tools	Cybersecurity_ Measures_in_ Financial_Institutions	Market_ Stability
Prevalence_of_ deepfake_ incidents	Pearson correlation			1	0.130	−0.088	0.016	−0.049
	Sig. (2-tailed)				0.055	0.196	0.814	0.472
	Sum of squares and cross-products			532.571	55.571	−47.429	7.143	−25.000
	Covariance			2.466	0.257	−0.220	0.033	−0.116
	N			217	217	217	217	217
	Bootstrap[b]	Bias		0	−0.002	−0.001	0.000	−0.002
		Std. Error		0	0.070	0.070	0.072	0.074
		95% Confidence interval	Lower	1	−0.006	−0.223	−0.123	−0.196
			Upper	1	0.262	0.054	0.156	0.097
Investor_ confidence	Pearson correlation			0.130	1	0.475[a]	0.536[a]	0.416[a]
	Sig. (2-tailed)			0.055		0.000	0.000	0.000
	Sum of squares and cross-products			55.571	341.410	204.636	190.724	170.000
	Covariance			0.257	1.581	0.947	0.883	0.787
	N			217	217	217	217	217
	Bootstrap[b]	Bias		−0.002	0	0.002	0.001	0.001
		Std. Error		0.070	0	0.052	0.056	0.062
		95% Confidence Interval	Lower	−0.006	1	0.373	0.426	0.290
			Upper	0.262	1	0.575	0.642	0.535

(Continued)

TABLE 15.4 (*Continued*)
Correlations

				Prevalence_of_Deepfake_Incidents	Investor_Confidence	Technological_Sophistication_of_Deepfake_Detection_Tools	Cybersecurity_Measures_in_Financial_Institutions	Market_Stability
Technological_sophistication_of_deepfake_detection_tools	Pearson correlation			−0.088	0.475[a]	1	0.528[a]	0.565[a]
	Sig. (2-tailed)			0.196	0.000		0.000	0.000
	Sum of squares and cross-products			−47.429	204.636	543.346	237.111	291.000
	Covariance			−0.220	0.947	2.515	1.098	1.347
	N			217	217	217	217	217
	Bootstrap[b]	Bias		−0.001	0.002	0	0.002	0.001
		Std. Error		0.070	0.052	0	0.058	0.054
		95% Confidence interval	Lower	−0.223	0.373	1	0.405	0.457
			Upper	0.054	0.575	1	0.634	0.669
Cybersecurity_measures_in_financial_institutions	Pearson correlation			0.016	0.536[a]	.528[a]	1	0.425[a]
	Sig. (2-tailed)			0.814	0.000	0.000		0.000
	Sum of squares and cross-products			7.143	190.724	237.111	370.995	181.000
	Covariance			0.033	0.883	1.098	1.718	0.838
	N			217	217	217	217	217
	Bootstrap[b]	Bias		0.000	0.001	0.002	0	0.004
		Std. Error		0.072	0.056	0.058	0	0.061
		95% Confidence interval	Lower	−.123	0.426	0.405	1	0.312
			Upper	0.156	0.642	0.634	1	0.544

(*Continued*)

TABLE 15.4 (*Continued*)
Correlations

				Prevalence_of_ Deepfake_ Incidents	Investor_ Confidence	Technological_ Sophistication_of_ Deepfake_Detection_Tools	Cybersecurity_ Measures_in_ Financial_Institutions	Market_ Stability
Market_stability	Pearson correlation			−0.049	0.416[a]	0.565[a]	0.425[a]	1
	Sig. (2-tailed)			0.472	0.000	0.000	0.000	
	Sum of squares and cross-products			−25.000	170.000	291.000	181.000	488.000
	Covariance			−0.116	0.787	1.347	0.838	2.259
	N			217	217	217	217	217
	Bootstrap[b]	Bias		−0.002	0.001	0.001	0.004	0
		Std. Error		0.074	0.062	0.054	0.061	0
		95% Confidence interval	Lower	−0.196	0.290	0.457	0.312	1
			Upper	0.097	0.535	0.669	0.544	1

Source: Authors.

[a] Correlation is significant at the 0.01 level (2-tailed).

[b] Unless otherwise noted, bootstrap results are based on 1,000 bootstrap samples.

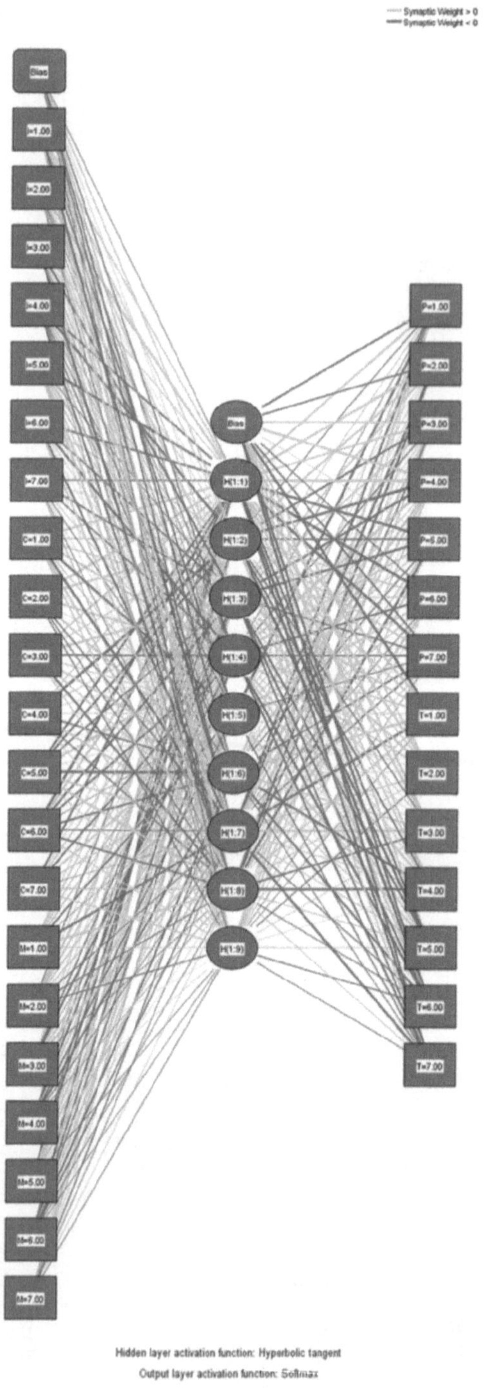

FIGURE 15.1 Neural network between variables. (Source: Authors.)

TABLE 15.5
Case Processing Summaryx

	Cases					
	Included		Excluded		Total	
	N	Percent (%)	N	Percent (%)	N	Percent (%)
Market_stability * prevalence_of_deepfake_incidents	217	100.0	0	0.0	217	100.0
Market_stability * investor_confidence	217	100.0	0	0.0	217	100.0
Market_stability * technological_sophistication_of_deepfake_detection_tools	217	100.0	0	0.0	217	100.0
Market_stability * cybersecurity_measures_in_financial_institutions	217	100.0	0	0.0	217	100.0

Source: Authors.

Table 15.8 presents a sophisticated statistical analysis of the correlation between "Market Stability" and varying levels of "Cybersecurity Measures in Financial Institutions," wherein each level of cybersecurity is associated with distinct central tendencies and deviations, rigorously adjusted for bias and standard error using bootstrapping, with confidence intervals offering a comprehensive view of the variability; this analysis underscores the critical influence of cybersecurity robustness on market stability, revealing that higher levels of cybersecurity measures correspond to greater market resilience, while lower levels introduce heightened volatility, all achieved through a methodologically rigorous framework ensuring the precision and reliability of the findings.

15.5 MANAGERIAL IMPLICATIONS OF THE STUDY

The managerial and practical implications of this research underscore the urgent necessity for financial institutions to integrate advanced deepfake detection tools, robust cybersecurity frameworks, and dynamic real-time threat intelligence systems into their strategic operations, not only to safeguard against the increasingly sophisticated nature of algorithmically driven disinformation and cyber fraud but also to preserve market integrity, maintain investor confidence, comply with regulatory demands, and ensure long-term stability, resilience, and sustainability in a financial ecosystem that is progressively more susceptible to destabilization through emergent AI-driven vulnerabilities and complex technological threats.

15.6 CONCLUSION

The overall analysis of the tables demonstrates that market stability is profoundly impacted by several interdependent factors including investor confidence, the prevalence of deepfake incidents, and the technological sophistication of deepfake

TABLE 15.6
Market_Stability * Prevalence_of_Deepfake_Incidents

Market_Stability

Prevalence_of_Deepfake_ Incidents		Statistic	Bootstrap[a] Bias	Std. Error	95% Confidence Interval Lower	Upper
1.00	Mean	2.5714	−0.0349[b]	0.5805[b]	1.4000[b]	3.7012[b]
	N	7	0[b]	3[b]	2[b]	13[b]
	Std. Deviation	1.51186	−0.15475[d]	0.34609[d]	0.00000[d]	1.73205[d]
2.00	Mean	4.0000	−0.0256[c]	1.0059[c]	2.0000[c]	6.0000[c]
	N	5	0[c]	2[c]	1[c]	10[c]
	Std. Deviation	2.23607	−0.34945[e]	0.68431[e]	0.50000[e]	3.05505[e]
3.00	Mean	3.4706	0.0014	0.3656	2.7503	4.1762
	N	17	0	4	9	25
	Std. Deviation	1.54587	−0.05495	0.19644	1.08373	1.88083
4.00	Mean	4.7907	0.0015	0.1709	4.4445	5.1199
	N	43	0	6	32	55
	Std. Deviation	1.14555	−0.01622	0.16929	0.79708	1.43870
5.00	Mean	4.2264	−0.0093	0.1765	3.8801	4.5555
	N	53	0	6	41	66
	Std. Deviation	1.26554	−0.02057	0.11772	1.01348	1.46902
6.00	Mean	3.8485	−0.0028	0.2896	3.2581	4.4371
	N	33	0	5	23	43
	Std. Deviation	1.69781	−0.03529	0.16117	1.32938	1.96795
7.00	Mean	3.6271	−0.0012	0.1863	3.2667	4.0134
	N	59	0	7	46	73
	Std. Deviation	1.48438	−0.01074	0.12465	1.22114	1.72240
Total	Mean	4.0000	−0.0034	0.0993	3.8203	4.1750
	N	217	0	0	217	217
	Std. Deviation	1.50308	−0.00121	0.05890	1.38767	1.61293

Source: Authors.

[a] Unless otherwise noted, bootstrap results are based on 1,000 bootstrap samples.

[b] Based on 998 samples.

[c] Based on 990 samples.

[d] Based on 993 samples.

[e] Based on 965 samples.

detection tools and cybersecurity measures in financial institutions while deepfake incidents pose a destabilizing risk by eroding trust and disseminating disinformation. Their negative effect is counterbalanced by strong investor confidence, advanced cybersecurity protocols, and robust detection mechanisms, which act as protective

TABLE 15.7
Market_Stability * Investor_Confidence

		Market_Stability				
			Bootstrap[a]			
					95% Confidence Interval	
Investor_Confidence		**Statistic**	**Bias**	**Std. Error**	**Lower**	**Upper**
1.00	Mean	1.6667	0.0061[b]	0.6195[b]	1.0000[b]	3.0000[b]
	N	3	0[b]	2[b]	1[b]	7[b]
	Std. deviation	1.15470	−0.38286[c]	0.54237[c]	0.00000[c]	1.41421[c]
2.00	Mean	2.8750	0.0004	0.6406	1.6667	4.3333
	N	8	0	3	3	14
	Std. deviation	1.72689	−0.16271[d]	0.41379[d]	0.70711[d]	2.33166[d]
3.00	Mean	3.2069	−0.0006	0.2494	2.7241	3.7388
	N	29	0	5	19	39
	Std. deviation	1.34641	−0.02289	0.15080	1.01003	1.61716
4.00	Mean	3.7213	−0.0091	0.1848	3.3540	4.0727
	N	61	0	7	48	75
	Std. deviation	1.39201	−0.01202	0.11160	1.16149	1.59897
5.00	Mean	4.2029	0.0016	0.1645	3.8573	4.5238
	N	69	0	7	56	83
	Std. deviation	1.37819	−0.01303	0.13038	1.09828	1.61328
6.00	Mean	4.9412	−0.0078	0.2110	4.5129	5.3226
	N	34	0	6	24	45
	Std. deviation	1.22947	−0.03291	0.19680	.81720	1.58617
7.00	Mean	4.7692	0.0068	0.4333	3.8953	5.6146
	N	13	0	3	7	20
	Std. deviation	1.53590	−0.07871	0.23619	0.95119	1.93629
Total	Mean	4.0000	−0.0034	0.0993	3.8203	4.1750
	N	217	0	0	217	217
	Std. Deviation	1.50308	−0.00121	0.05890	1.38767	1.61293

Source: Authors.

[a] Unless otherwise noted, bootstrap results are based on 1,000 bootstrap samples.

[b] Based on 959 samples.

[c] Based on 816 samples.

[d] Based on 998 samples.

barriers to safeguard market integrity. Investor confidence emerges as a critical psychological driver that sustains liquidity and price equilibrium, whereas the deployment of sophisticated deepfake detection technologies and fortified cybersecurity frameworks ensures the preservation of information authenticity, which mitigates the risks posed by cyber fraud and enhances overall market resilience. The statistical data underscores that the synergy between high investor confidence and the adoption of advanced technological tools is crucial to stabilizing market dynamics protecting

TABLE 15.8
Market_Stability * Cybersecurity_Measures_in_Financial_Institutions

Market_Stability							
			Bootstrap[a]				
					95% Confidence Interval	
Cybersecurity_Measures_in_Financial_Institutions		Statistic	Bias	Std. Error	Lower	Upper
1.00	Mean	2.2000	0.0083[b]	0.7867[b]	1.0000[b]	3.8712[b]
	N	5	0[b]	2[b]	2[b]	10[b]
	Std. deviation	1.78885	−0.33151[d]	0.69081[d]	0.00000[d]	2.82843[d]
2.00	Mean	2.0000	−0.0076[c]	0.3206[c]	1.3750[c]	2.6667[c]
	N	8	0[c]	3[c]	3[c]	14[c]
	Std. deviation	0.92582	−0.07655[e]	0.16851[e]	0.48795[e]	1.09545[e]
3.00	Mean	2.6897	.0053	.2342	2.2145	3.1706
	N	29	0	5	19	40
	Std. deviation	1.28462	−0.02692	0.16953	0.93743	1.58810
4.00	Mean	4.1250	−0.0037	0.1658	3.7834	4.4575
	N	56	0	6	44	68
	Std. deviation	1.25136	−0.01673	0.11662	1.01674	1.46746
5.00	Mean	4.4762	0.0001	0.1636	4.1251	4.7965
	N	63	0	7	49	76
	Std. deviation	1.33026	−0.01238	0.12356	1.08410	1.55285
6.00	Mean	4.5217	0.0012	0.1994	4.1429	4.9037
	N	46	0	6	35	57
	Std. deviation	1.34560	−0.02134	0.14400	1.03148	1.60870
7.00	Mean	4.2000	0.0012	0.4639	3.2500	5.0000
	N	10	0	3	4	17
	Std. Deviation	1.47573	−0.12863[c]	0.39644[c]	0.54772[c]	2.06828[c]
Total	Mean	4.0000	−0.0034	0.0993	3.8203	4.1750
	N	217	0	0	217	217
	Std. deviation	1.50308	−0.00121	0.05890	1.38767	1.61293

Source: Authors.

[a] Unless otherwise noted, bootstrap results are based on 1,000 bootstrap samples.

[b] Based on 995 samples.

[c] Based on 999 samples

[d] Based on 974 samples.

[e] Based on 997 samples.

against financial disarray and ensuring smooth capital flows; ultimately, the key to maintaining long-term market stability in a volatile and increasingly digitalized economy lies in the integration of both psychological and technological safeguards with advanced cybersecurity and deepfake detection systems forming the backbone of a resilient financial system.

REFERENCES

Abbas, N. N., Ahmad, R., Qazi, S., & Ahmed, W. (2022). Impact of deepfake technology on FinTech applications. In *Handbook of Research on Cybersecurity Issues and Challenges for Business and FinTech Applications*. IGI Global. https://doi.org/10.4018/978-1-6684-5284-4.ch012

Afanasyeva, T., & Yumasheva, I. (2022). Research on the effects of "deepfake" technology for the modern digital space. In Rumyantseva, A., Plotnikov, V., Minin, A., & Anyigba, H. (Eds.), Springer *Proceedings in Business and Economics* (pp. 57–65). Springer Nature. https://doi.org/10.1007/978-3-031-14410-3_7

Al-Khazraji, S. H., Saleh, H. H., Khalıd, A. I., & Mıshkhal, I. A. (2023). Impact of deepfake technology on social media: detection, misinformation and societal implications. In Mahmud, O. (Ed.), *Eurasia Proceedings of Science, Technology, Engineering and Mathematics* (Vol. 23, pp. 429–441). ISRES Publishing. https://doi.org/10.55549/epstem.1371792

Arora, M., Mishra, K. K., Singh, M., Singh, P., & Tripathi, R. (2024). Deepfake technology and its implications for influencer marketing. In *Navigating the World of Deepfake Technology*. IGI Global. https://doi.org/10.4018/979-8-3693-5298-4.ch005

Hasani, A., Rasheed, J., Alsubai, S., & Luma-Osmani, S. (2024). Personal rights and intellectual properties in the upcoming era: the rise of deepfake technologies. *Lecture Notes in Networks and Systems, 1036 LNNS*, 379–391. https://doi.org/10.1007/978-3-031-62881-8_32

Kerner, C., & Risse, M. (2021). Beyond porn and discreditation: epistemic promises and perils of deepfake technology in digital lifeworlds. *Moral Philosophy and Politics*, *8*(1), 81–108. https://doi.org/10.1515/mopp-2020-0024

Lakhera, G., Taneja, S., Ozen, E., Kukreti, M., & Kumar, P. (2024). *Navigating the World of Deepfake Technology*. IGI Global. https://doi.org/10.4018/979-8-3693-5298-4

Matli, W. (2024). Extending the theory of information poverty to deepfake technology. *International Journal of Information Management Data Insights*, *4*(2), 100286. https://doi.org/10.1016/j.jjimei.2024.100286

Roy, S., & Bag, A. (2023). Dangers of deepfake technology. *Economic and Political Weekly*, *58*(50), 4–6.

Sardana, F., Mishra, K. K., Singh, A., & Saini, N. (2024). Transforming social media marketing through deepfake technology. In *Navigating the World of Deepfake Technology*. IGI Global. https://doi.org/10.4018/979-8-3693-5298-4.ch022

Sharma, K., Taneja, S., & Monga, S. (2024). Deepfake technologies: concepts, methods, and applications. In *Navigating the World of Deepfake Technology*. IGI Global. https://doi.org/10.4018/979-8-3693-5298-4.ch004

Westerlund, M. (2019). The emergence of deepfake technology: a review. *Technology Innovation Management Review*, *9*(11), 39–52. https://doi.org/10.22215/TIMREVIEW/1282

Zendran, M., & Rusiecki, A. (2021). Swapping face images with generative neural networks for deepfake technology – Experimental study. In Jakóbczak, W., Wierzchoń, S., Czachórski, T., Chao, Z.-M., Jung, H. R., & Cieśliński, J. L. (Eds.), *Procedia Computer Science* (Vol. 192, pp. 834–843). Elsevier B.V. https://doi.org/10.1016/j.procs.2021.08.086

Index